Brickwork Level 3

For CAA Construction Diploma and NVQs

Malcolm Thorpe

ELSEVIER

AMSTERDAM • BOSTON • HEIDELBERG • LONDON • NEW YORK • OXFORD
PARIS • SAN DIEGO • SAN FRANCISCO • SINGAPORE • SYDNEY • TOKYO

Butterworth-Heinemann is an imprint of Elsevier

B H

Butterworth-Heinemann is an imprint of Elsevier
The Boulevard, Langford Lane, Kidlington, Oxford OX5 1GB, UK
30 Corporate Drive, Suite 400, Burlington, MA 01803, USA

First edition 2010

Notice
No responsibility is assumed by the publisher for any injury and/or damage to persons or property as a matter of products liability, negligence or otherwise, or from any use or operation of any methods, products, instructions or ideas contained in the material herein. Because of rapid advances in the medical sciences, in particular, independent verification of diagnoses and drug dosages should be made

British Library Cataloguing in Publication Data
A catalogue record for this book is available from the British Library

Library of Congress Cataloging-in-Publication Data
A catalog record for this book is available from the Library of Congress

ISBN: 978-1-85617-764-1

For information on all Butterworth-Heinemann publications visit our web site at books.elsevier.com

Printed and bound in China

10 11 12 13 14 15 10 9 8 7 6 5 4 3 2 1

Contents

Preface

The changes in construction training have led to the need to produce this series of books which incorporate both National Vocational Qualifications (NVQs) and Diplomas.

The content of each book follows both routes and provides the necessary information for the various job knowledge tests.

After the initial chapter, which gives the construction student an insight into the industry they are entering, each chapter follows very closely the NVQ and Diploma units.

The aim of each book is to provide an information resource and student workbook for all building craft students. It can be used to provide teaching and assessment material, or used simply to reinforce college lectures.

Each chapter has a set of multiple-choice questions designed to test your level of knowledge before moving on to the next chapter.

Malcolm Thorpe

CHAPTER 1

The Construction Industry

This chapter will cover the following NVQ and Diploma units:

- NVQ All
- CC 3003K

This chapter is about:

- Types of industrial building
- The construction industry

The following NVQ performance criteria will be covered:

This chapter has no comparable Level 3 NVQ units but it gives the student an early introduction to the construction industry.

The following Diploma outcomes will be covered:

- New technology and building methods
- Energy conservation
- Sustainable materials

Introduction

The construction industry is one of the most important industries in the UK.

Construction means creating, not only the houses we need to live in but many other buildings such as schools, hospitals and shopping centres.

The majority of buildings and structures are designed and constructed for a specific purpose. The use of the building will determine the size, shape, style and ultimately the cost.

There are also several different types of construction work attached to the building industry.

Types of building

Many different types of construction are required to fulfil the needs of today's ever-demanding society.

Industrial buildings

For the purpose of Level 3, only buildings for industrial purposes will be discussed.

These buildings are constructed as a place of work. Examples are shown in Figure 1.1. Buildings in this category vary a great deal in their design. Their physical size and type of construction depend on the type of business being conducted and the size of the form.

The production of most consumer items is best achieved under cover of a building. The design of the building depends on the consumer item being produced. For example, car manufacture requires large production-type factory units with open floor area, whereas small electrical items could be produced in small individual factory units.

Buildings are also constructed simply to store materials, for example warehouses.

Other types of buildings constructed for work are offices, shops, banking establishments, etc. These buildings need other aspects not required in

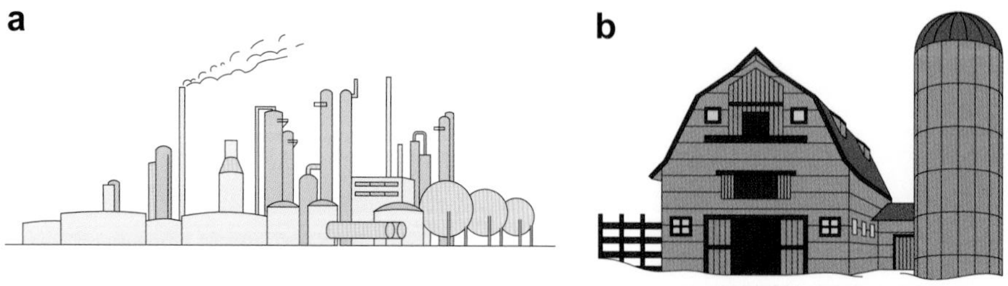

FIGURE 1.1
Types of industrial building: (a) factory; (b) agricultural

physical work areas. Offices have light, ventilation and comfort as their main criteria, which are normally found in small units.

There are many buildings in this category that could be considered for other categories, such as shops. These could be classified as either industrial or community: are they a place of work for the shopkeepers, or a place for the community to relax while browsing through the shops?

TYPES OF INDUSTRIAL BUILDING

These buildings are required for people to work in the manufacture of goods. A surplus of these goods can then be exchanged for other goods produced by other groups of workers. These goods are sold in shops, stores and markets.

The organization of the workforce and exchange of goods is carried out in offices.

Storage

All living, communal and working accommodation has storage areas designed into it, but some buildings are designed solely for the storage of manufactured goods. These are classed as warehouses. Examples are shown in Figure 1.2.

Most storage accommodation requires large unobstructed areas, to facilitate modern mechanical handling of goods. Other storage accommodation includes reservoirs, water towers and oil tanks.

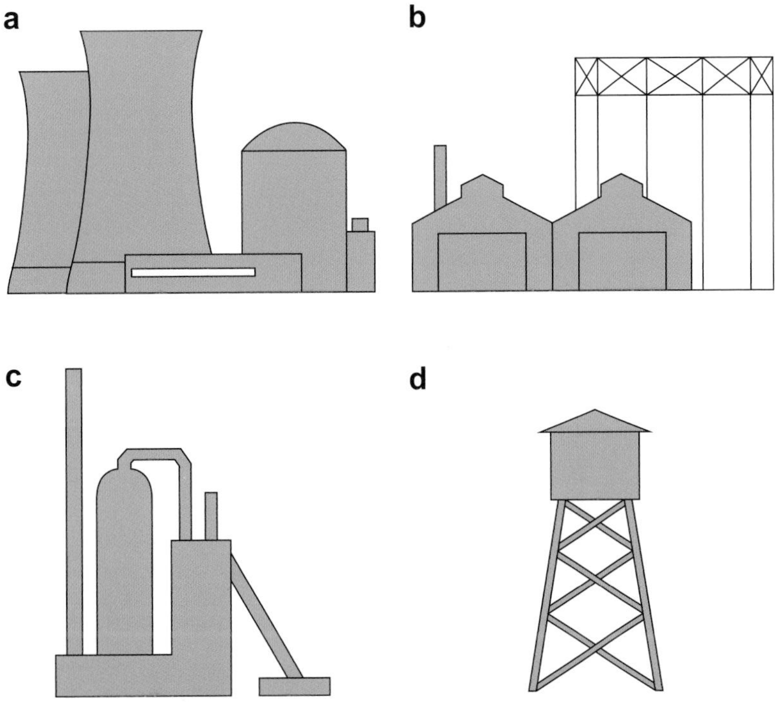

FIGURE 1.2
Types of storage building: (a) nuclear power station; (b) warehouse; (c) chemical storage; (d) water tower

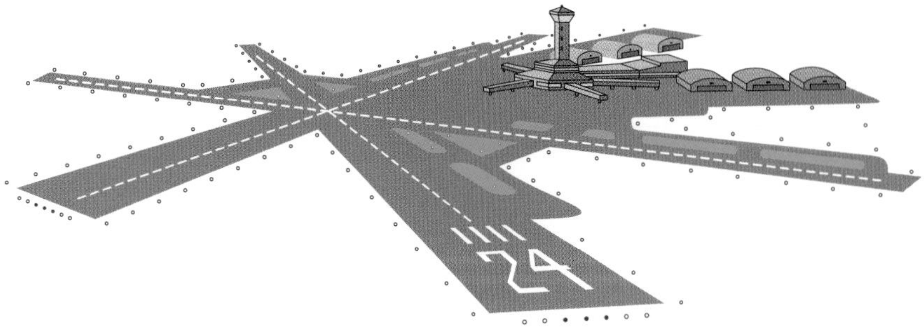

FIGURE 1.3
Type of transport buildings: airport

Each construction must be specially designed to accommodate the material to be stored.

Transport

In our modern society there is a need for many facilities for transport. To facilitate this there are national networks of roads, rail, rivers, etc.

These facilities, in turn, require ancillary buildings such as terminals, stations, waiting rooms and storage space. An example is shown in Figure 1.3.

The construction industry

The construction industry is the term given for the two main operations within the industry: building and civil engineering. Figures 1.4 and 1.5 show examples of each.

- Building – This is the term applied to work that relates to accommodation.
- Civil engineering – This is the term applied to work that relates to features and/or services around the accommodation.

FIGURE 1.4
Building

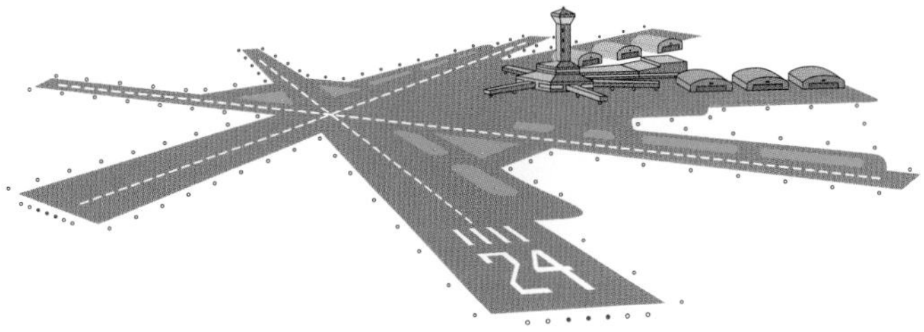

FIGURE 1.5
Engineering

New technology and methods used in construction

As the industry advances with new technology the design of buildings changes accordingly.

The majority of buildings are constructed for a specific purpose. This will determine their shape, size and style, and will also affect the quality and the eventual cost of the building.

The actual structure of the building is termed the external envelope, and protects the internal environment from all outside elements.

A structure can be defined as an organized combination of connected elements, which are constructed to perform some function, e.g. office block, factory unit or domestic building. Each building will be constructed with different structures to ensure that the correct facilities have been included and used to their full potential.

Structural forms

There are many structural forms in present-day use, each being modified from time to time to gain the best possible benefits from new materials and new techniques.

There are two basic forms of structure used in the construction of buildings:

- solid structure
- framed structure.

SOLID STRUCTURE

The walls of this type of construction support the loads, protect the internal environment and enclose space. The walls are therefore load-bearing, transferring the loads from the building down to the foundations.

The characteristic of this particular form is the thickness of the wall, owing to the materials and the manner in which they are used. Solid construction in the form of brick, stone and concrete is an easily erected structure. It is gradually built up from small units, block by block. When concrete is used some form of mould has to be provided to support the wet concrete until it has hardened.

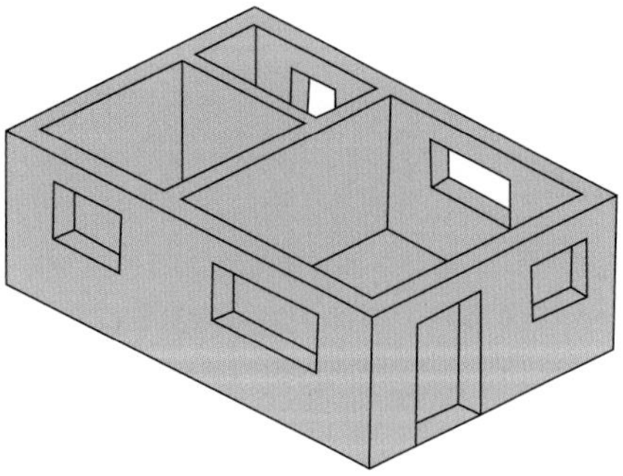

FIGURE 1.6
Cellular structure

There are two main types of sold wall construction: cellular and cross-wall construction.

In cellular construction (Figure 1.6), the structure consists of walls, connected together to form the rooms of the building. The result is a very rigid, load-bearing structure. The external walls form the external envelope and the internal walls divide and build up the rooms.

Cross-wall construction (Figure 1.7) consists of a series of walls parallel to each other and at right angles to the front of the building. These walls take the main structural loads and transmit them to the foundations.

Both methods can be built in cavity wall construction. It is very rare for either domestic or commercial buildings to be built with solid walls. Cavity walls are used in modern construction and provide better insulation to the property. They can be filled with various insulation materials to prevent heat loss.

FRAMED STRUCTURES

A framed structure is a network of beams and columns connected together to form the skeletal frame of the building (Figure 1.8).

The interconnecting members also have a supporting function and they transfer the loads to the foundations. Protection is provided by either cladding fixed over the framework or infill panels fixed between the members.

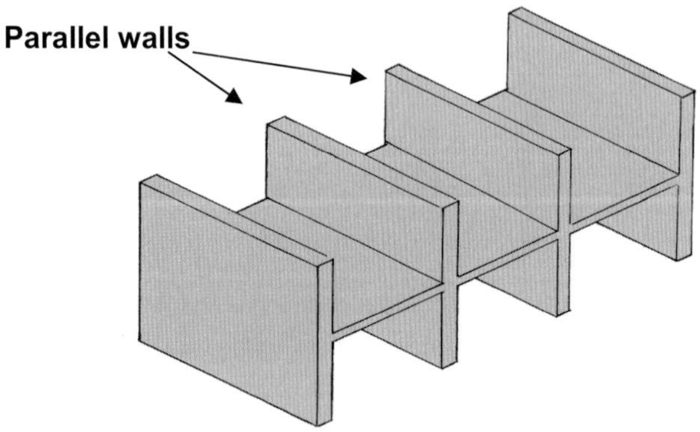

FIGURE 1.7
Cross-wall structure

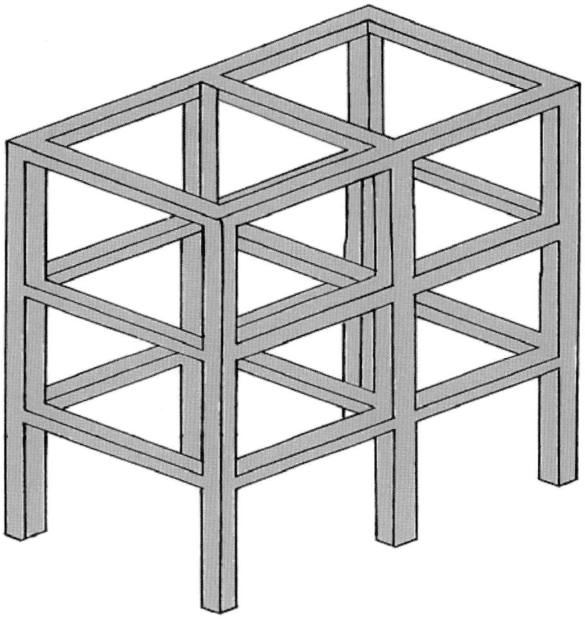

FIGURE 1.8
Framed structure

Framed construction is suitable for a very wide range of buildings from low rise to high rise, and is easily erected from prefabricated members. The members are simply connected together in a certain sequence to form the structural framework.

Concrete and steel are the two main materials used, with a few low-rise buildings having a timber-framed structure.

The rectangular framed structure is the most common type. This consists of a series of upright and horizontal members. The resulting frame provides the bearing for the floors, walls and roof.

The structure can be cladded with brick when the building has been made watertight. The whole frame is insulated before the brick cladding is added, to prevent heat loss from the building.

Design aspects

The appearance of the building should be pleasing to the eye. It should also be pleasing to the eventual owner and the public, and blend in with the surrounding buildings or area.

Technological aspects take into account the choice of the most suitable material for a specific purpose and the use of sustainable materials.

Good buildings are naturally those that are constructed with suitable materials and appropriate methods. In the past, buildings were constructed of materials that were local to the area to save on transportation costs, but nowadays materials can be used from anywhere in the world. The designer must be aware of the capabilities of all materials and the constructional methods to be adopted.

Buildability

It is impossible to design buildings to be used indefinitely, even if that were thought desirable. All building materials deteriorate eventually, and the cost of maintenance becomes prohibitive. Most buildings are designed to have a life-span of 70 years, or even less in some instances. The aim, though, is a building that will require as little maintenance as possible during its planned life.

The materials selected must be ones which, subject to conscientious maintenance by the occupiers of the building, will not show serious deterioration during the building's life.

Sometimes architects let their fantasies take over, resulting in a building that seems impossible to build.

Structural stability

Given determination, even the most apparently unstable structures can be satisfactorily built. Clearly, every building must be stable, whether it looks it or not.

Durability

Most architects will need to know the required life of the building. Most clients would not have any idea, but this aspect is crucial to the design team. It could determine the structural form and type, the selection of materials and the eventual building cost.

The architect will explain these reasons to the client, and between them they will decide on the life expectancy of the building.

Energy efficiency

A great deal of energy is used in the construction of a building.

Many years ago the whole of the construction process was carried out by hand. All the materials were produced and formed manually, the excavation was dug by hand, materials were mixed and transported by hand. A great deal of manual energy was expended.

Things have changed over the years and machines are used constantly on building sites to save on manpower and produce a more cost-effective building.

Construction process

When commencing building on a new site it is important to clear the site of all debris, old buildings and any other items.

The next operation is to remove the vegetable soil. Clearance of vegetable soil or top soil from the construction site area is mandatory, unless the work consists of alterations or amendments to existing buildings.

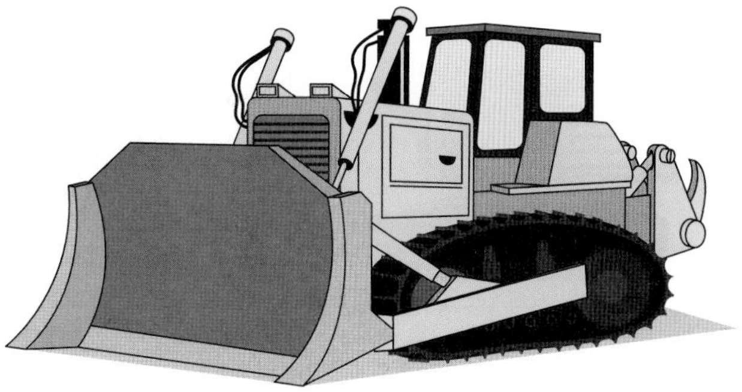

FIGURE 1.9
Bulldozer

Vegetable soil is soft, easily compressed and contains plant life, properties excluding its use as a basis for construction. It is found in variable depths, of approximately 150–300 mm, and has been built up by nature over many years.

This layer is excavated separately from the other excavation work and the material is usually stockpiled on a part of the site where it will not be in the way of any other building work until it is required for landscaping. If it is not required it can be sold. The bulldozer is ideal for stripping the top soil and stockpiling it ready for later use (Figure 1.9).

Setting out

Once the area is clear the proposed building can be set out. It is important that this procedure is carried out accurately, as it is very expensive to correct later.

When contractors take possession of a building site they will receive a full set of contract drawings, including a site plan of the proposed building.

The production drawings produced by the architect will show the position of the building on the site, and it is important to ensure that the building is constructed as shown, as this has been approved by the local authority.

The front of the building will be a certain distance back from the road, which will have been prescribed by the Highway Authority, and the distance from the other site boundaries will be controlled by the planning authority.

Before the foundations can be constructed it is necessary to establish the exact position of the building in relation to the site boundaries. This is done by a process known as 'setting out'. Setting-out procedures are dealt with in Chapter 6.

Excavations

On small sites, hand-held tools such as picks, shovels and wheelbarrows are used. However, if the depth of excavation exceeds 1.2 m some method of removing spoil from the excavation will have to be used.

FIGURE 1.10
Backactor

On all sites mechanical methods could be used, depending on factors that are different on each site. These include the volume of soil involved, the nature of the site and any cost constraints.

The most common machine for digging site trenches is the backactor (Figure 1.10).

Foundations

The foundation of a building is that part which is in direct contact with the ground.

The current Building Regulations require that the foundations of a building shall safely sustain and transmit to the ground the combined dead and superimposed loads in such a way as not to impair the stability or cause damage to any part of the building.

The ground or subsoil on which a building rests is called the natural foundation or subfoundation, and has a definite load-bearing capacity, according to the nature of the soil (Figure 1.11).

Subsoils are of many varieties and may generally be classified as rock, compact gravel or sand, firm clay and firm sandy clay, silty sand and loose clayey sand (Table 1.1).

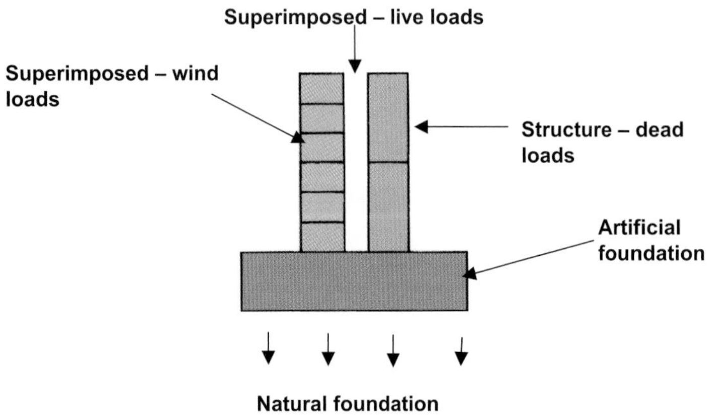

FIGURE 1.11
Loads

Table 1.1 Types of subsoil

Type of subsoil		Condition of subsoil
Class 1	Rock	Sandstone, limestone or firm chalk: requires pneumatic pick or similar appliance to excavate
Class 2	Compact gravel or sand subsoil	Requires a pick for excavation; a 50 mm wood peg; hard to drive more than about 75 mm
Class 3	Clay and sandy clay	Stiff, cannot be moulded in the fingers and requires a pick or pneumatic spade for removal
Class 4	Clay or sandy clay	Firm, can be moulded with the hand but can be excavated with a spade
Class 5	Sand, silty sand and clayey sand	Loose, can be excavated with a spade
Class 6	Silt, clay, sandy clay and silty clay	Soft, can be moulded with the hand and easily excavated
Class 7	Silt, clay, sandy clay and silty clay	Very soft and squeezes through fingers when squashed

Based on information found in Approved Document A in the current Building Regulations

It is possible to erect a wall on rock with little or no preparation, but on all soils it is necessary to place a continuous layer of in situ concrete in the trench, called the building foundation. (The term 'in situ' means cast in place in its permanent position; unlike 'precast' which is made elsewhere, lifted and transported later to the place where it is required for use.) This cast, in situ concrete is made from Portland cement with coarse aggregate, plus sharp sand or ballast, graded from 40 mm to fine sand, and mixed in the proportion of 1:6.

The design of a foundation is an important subject for the architect's consideration, particularly where the building structure is heavy, with possible concentrated loading, and the ground on which the building rests is of poor load-bearing capacity or is affected by other conditions such as seasonal change.

Even buildings of the small domestic and industrial type may require careful site exploration so that a suitable foundation is constructed and the future stability of the building assured.

Where buildings are extremely heavy it may be necessary to construct foundations of a special type, such as piles, where the site consists of deep beds of soft soil overlaying a hard soil, but it is the intention here to deal only with buildings of the smaller domestic and industrial range.

FOUNDATION TYPES

This chapter will deal with the four main types of foundations:

- strip foundations – narrow strip and wide strip foundations
- pad foundations
- raft foundations
- piled foundations.

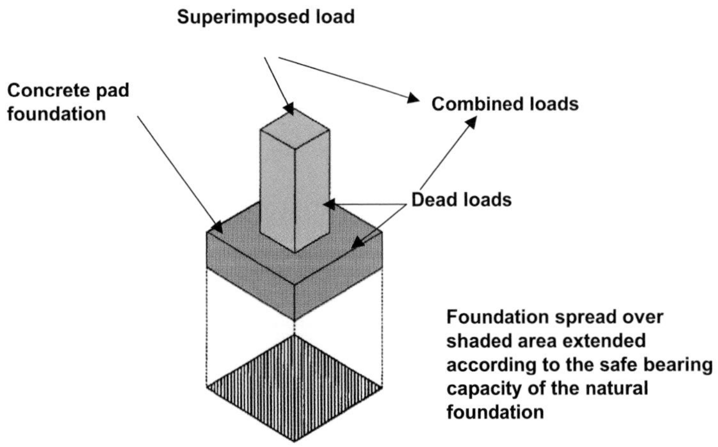

FIGURE 1.12
Transfer of building load onto subsoil under pad foundation

FOUNDATION DESIGN

When a load is placed on soil it is necessary to spread or extend the foundation base to ensure stability. This extended or spread foundation is referred to as a strip foundation in the case of a continuous wall structure, or a pad foundation in the case of an isolated pier.

The important loads to be resisted by a foundation are the dead loads, or the weight of the building structure, and the superimposed loads, or the weight that may be placed on the building structure. The area or spread of the foundation base should be sufficient to resist the downward thrust or bearing pressure of these combined loads (Figure 1.12).

The spread of a pad foundation required for a pillar or isolated pier can be determined by dividing the combined loads by the safe bearing capacity of the soil, i.e.

$$\frac{\text{Combined load of pier}}{\text{Safe bearing capacity of soil per square metre}}$$

or

$$\frac{\text{Total load of pier}}{\text{Safe load on soil}}$$

These calculations are published in the current Building Regulations for the more common strip foundations on subsoils up to a maximum load of 70 k/N (Table 1.2). Any calculations outside this table should be carried out professionally.

The foundation required for a continuous wall structure (Figure 1.13) is obtained by similar methods; in this case a 1 m length of wall is taken for the purpose of calculation.

Having determined the spread of the foundation base it is now necessary to consider the depth. The base, in spreading the load, is subjected to stresses known as tension and punching shear, and must be of sufficient depth to resist them.

Table 1.2 Minimum width of strip foundations

	Type of subsoil	Condition of subsoil	Total load of load-bearing walling not more than: (kN/m) 20	30	40	50	60	70
Class 1	Rock	Sandstone, limestone or firm chalk	In each case equal to the width of the wall					
Class 2	Gravel Sand	Compact Compact	250	300	400	500	600	650
Class 3	Clay Sandy clay	Stiff Stiff	250	300	400	500	600	650
Class 4	Clay Sandy clay	Firm Firm	300	350	450	600	750	850
Class 5	Sand Silty clay Clayey sand	Loose Loose Loose	400	600	Note: If the total load exceeds 30 N/m then this table does not apply to types 5, 6 and 7			
Class 6	Silt Clay Sandy clay Silty clay	Soft Soft Soft Soft	450	650				
Class 7	Silt Clay Sandy clay Silty clay	Very soft Very soft Very soft Very soft	600	850				

Based on information found in Approved Document A in the current Building Regulations

Tension is due to the bending tendency and punching shear to the tendency of the wall or pier structure to punch a hole through the foundation base (Figure 1.14).

The thickness of the concrete foundation should not be less than the projection of the strip either side of the wall, but in no case less than

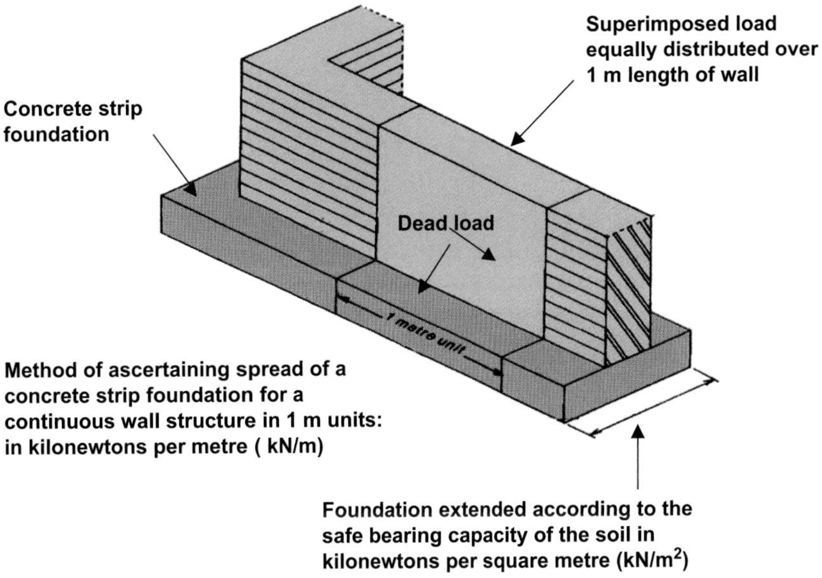

FIGURE 1.13
Loading under strip foundation

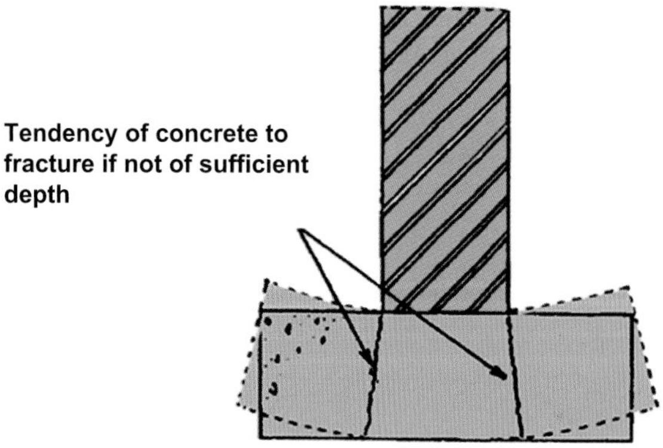

FIGURE 1.14
Potential failure of a strip foundation due to punching shear

150 mm. One method of ascertaining the depth of concrete is shown in Figure 1.15.

Design of simple strip foundations

Using the information given, it is now possible to calculate the spread and depth of simple concrete foundations.

You need to know the following:

- the total load per metre run of wall
- the type of subsoil
- the width of the wall.

Example 1

Design a strip foundation to carry a 275 mm wide cavity wall, if the total load is 60 kN per metre run of wall and the ground is stiff sandy clay (Figure 1.16).

Remember: the wall should sit in the middle of the strip foundation.

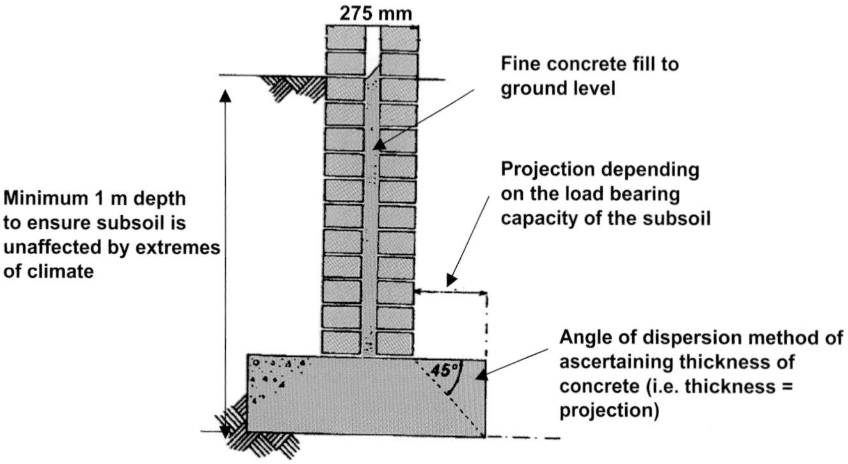

FIGURE 1.15
Establishing thickness of strip foundation concrete

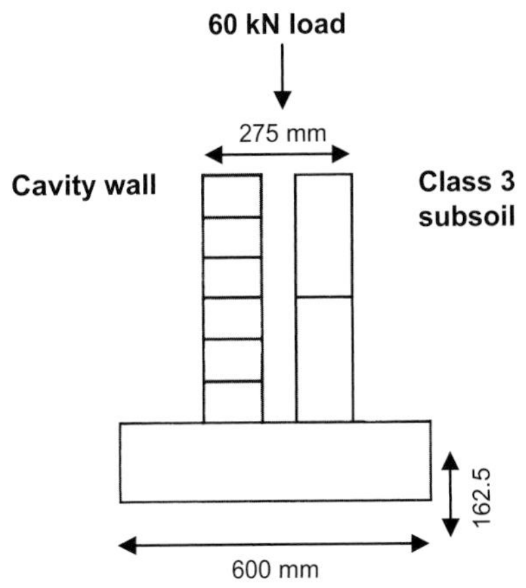

FIGURE 1.16
Designing a strip foundation

The first thing to check out is Table 1.2.

Find the column with the correct loading of 60 kN and trace down the column until you reach the correct class of subsoil – class 3. The two join at 600 mm.

This gives the minimum width of the strip foundation.

The next calculation is to find the projection of the concrete.

The wall is 275 mm wide and the total width of the foundation is 600 mm.

If we deduct 275 from 600 and divide by 2, it will give the projection:

$$\frac{600 - 275}{2} = 162.5 \text{ mm projection}$$

According to the rules for the projections the depth must be at least equal to the projection but in no case less than 150 mm. Therefore the depth of the foundation concrete should be 162.5 mm.

This is the mathematical method of finding the depth; the angle of dispersion method is shown in Figure 1.15.

Example 2
Design a strip foundation to carry a 275 mm wide cavity wall, if the total load is 30 kN per metre run of wall and the ground is sandy gravel (Figure 1.17).

Again the first thing to check out is Table 1.2.

Find the column with the correct loading of 30 kN and trace down the column until you reach the correct class of subsoil – class 2. The two join at 300 mm.

This gives the minimum width of the strip foundation.

The next calculation is to find the projection of the concrete.

The wall is 275 mm wide and the total width of the foundation is 300 mm.

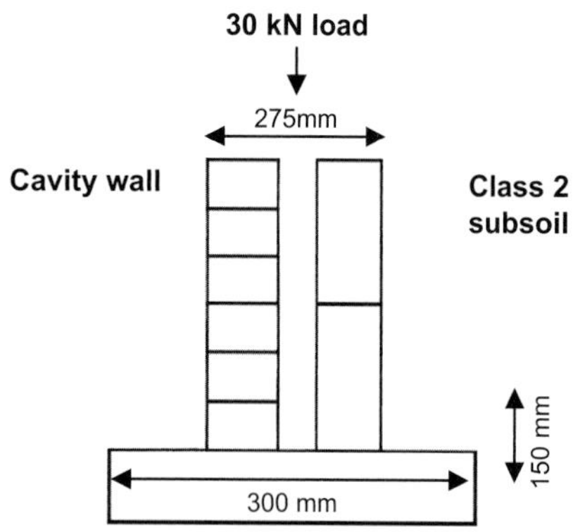

FIGURE 1.17
Designing a strip foundation

If we deduct 275 from 300 and divide by 2, it will give the projection.

$$\frac{300 - 275}{2} = 12.5 \text{ mm projection}$$

According to the rules for the projections the depth must be at least equal to the projection but in no case less than 150 mm. Therefore the depth of the foundation concrete in this example should be 150 mm.

These narrow foundations, although acceptable by the local authority, are practically impossible to build. They are usually extended in width to 150 mm to allow the bricklayer to stand in the trench when foundation walling is being built.

ATMOSPHERIC DEPTH

This is the depth below ground level to which foundations should be taken.

It depends on the type of soil and is the depth at which the subfoundation ceases to be affected by the weather. This is between 600 and 1500 mm, decreasing as the proportion of gravel increases (Figure 1.15).

In buildings of the smaller domestic and industrial type, or where the loads are not excessive, standard methods and rules are applied to the design of the foundations. In general, the resultant construction complies with the Building Regulations, but in all cases the decision as to adequacy rests with the local authority building control office.

OTHER FOUNDATION TYPES

Narrow strip

As in the last example, the working space required to build on top of the concrete strip foundation would make the strip wider than it needs to be to carry the load. In these circumstances, an economical alternative is the narrow strip, or trench fill as it is sometimes known (Figure 1.18).

A high standard of accuracy is required in constructing such a foundation.

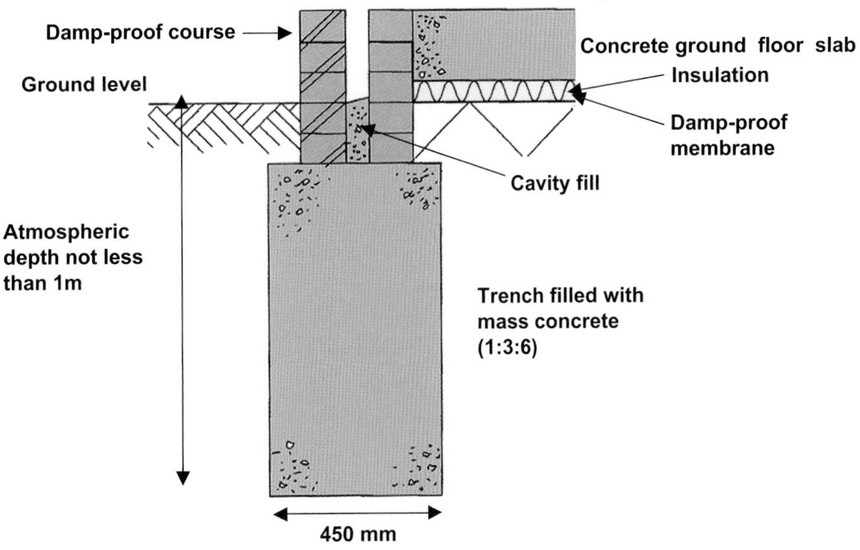

FIGURE 1.18
Narrow strip foundation

A narrow strip is excavated by the mechanical excavator and backfilled with mass concrete up to a level just below the finished ground level. It is cheaper and quicker to fill the trenches with mass concrete than to excavate a wider trench. There is less excavated material to be removed and backfilling is eliminated. A wider trench for a strip foundation requires timbering for safety purposes, foundation concrete to be laid, foundation brickwork constructed and the cavity filled with weak concrete up to ground level. This work could be reduced by using foundation blocks, but would still be more expensive than trench fill.

Wide strip foundations

Where the structural loads are very heavy or the safe bearing capacity of the soil is low, the spread of the foundation base increases. This is normally referred to as a wide strip foundation. It follows that the required depth of the concrete foundation base may be considered excessive; it can be reduced by the introduction of steel reinforcement, but the foundation must always be of sufficient depth to ensure that, in combination with the steel, it will resist the stresses of tension and shear. Figure 1.19 shows a simple example of a reinforced concrete strip foundation.

Pad foundations

Loads are not always evenly distributed along the wall, but may at times be concentrated at various points.

Figure 1.20 shows part of a wall with an attached pier, which may, for instance, bear a roof truss or beam. To obtain even pressure over the soil the foundation would be extended as shown.

For single loads that are transmitted down a column, the most common foundation is a square or rectangular block of concrete of uniform thickness known as a pad foundation (Figure 1.21).

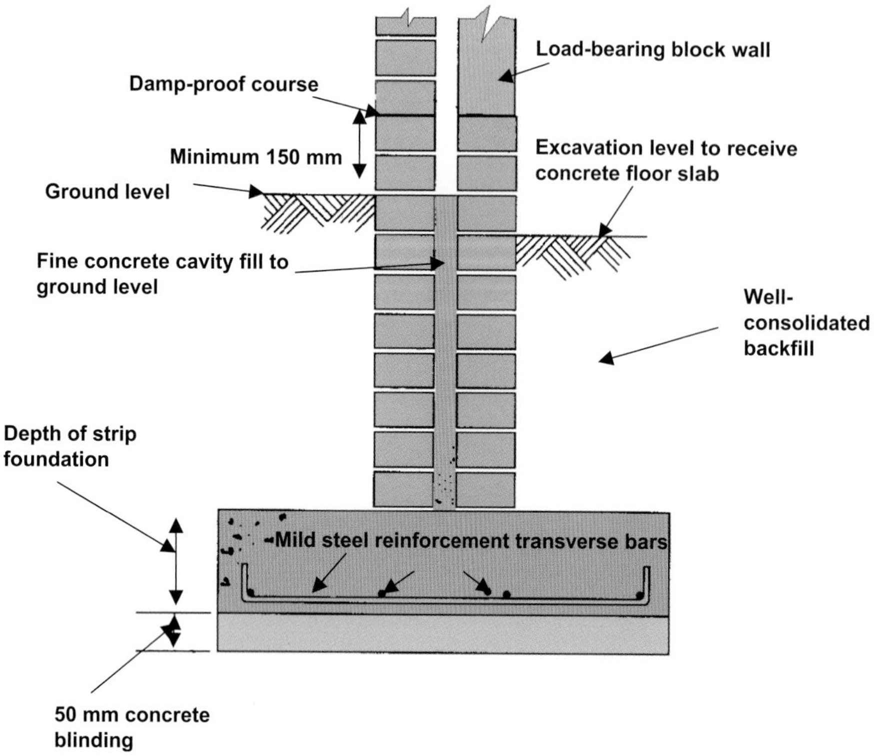

FIGURE 1.19
Reinforced wide strip foundation

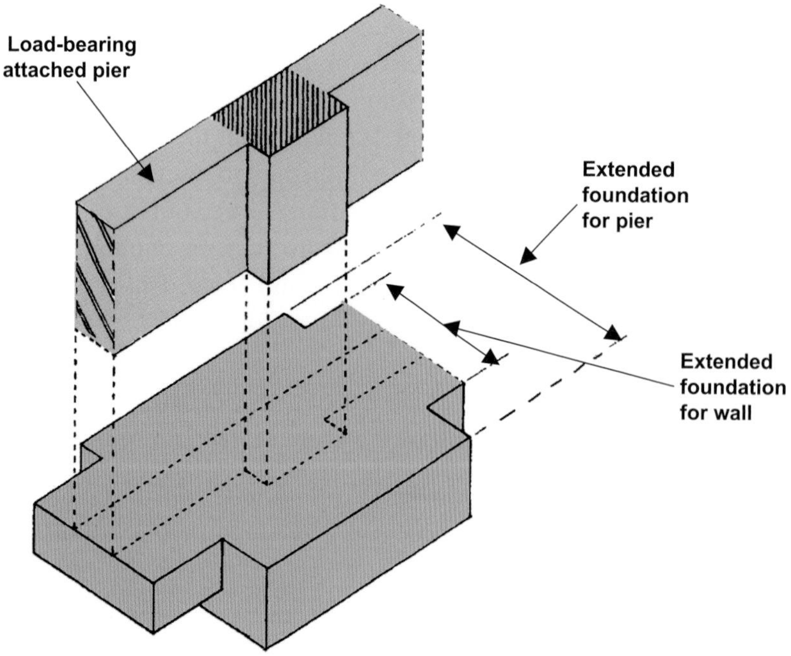

FIGURE 1.20
Extension of strip concrete foundation around attached pier

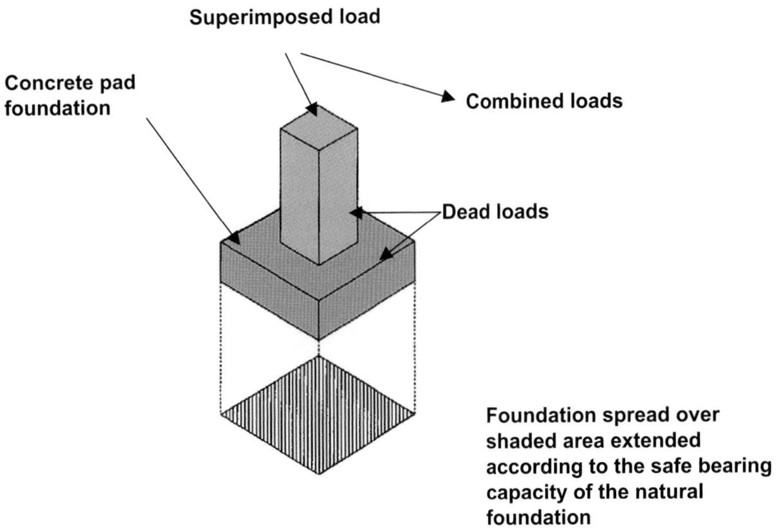

FIGURE 1.21
Pad foundation

It is sometimes more economical to construct a foundation of isolated pads with pillars of brick or concrete, which in turn support concrete ground beams and concrete floor slabs, which then support the walls of the building.

This method of construction avoids total trench excavation, timbering to the trenches and foundation brickwork all the way around the perimeter of the building.

To spread the load over a greater area it is necessary either to make the pad thicker or to use reinforced concrete.

Raft foundations

These foundations consist of a raft of reinforced concrete under the whole of the building.

Raft foundations are often used on poor subsoils for lightly loaded buildings and are considered capable of accommodating small settlements of the subsoil.

The simplest and cheapest form of raft is the thick reinforced concrete raft (Figure 1.22). Its rigidity enables it to minimize the effects of differential settlement.

SHORT BORED PILED FOUNDATIONS

If, instead of spreading the load from the wall over a wide area, it is decided to transfer it to a greater depth, an economical solution is the use of a short bored pile foundation.

Short bored piles are formed by boring circular holes 300 mm in diameter to a depth of about 3 m by means of an auger. This depth is governed by the level of suitable bearing capacity ground. The holes are filled as soon as possible with mass concrete.

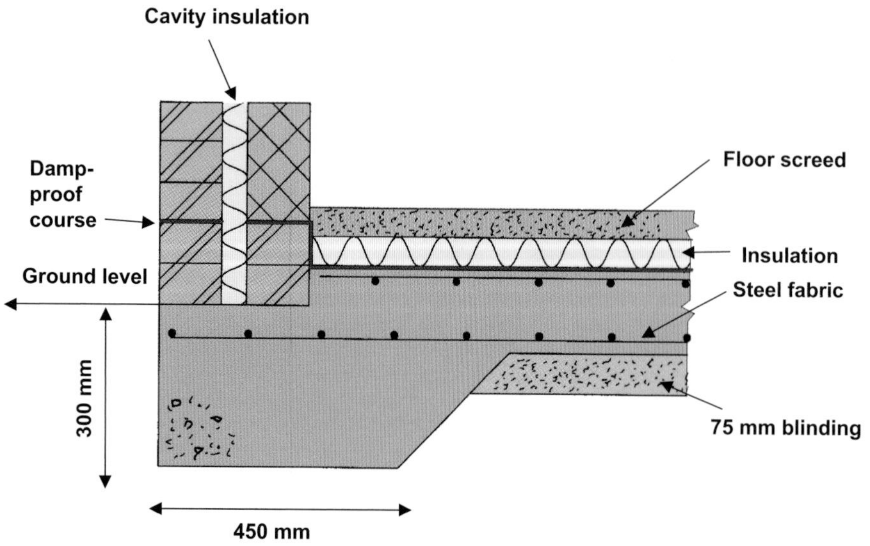

FIGURE 1.22
Raft foundation

The piles are placed at the corners of the building and at intermediate positions along the walls. They support reinforced concrete beams which are cast in place in the ground (Figure 1.23).

BRICK FOOTINGS

Many authorities consider brick footings to be obsolete, mainly because of the quality of concrete used in modern construction. However, the

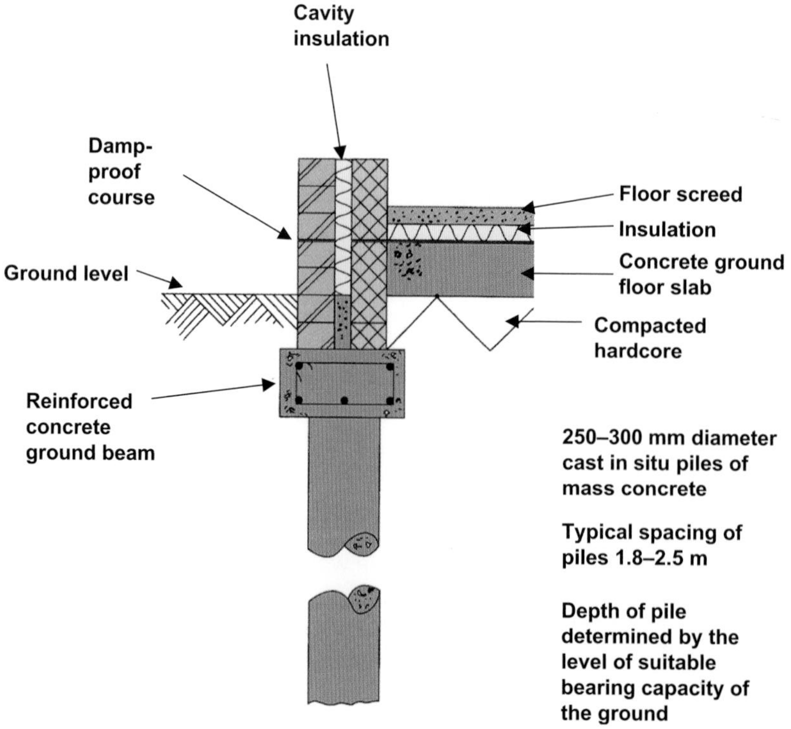

FIGURE 1.23
Piled foundation

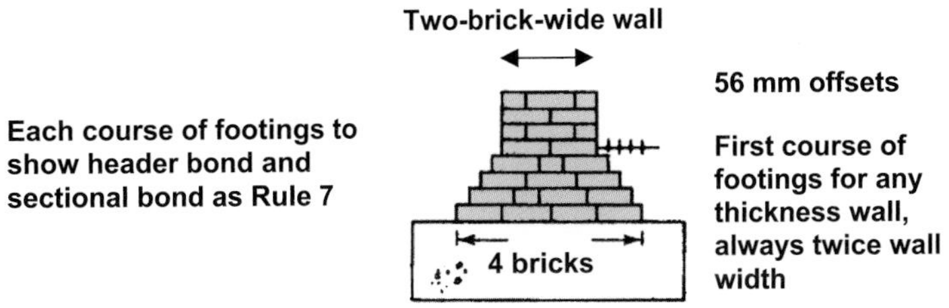

FIGURE 1.24
Cross-section of footing courses to a two-brick wall

apprentice should be aware of the principles involved when this form of construction is adopted and of the necessary bonding arrangements.

The stresses of tension and shear in a foundation base may be eased by the addition of a construction known as a footing. Where brick is used this is achieved by regular offsets at the base of the wall or pier structure. The footings spread the weight of the wall or pier and superimposed loads over the concrete, which in turn distributes the combined loads over the soil.

Thus, the first course of footings is always double the wall width, and the second and each subsequent course of footings is offset 56 mm each side. Every course of footings should be header bond and sectional, following rule 7 of bonding (Figure 1.24).

Strip foundations in Victorian or Edwardian buildings may not be concrete but rammed ‘hoggin’ instead (that is, ‘as dug’ gravel, before clay has been washed out).

STEPPED FOUNDATIONS

These are constructed on sloping sites to ensure a horizontal bearing on the natural foundations (Figure 1.25). Note the overlap of concrete at the change of levels. The height of the steps must be maintained at not more than 450 mm; where this dimension is exceeded, special precautions may be necessary.

Walls

The standard form of construction for the external walls of domestic and communal brick buildings is called cavity walling. This means that the bricklayer builds the two separate leaves or ‘skins’ of brick masonry (a general term indicating brickwork and/or blockwork), with a 50–75 mm wide space between them.

The outer skin is usually 102.5 mm thick face brickwork, but may be constructed from facing quality blocks.

The inner skin is usually 100 mm thick common blocks that are later plastered to receive internal decoration.

Both skins of brick masonry are joined together with a regular pattern of corrosion-resistant ties, so that they behave as a single wall.

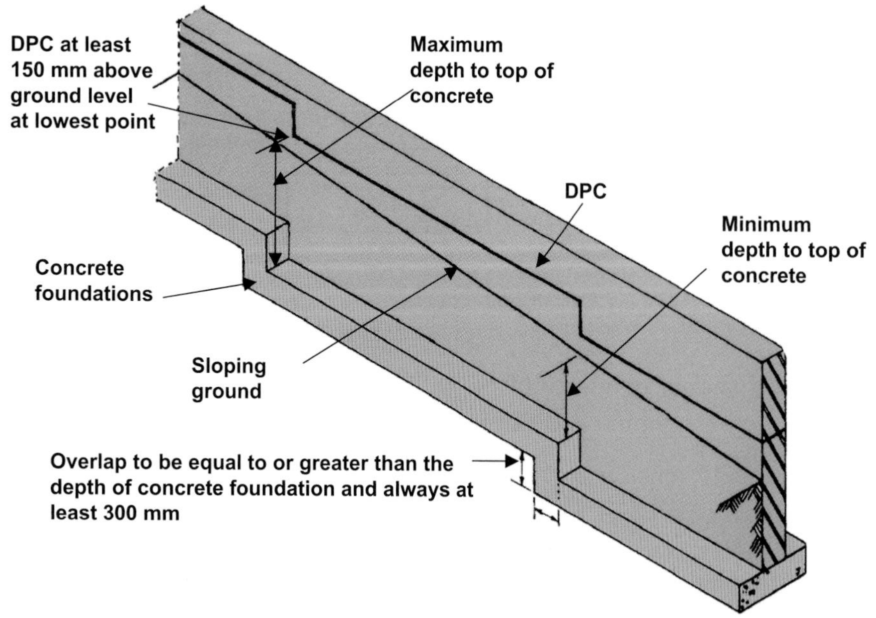

FIGURE 1.25
Stepped foundations on sloping sites

The main objectives of cavity walls are to prevent rain penetration and provide a greater degree of thermal insulation than solid wall construction.

New developments in the design of cavity walls have allowed for greater insulation values to prevent heat loss through the walls.

The external walls of domestic and communal buildings can also be designed in a framed construction. Frames can be timber, steel or concrete.

One of the great advantages of framed buildings is that the main construction can be almost finished and watertight before the external skin of brickwork is added. This is often considered as a faster method of building and allows other trades to continue while the external cladding is being fixed.

When designing to meet the current Building Regulations for heat loss through walls, it is important to consider which method of construction will be used, as the cost can vary according to which method is used.

The current Building Regulations require that external walls have a maximum U value of $0.35\,W/m^2\,K$. The U value denotes the thermal transmittance, which is the rate of heat transfer through a wall from air to air.

The most common method of insulating cavity walls, whether they are brick and block or framed, is by inserting slabs of polyurethane foam, expanded polystyrene or bonded glass fibre into the cavity space.

Figures 1.26 and 1.27 show insulation placed in the cavity of a brick and block wall and insulation to timber-framed construction, respectively.

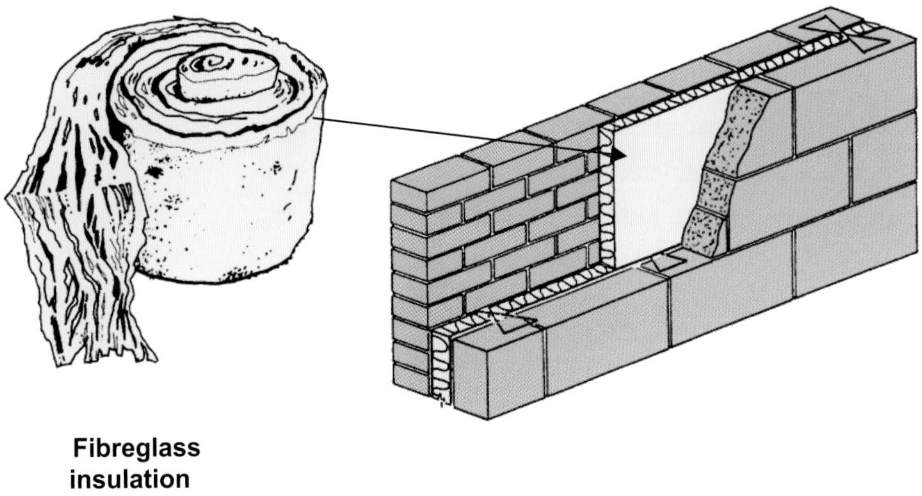

FIGURE 1.26
Insulation to cavity wall

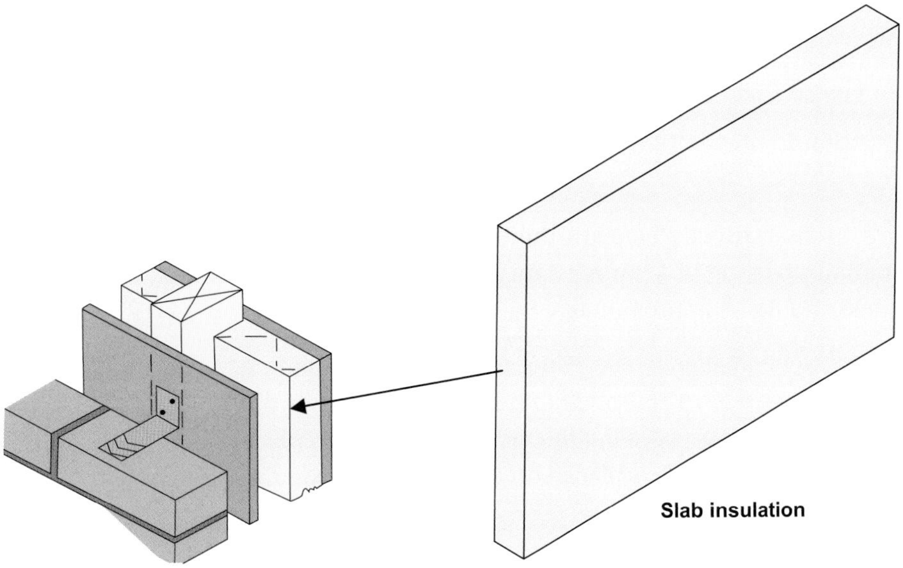

FIGURE 1.27
Insulation to timber frame cladding

Floors

GROUND FLOORS

There are two types of ground floor construction found in both domestic and communal buildings:

- solid
- hollow.

The functions of ground floors are to provide a level surface with sufficient strength to support all the loads, and to prevent dampness entering the building and heat loss from the building.

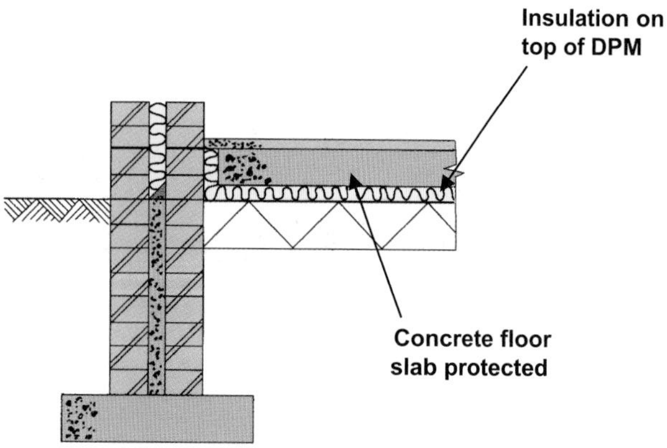

FIGURE 1.28
Damp-proofing a solid ground floor slab with the DPM under the concrete floor slab

SOLID GROUND FLOORS

Solid ground floors consist of a compacted hardcore with sand blinding on top. This protects the damp-proof membrane (DPM). Insulation is laid on the membrane on which a slab of concrete is laid.

Details are shown in Figure 1.28.

The horizontal damp-proof course (DPC) level must never be above the floor level.

Brick rubble or hardcore laid directly beneath the concrete floor will not only prevent settlement, but, being of a porous nature, also help to prevent dampness. Its thickness should be approximately that of the concrete floor.

Solid ground floors have to be insulated, according to the current Building Regulations, to provide resistance to unacceptable heat loss through the floor. This can be achieved in various ways but the most common is to place the insulation on top of the DPM, which is placed on a blinding layer on top of the hardcore.

This method of construction protects the concrete floor slab from any moisture or harmful salts. The only problem with this method is the risk of damage to the DPM and insulation when laying the concrete floor slab.

An alternative method is shown in Figure 1.29. This method is easier but the concrete floor slab is not protected against the ingress of moisture or harmful salts.

Materials suitable for floor insulation are dense resin bonded mineral or glass fibre slabs, and polystyrene and cork slabs. These should be placed above the DPM and turned up at the edges of the floor slab to prevent heat loss through the external wall.

Remember

The DPM *must* be continuous with the horizontal DPC in the external and internal walls.

HOLLOW GROUND FLOORS

Suspended timber floors need to have a well-ventilated space beneath the floor construction to prevent the moisture content of the timber rising

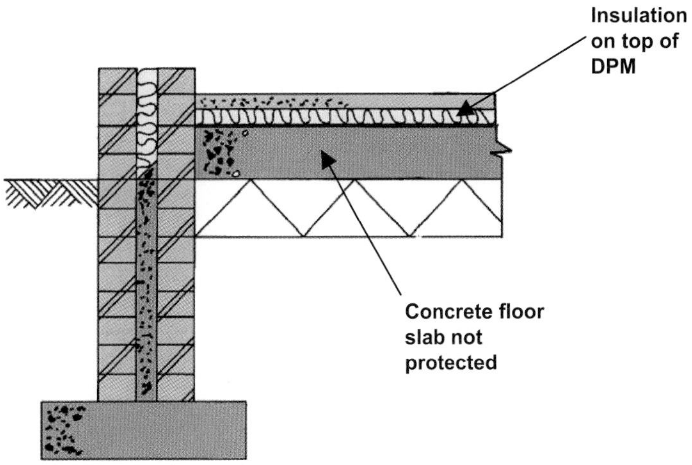

FIGURE 1.29
An alternative method with the DPM on top of the concrete floor slab

above an unacceptable level, i.e. above 20 per cent, which would create the conditions for possible fungal attack.

Hardcore and oversite concrete are still required for these floors, but in this case the concrete does not require a waterproof membrane.

Hollow sleeper walls are constructed on the oversite concrete to receive the wall plate, which in turns supports the floor joists (Figure 1.30). A horizontal DPC is inserted under the wall plate to resist rising damp.

Underfloor ventilation against dry rot

Having placed the oversite, the floor must be supported, so as to allow free passage of air to prevent the floor timbers from rotting. This is achieved by building honeycomb sleeper walls on the oversite concrete and air bricks built into the external wall.

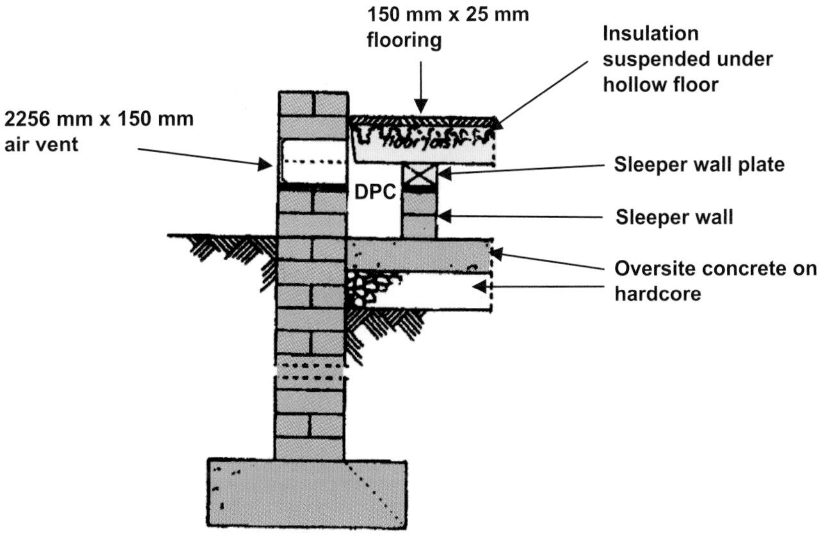

FIGURE 1.30
Solid wall: location of sleeper wall

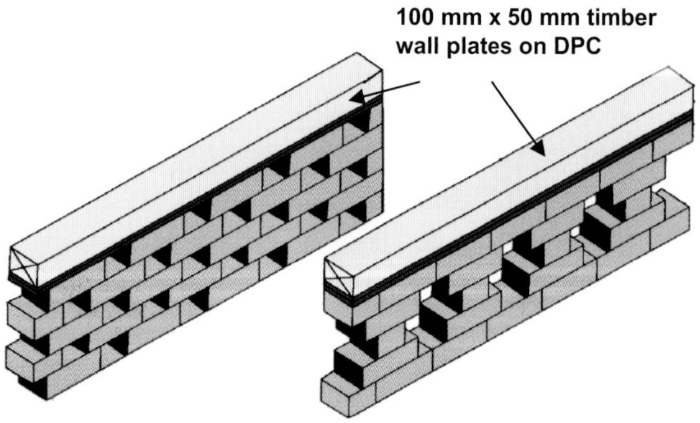

FIGURE 1.31
Types of sleeper wall construction

Two types are illustrated in Figure 1.31.

The sleeper walls nearest to the main wall are positioned as in Figure 1.30; this provides freedom of movement when the bricklayer is building the wall and reduces the possibility of mortar droppings collecting between the main wall and the sleeper wall.

Figure 1.32 shows a construction that will provide efficient support and very important ventilation to the floor timbers.

Damp, stagnant air provides ideal growing conditions for a fungus commonly called dry rot to take root in constructional timbers. Although needing moisture to grow, the fungus roots destroy the timber, leaving it split, broken and 'dry', thereby explaining the name.

Hollow floors have to be insulated to prevent cold air rising into the building (Figures 1.30 and 1.32).

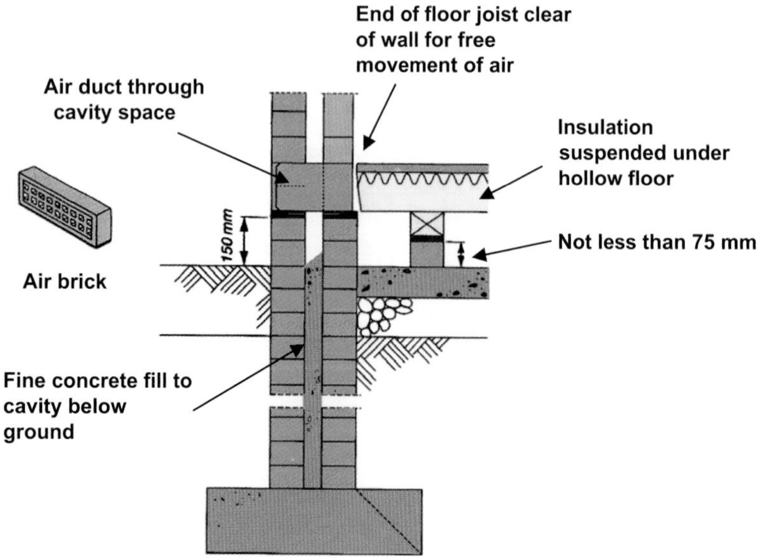

FIGURE 1.32
Cavity wall, showing underfloor ventilation

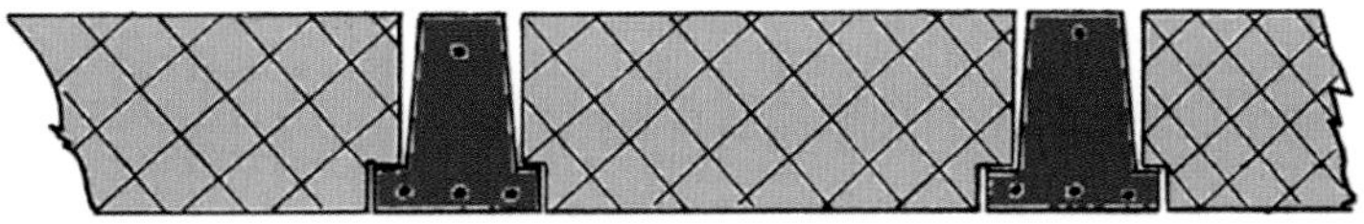

FIGURE 1.33
Precast concrete ground floors: concrete blocks resting on prestressed precast concrete beams

Precast beam and pots

Prestressed precast concrete beams were first used in commercial ground floor construction and are now common in domestic construction. They have been designed as an alternative to suspended timber floors. The same preparation is required as for suspended timber floors (Figure 1.33).

When precast floors have been constructed they provide a safer work area than suspended timber floors at both ground and first floor level.

Suspended timber floors require boarding out before work can proceed, whereas the complete floor area of precast floors is ready to work on and can take full loads.

SUSPENDED UPPER FLOORS

Timber floors are the most common type in domestic buildings and precast concrete floors are more common in commercial buildings.

A suspended timber floor consists of a number of joists either built into the walls (Figure 1.34) or resting on joist hangers (Figure 1.35).

Suspended floors that belong to the same dwelling unit are not required to be insulated, but when they are between two properties then insulation for sound, fire and heat loss is required.

Galvanized steel restraint straps (Figure 1.35) are used to restrict movement. Herringbone strutting is fixed between the joists to prevent movement and twisting (Figure 1.36).

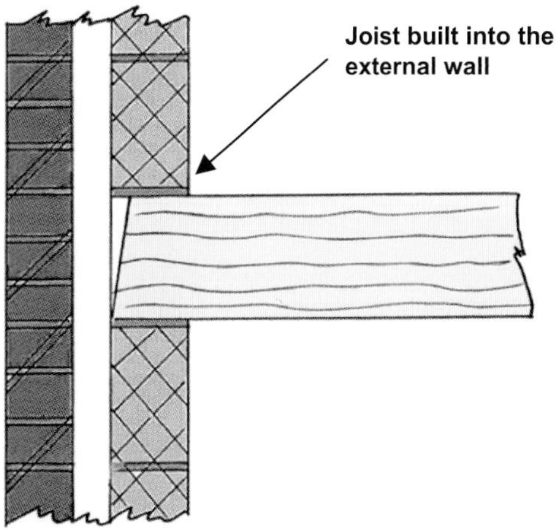

FIGURE 1.34
Suspended timber upper floors

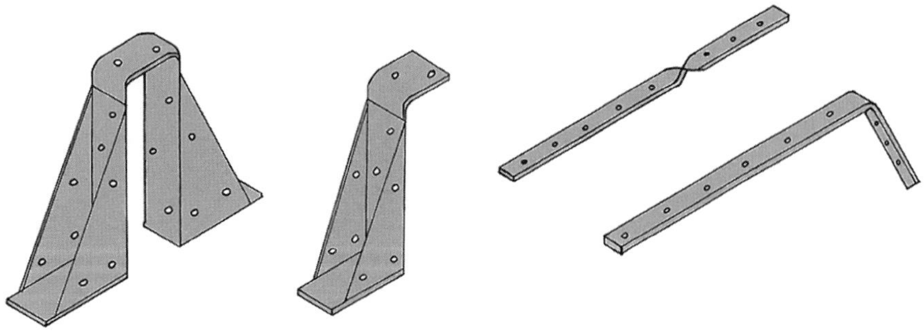

FIGURE 1.35
Double and single joist hangers and restraint straps

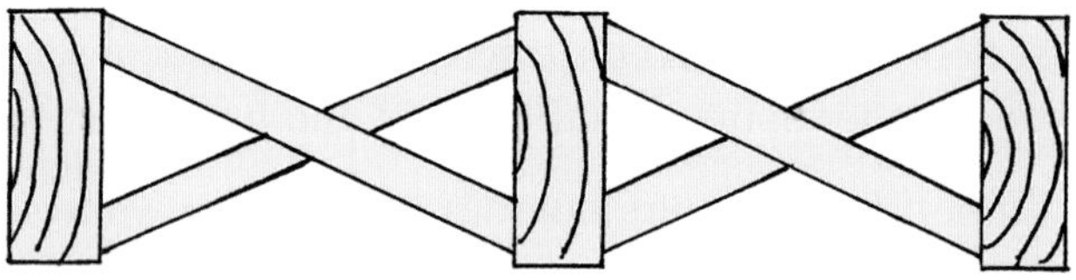

FIGURE 1.36
Herringbone strutting

Concrete suspended floors

Concrete beam and block floors are the most common in domestic buildings, but there are many other designs for commercial buildings where greater loading is required.

Concrete suspended floors fall into two categories: those that require formwork and those that do not.

Precast beams with concrete blocks or clay pots do not require formwork and can be used immediately the floor is completed. A typical design is shown in Figure 1.37.

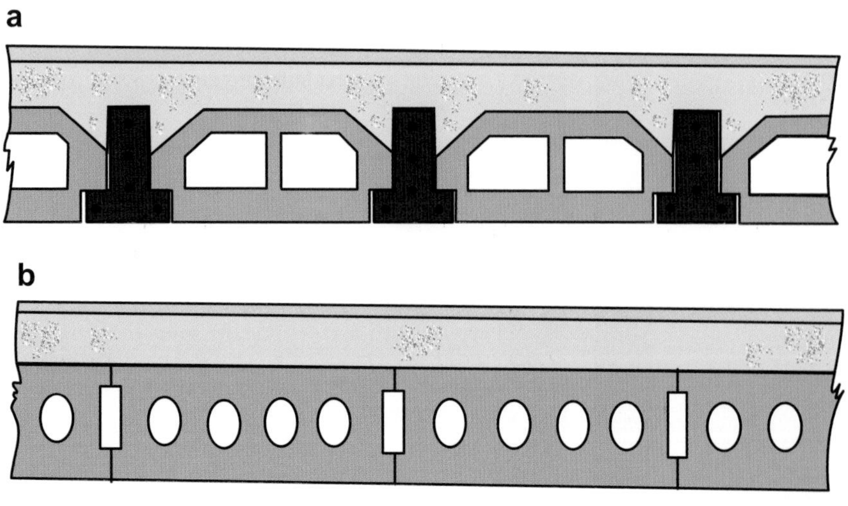

FIGURE 1.37
Precast first floors: (a) precast concrete beam and pot composite floor; (b) precast concrete cored slabs

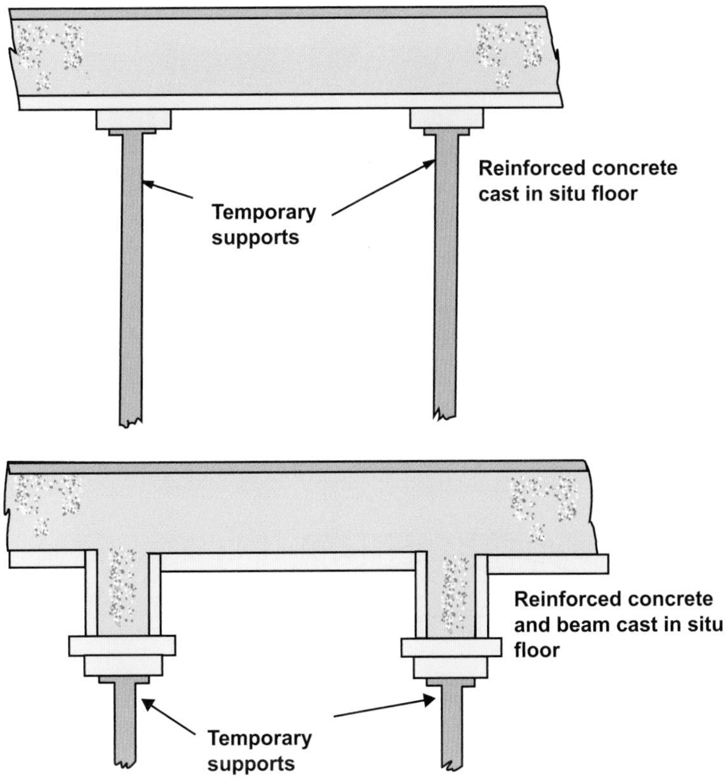

FIGURE 1.38
Cast in situ concrete floors

Cast in situ concrete floors require temporary formwork while the concrete sets (Figure 1.38).

There are many methods available but all consist of a slab of reinforced concrete up to a spans of 5 m. When larger spans are required, beams are incorporated into the design.

There are several disadvantages of cast in situ floors over precast floors, the main one being the time involved. With precast floors, there is no waiting for the concrete to cure and the working area is available immediately.

FLOOR FINISHES

There are numerous floor finishes available, some of which form an integral part of the construction, such as floor boarding and screeds.

Screeds

These are used to give the concrete floor, at either ground floor or upper floor level, a durable finish to receive the final floor finish, such as tiles.

Cement and fine aggregate screed are laid up to a thickness of 75 mm directly on top of the concrete floor (Figure 1.39).

Floor boarding

Timber suspended floors can be covered with either floor boarding or sheets (Figure 1.40).

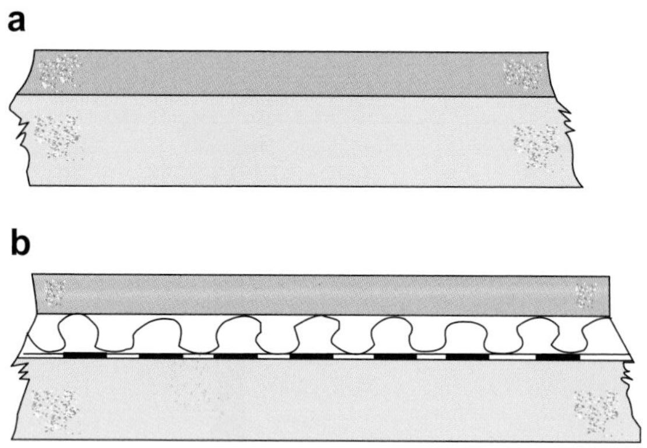

FIGURE 1.39
Screeds to concrete floors: (a) sand and cement screed laid on concrete slab; (b) sand and cement screed laid on insulation, DPM and concrete slab

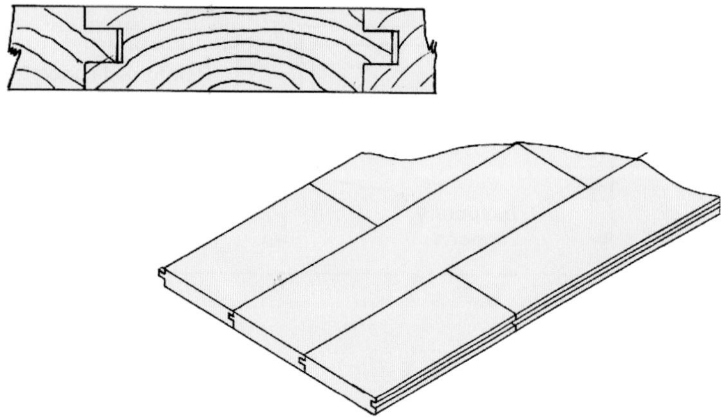

FIGURE 1.40
Covering to timber joists: tongue-and-groove floor boarding

Tongue-and-groove boarding is laid at right angles to the floor joists and should be fastened using floor brads.

Sheeting is available in 600 mm × 19 mm sheets and is laid at right angles to the floor joists in a chequerboard design. Sheeting should be screwed into position to prevent movement.

FLOATING FLOORS

When quiet floors are required, such as for a library, floating floors can be constructed (Figure 1.41). They require the final floor to be separated from the floor structure to prevent the passage of sound up through the structure and into the room above.

Windows

The primary function of a window is to provide a means for admission of natural daylight to the interior of the building. A window can also serve as a means of providing the necessary ventilation of dwellings.

Window frames can be made from a variety of materials, such as timber, plastic or metal, e.g. aluminium.

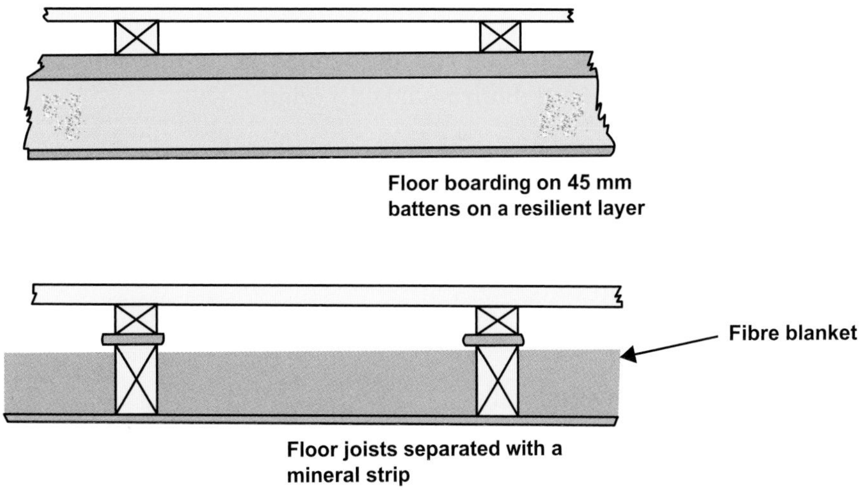

FIGURE 1.41
Floating floors

They can be designed to open in various ways by arranging for the sashes to slide, pivot or swing.

DOUBLE GLAZING

The window when fixed forms part of the external envelope, so it has to function in the same way, for weather exclusion, sound and thermal insulation, and fire resistance. A typical design is shown in Figure 1.42.

Glass is a material that is poor in preventing the transfer of heat and sound and the spread of fire. Many windows are now designed with double-glazed units to achieve the levels of heat transfer prevention required by the current Building Regulations.

Double glazing refers to the use of two sheets of glass in a window or door with air trapped between. This serves as thermal insulation to reduce the transfer of heat through the window.

The building can have a larger area of glass if it is double glazed than if only single glazing were used.

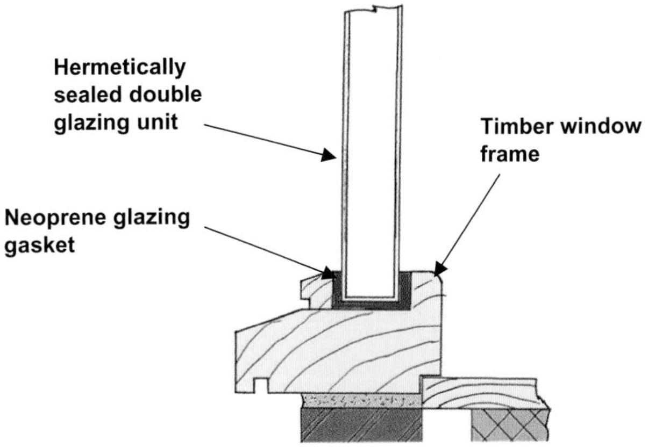

FIGURE 1.42
Double-glazed unit set in timber frame

Superstructure

The superstructure of a building can be defined as all elements of the building above ground level (horizontal DPC) (Figure 1.43).

The substructure of a building can be defined as all elements of the building below ground level.

The main elements of superstructure include:

- framed structures
- claddings
- external and internal walls
- ground and upper floors
- windows and doors
- roofs.

EXTERNAL ENVELOPE

The external envelope of the building should be designed to:

- give protection from the elements
- enclose inside space
- provide a suitable internal environment.

The external envelope is generally a combination of walls and roofs, with openings for light and access.

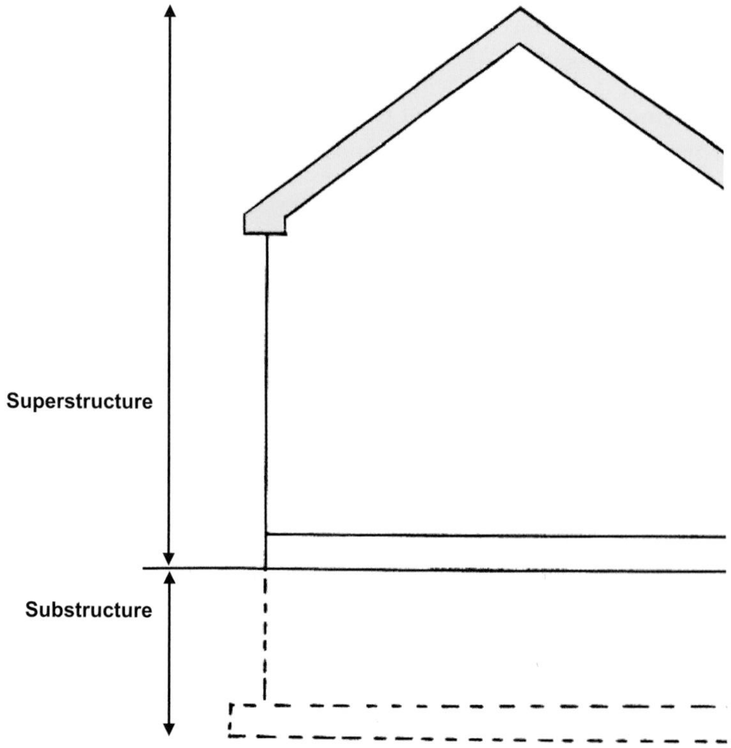

FIGURE 1.43
Substructure and superstructure

Performance

The functions of the external envelope are as follows:

- strength and stability
- weather exclusion
- thermal insulation
- sound insulation
- durability
- fire resistance
- appearance
- access and egress.

Strength and stability

The envelope has to be strong enough to carry the loads which may be imposed upon it without excessive deformation and to transfer these safely to the structural frame or foundation.

The loading may occur in the form of:

- the weight of the external envelope
- any loads transferred to the envelope from the internal construction, such as floors, walls and fixtures
- any external loads caused by the wind, snow and rain.

Weather exclusion

Keeping out the weather is an extremely important consideration if the internal environment is to remain constant.

The problem depends greatly on the position of the building, its height and the amount of exposure to the weather.

Thermal insulation

Heat will flow from a high temperature inside the building to a low temperature on the outside.

During the winter there will be a constant flow of heat from the building, which means that extra heating will be required to counteract the heat loss through the external envelope.

This rate of heat loss can be reduced by insulating the external envelope, thereby saving on the amount of fuel required and the size of the heating system.

In the summer this process can be reversed and the external envelope should be able to resist the heat flow into the building, thereby providing a cooler internal environment.

Sound insulation

Noise is defined as 'unwanted sound'.

Extremely loud noises can be detrimental to the health of the users of the building, while small noises can be irritating and could cause loss of concentration.

The transfer of sound can take place in one of two ways:

Air-borne sound is sound that has been produced as waves in the air by a vibrating object such as a loudspeaker or a musical instrument

Structure-borne sound, also known as impact sound, occurs when a structural unit is set vibrating owing to impact such as by footsteps, vibrating machinery, etc., and the sound is transmitted through the structure of the building.

When sound waves hit the envelope surface, some will be reflected and some absorbed, while others will be apparently transmitted through the envelope.

Since the majority of people do not have control over the sounds outside the building, such as road traffic and aircraft noise, the external envelope must reduce that noise to an acceptable level.

Durability

The material from which the external envelope is constructed must have sufficient resistance to the damaging effects of the climate – in the form of erosion, atmospheric pollution, frost, rain, and chemical and solar degradation – to provide a building that will be relatively free of maintenance for its anticipated useful life.

Fire resistance

Many million pounds worth of damage occur to buildings every year through fire. It is, therefore, necessary to keep the effects of fire to a minimum. These effects can be minimized if the building itself can contain the fire until the arrival of fire-fighting appliance.

The prime consideration, in the event of an outbreak of fire, is the safety of human life, hence the external envelope must be strong enough to withstand the ravages of fire for such time as it will take to evacuate the building of all the occupants and allow them to reach places of safety some distance away.

The external envelope should be able to contain the fire and to prevent the spread of fire to other buildings in the vicinity.

Appearance

This is the aesthetic requirement which should be considered at the design stage.

The external envelope should be compatible with others in the vicinity, by either blending in with them or providing a contrast which is not too stark.

Remember that in the rural areas the envelope may be required to blend in with the landscape so that it becomes as inconspicuous as possible.

Access and egress

The external envelope has to have openings formed to allow the users in and out of the building, and allow for light to enter.

Internal walls

These usually act as partitions or dividers for the internal space of the building, although occasionally they also support the upper floor and roof loads.

The most common types of internal walls (Figure 1.44) are:

- block partitions
- timber frame or stud partitions
- brick partitions
- demountable partitions.

Block partitions

These can be built in either concrete or clay blocks (Figure 1.45).

The materials available are:

- crushed slag
- foamed slag
- aerated concrete
- sintered pulverized fly ash
- clinker
- expanded clays and shales.

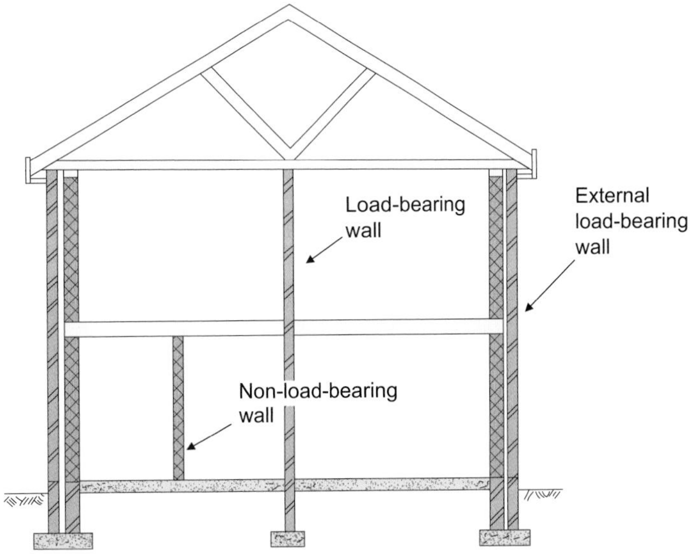

FIGURE 1.44
Types of wall

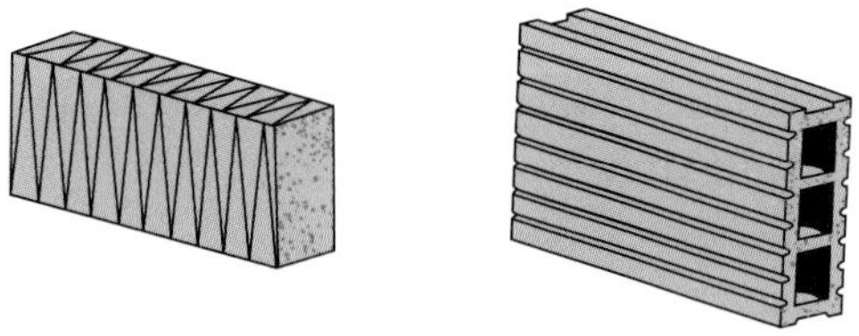

FIGURE 1.45
Types of lightweight internal block

These are one of the strongest of the internal walls and will support the loads subjected by the floors, roofs and superimposed loadings from the inhabitants and furniture.

They are built exactly as external walls, without the finish to the joints as external walls require.

The mortar is usually weaker than the material being used, to allow any shrinkage to take place in the joint rather than in the block.

Timber frame or stud partitions

Timber studding is a non-load-bearing partition.

These partitions must be designed and constructed to carry their own weight and any fittings that may be attached to them.

They must be strong enough to withstand impact loadings on their faces and also any vibrations set up by doors being closed or slammed.

Details are shown in Figure 1.46.

Timber studded walls are lighter than brick or block partitions, but are less efficient as sound or thermal insulators. They are easy to construct and provide a good fixing background. Because of their lightness they are suitable for building off suspended timber floors.

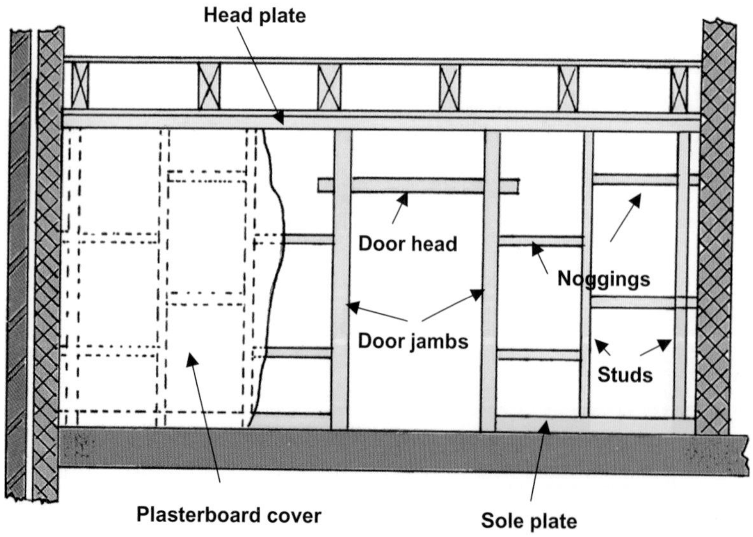

FIGURE 1.46
Timber studding

The basic principle is to construct a simple framed grid of timber to which a dry lining such as plywood or plasterboard can be attached. The lining material will determine the spacings of the uprights or studs to save undue wastage in cutting the boards to terminate on the centre of the studs. Openings can be formed and linings inserted to receive the door.

There are also numerous preformed partition units available.

One particular unit consists of plasterboard bonded on either side of a strong cardboard cellular core to form rigid panels. These are fixed to ceiling and wall battens and supported on a timber wall sole plate.

These units can also be provided with suitable facing for direct decoration or even with self-finish applied.

Brick and block partitions

Brick internal walls provide the strongest method but are not very common in modern buildings.

Their advantage over other materials is they provide very high fire resistance.

The walls are built in exactly the same way as for external walls without the joint finish.

When building blockwork partitions, lightweight blocks are used.

Demountable partitions

A wide range of lightweight, non-load-bearing and demountable partitions is available (Figure 1.47). The construction is usually of metal, wood or plastic sections or trim with various infill panels in a variety of materials.

Their main advantages are their versatility and non-permanent nature. They can be used to alter the floor layout of existing properties with the minimum of disruption.

Roofs

The roof is that part of the external envelope which spans the external walls at their highest level and, being part of the envelope, it must fulfil the same functions.

BASIC ROOF FORMS

These may be either flat or pitched, as shown in Figure 1.48, along with some basic roof terminology.

Performance

The performance requirements for roofs will vary depending on:

- the type of building
- the span to be covered
- the nature and magnitude of the loadings.

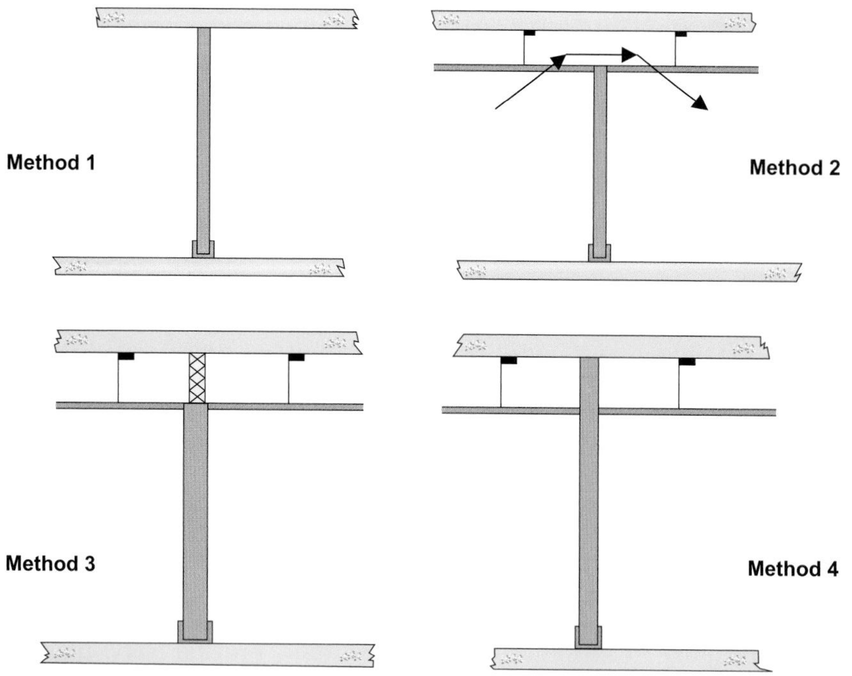

FIGURE 1.47
Types of demountable partition. Method 1: easily demounted and re-erected, provides good sound insulation; method 2: demountable partition between suspended ceiling and floor; sound and fire could pass over partition; method 3: demountable partition from floor to ceiling; fire and sound baffle inserted above partition; method 4: demountable partition through suspended ceiling

For some types of building there will only be a few solutions.

Weather exclusion

It is an essential requirement of the roof to be watertight.

Structural strength and stability

A roof structure must be capable of bearing its own dead weight together with the superimposed loading of the wind, rain and snow.

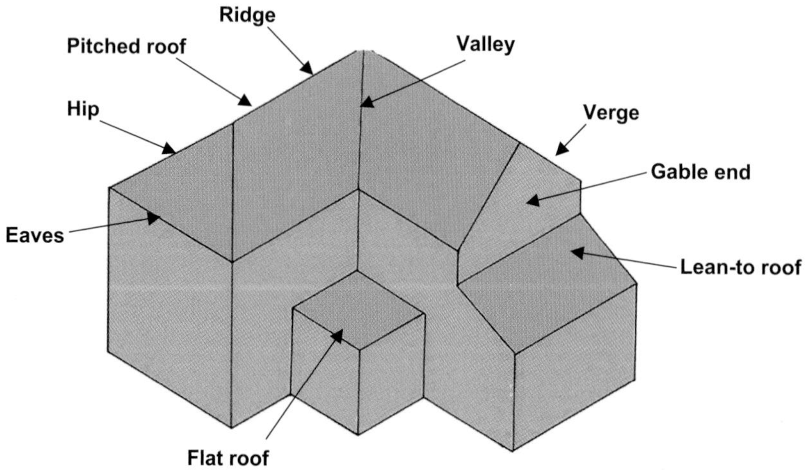

FIGURE 1.48
Basic roof terminology

Drainage

When the rain falls on the roof it must be directed down to the ground via the gutter system.

Durability

The roof is the most inconvenient part of the building to reach. Therefore, the roof must resist atmospheric pollution, frost, etc.

Thermal insulation

The greatest heat loss can be through an uninsulated roof.

Sound insulation

Unless it is a special building. e.g. a concert hall, the roof is not normally required to be soundproof.

Fire resistance

The fire resistance of a roof largely depends on its proximity to other buildings, and the primary function is to prevent the spread of fire to and from buildings via the roof.

Appearance

The roof should harmonize with its surroundings; this depends on the shape of the roof together with the shape and colour of the roof coverings.

Ventilation of roof space

A high level of insulation can cause problems with condensation in the roof space. This could lead to rotting of the roof timbers and cause structural problems.

To overcome this problem, ventilation should be provided by introducing a 10 mm air space along the eaves of the roof where the roof has a pitch of over 15 degrees. In roofs with less than 15 degree pitch a 25 mm air space is required.

Details are shown in Figure 1.49.

FLAT ROOFS

Flat roofs on domestic buildings are mainly of timber construction, with concrete flat roofs being used for industrial buildings.

Roofs with less than 10 degree slope are termed flat roofs. Most flat roofs have sufficient slope to prevent water standing on the roof surface.

Typical details of flat roofs are shown in Figure 1.50.

PITCHED ROOFS

Pitched roofs include the monopitch roof, which slopes in one direction only; the ridge roof, which has two slopes from a central centre ridge; and the lean-to, which abuts a wall. Details are shown in Figure 1.51.

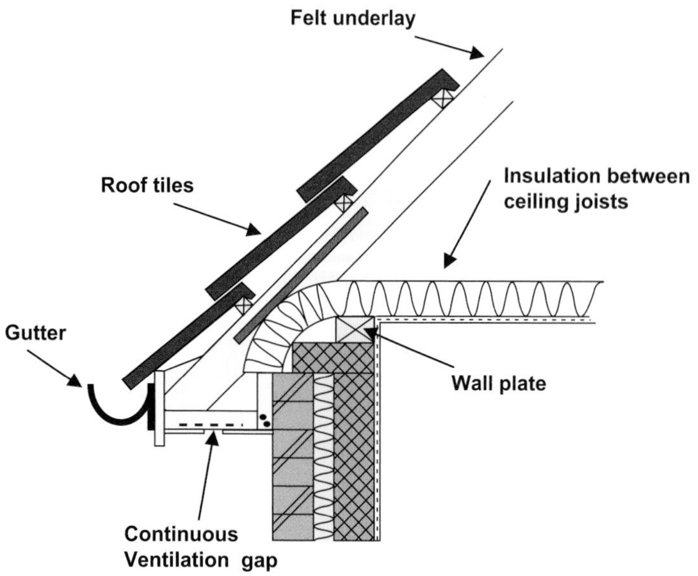

FIGURE 1.49
Eaves detail

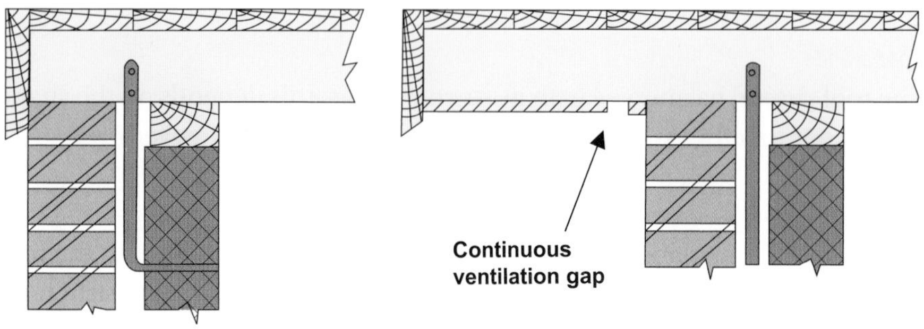

FIGURE 1.50
Types of flat roofs showing restraint fixings

ROOF COVERINGS

The covering for pitched roofs starts with the underfelt. This is a watertight membrane that is laid with overlapping joints.

Roof battens are then nailed onto the roof joists to provide a bearing for the slates or tiles. The tiles or slates are nailed to the battens to secure them in position.

The main covering for flat roofs is a watertight membrane such as bituminous felt with sealed watertight joints.

There are numerous types of tiles available, including double and single lap tiles. All are available with matching ridge, eave and valley tiles.

Details of roof coverings are shown in Figure 1.52.

Sustainable methods and materials

It is well known that the construction industry has a greater effect on the external environment than any other industry.

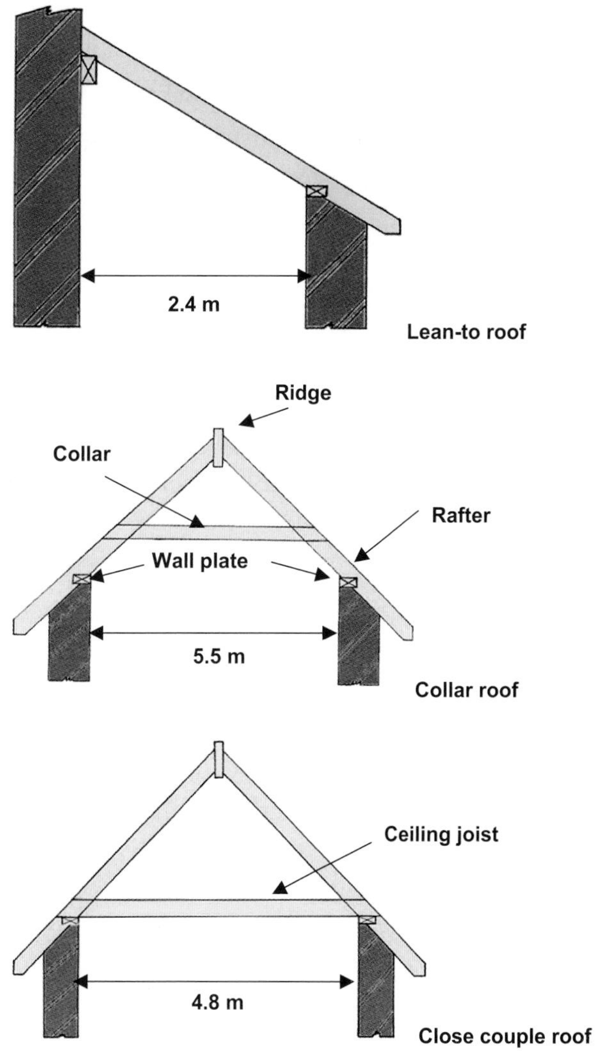

FIGURE 1.51
Types of pitched roofs

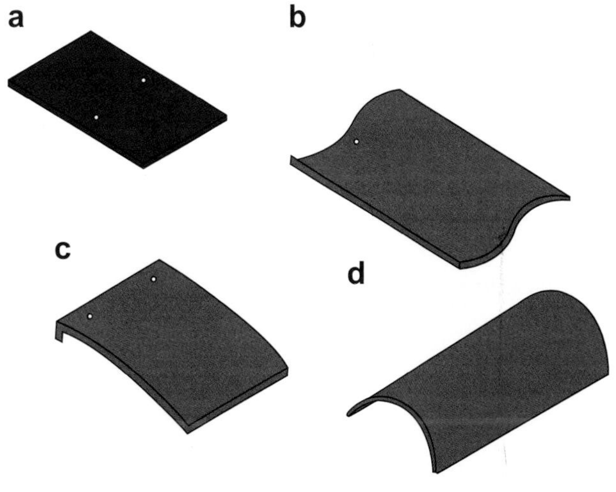

FIGURE 1.52
Types of roof coverings: (a) slate; (b) pantile; (c) rosemary tile; (d) semi-circular ridge tile

Over the years a tremendous amount of construction projects are finished in the UK, from houses to factories, bridges to roads, and small shops to large shopping centres, and these are increasing every year.

This adds work to another area of the industry that has the responsibility to maintain, refurbish and even rebuild where applicable.

All this work in the construction industry has an adverse effect on the external environment as the landscape has been irreversibly changed.

Buildings produce nearly 30 per cent of the UK's carbon emissions, which is almost twice the carbon emissions from transport.

The way the buildings are constructed, insulated, heated and ventilated, and the type of fuel used, all contribute to their carbon emissions. The industry is committed to protecting and enhancing the environment and tackling climate change, which is one of the most serious threats facing us today.

The government has a long-term plan to reduce carbon emissions by 60 per cent by 2050. If the construction industry is to meet this target then the design of our homes and the external environment clearly needs to change. Sustainable building must be the way forward.

A building that is sustainable should be constructed using locally sustainable materials, i.e. materials that can be used with minimal adverse effects on the environment, and which are produced locally, reducing the need to travel.

New national standards

The Code for Sustainable Homes has been developed to enable a change in sustainable building practice for new homes. It has been prepared by the government in close working consultation with the Building Research Establishment (BRE) and Construction Industry Research and Information Association (CIRIA).

The Code is intended as a single national standard to guide the construction industry in the design and construction of sustainable homes.

The Code will complement the system of energy performance certificates (EPCs), which were introduced in June 2007 under the Energy Performance of Buildings Directive (EPBD). EPCs were introduced to improve the energy efficiency of buildings. If you are buying or selling a home you need a certificate by law. Since October 2008 EPCs have been required whenever a building is built, sold or rented out.

The certificate provides A to G ratings for the building, with A being the most energy efficient and G the least, with the average up to now being D. The rating chart is shown in Figure 1.53.

Accredited energy assessors produce EPCs alongside an associated report, which suggests improvements to make a building more energy efficient.

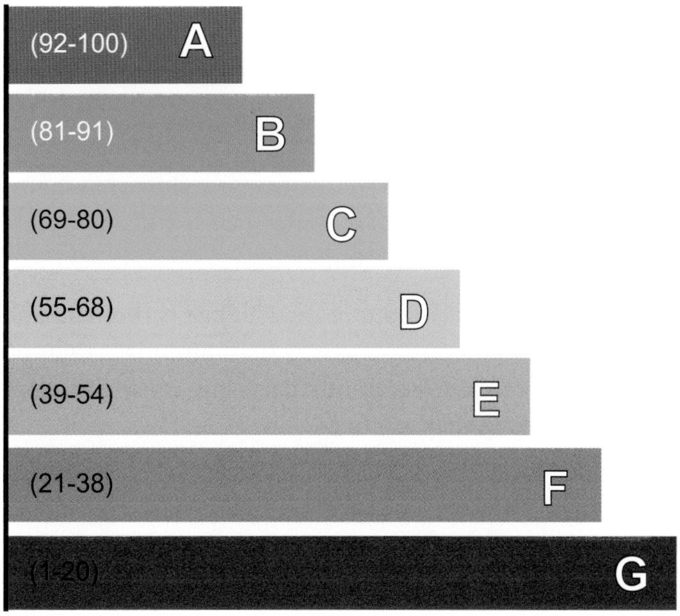

FIGURE 1.53
Energy rating chart

The Code for Sustainable Homes

The Code measures the sustainability of a new home against nine categories of sustainable design, rating the 'whole home' as a complete package.

The design categories included in the Code are:

- energy/carbon dioxide (CO_2)
- pollution
- water
- health and well-being
- materials
- management
- surface water run-off
- ecology
- waste.

The Code uses a 1 to 6 star rating system to communicate the overall sustainability performance of a new home. The Code sets minimum standards for energy and water use at each level and, within England, replaces the Eco Homes scheme, developed by the BRE.

The Code provides valuable information to home buyers, and offers builders a tool with which to differentiate themselves in sustainability terms.

Climate change

The construction industry, as well as the general public, has a duty to act on climate change. Evidence shows that it is real and happening already, and that urgent action is needed now.

At the same time we need to build more houses, so it is vital that we ensure these homes are built in a way that minimizes the use of energy and reduces their harmful carbon dioxide emissions.

Building sustainable homes is about more than just carbon dioxide. We also need to build and use our homes in a way that minimizes their other environmental impacts, such as the water they use, the waste they generate and the materials from which they are built.

The future

The Code is closely linked to Building Regulations, which are the minimum building standards required by law.

Minimum standards for Code compliance have been set above the requirements of Building Regulations.

It is intended that the Code will signal the future direction of Building Regulations in relation to carbon emissions from, and energy use in homes, providing greater regulatory certainty for the house-building industry.

Materials

It is important to encourage the use of materials with lower environmental impacts over their life cycle.

The production, use and disposal of building materials accounts for a significant amount of energy and resources being used.

Many firms now specialize in recycled or reused materials.

Reused materials are materials that can be extracted from the demolition of buildings and used again without further processing, or with minor processing that does not alter the nature of the material (e.g. cleaning or cutting). Examples of reused materials are bricks and roofing tiles.

Recycled materials are materials extracted from the demolition of buildings that require significant processing (which alters the nature of the material) before they can be used again (e.g. crushing, grinding, reprocessing). A good example is hardcore.

To be classed as a sustainable material it is important that the supply chain is stated. This covers all of the major aspects of processing and extraction involved in the supply chain for the end product (Table 1.3).

SOURCE OF MATERIALS

All construction operations involve alterations and modifications to the environment. This activity consumes a great deal of energy.

Note

Recycled materials are not required to demonstrate a supply chain.

Table 1.3 Supply chain processes

Materials	Manufacture	Processing
Brick and clay products	Product manufacture	Clay extraction
Composites	Composite manufacture	Glass fibre production, polymer production
Concrete (including blocks, tiles, precast and in situ)	Concrete production or concrete product manufacture	Cement production, aggregate extraction
Glass	Glass production	Sand extraction, soda ash production or extraction
Plastics	Plastic product manufacture	Plastic polymer production
Metals (steel, aluminium, etc.)	Metal product manufacture, e.g. metal production, e.g. for steel: cladding production, steel section production	EAF/BOF process, for aluminium ingot production, copper ingot or cathode production
Plasterboard	Plasterboard manufacture	Gypsum extraction
Stone	Stone product manufacture	Stone extraction
Timber	Only if it is 100% recycled	Only if it is 100% recycled

Selecting materials for the construction process will also raise questions about the amount of damage caused in these processes.

Nearly all material production involves extracting, processing and transporting, all of which use energy. As you can see from Table 1.3, many of the materials used in the production of construction materials require the extraction of raw materials.

EXTRACTION

Materials that are extracted from the ground include clays, gravels and sands, which are involved in the manufacture of bricks, blocks, plaster, metals and glass.

These require specialized machinery and produce noise, dust and waste. They are all extracted from a quarry which leaves great depressions in the ground. All the raw materials require some amount of processing before they are transported to where they are required.

Clays are extracted and used in the manufacture of bricks. The energy used in the making of clay bricks can be greatly reduced if materials with a high fuel content are used, such as shale.

Some companies involved in the extraction of raw materials attempt to return the ground to its original state by infilling or landscaping. In many cases it is possible to improve the ecological value of the site. However, this requires careful consideration of the existing and neighbouring features, in addition to careful selection of plant species and habitats. This is an area of specialist expertise and requires input from experts at both planning and detailed design stages.

Timber

This material can have a minimal effect on the environment. Trees in their natural environment perform an important function, as each tree consumes on average 9 kg of carbon dioxide and gives out approximately 7 kg of oxygen each year.

Timber is obtained by cutting down forests but unlike extraction, which leaves huge holes in the ground, the forest be replanted for future generations.

The growing cycle is approximately 40 years, and therefore timber is classed as a renewable resource.

Insulation

Reducing the amount of energy used from fossil fuels is the most important factor in promoting sustainability.

Insulation has the greatest potential for reducing carbon dioxide emissions. The energy conserved through insulating buildings and reducing heat loss far outweighs the energy used in the manufacture of the various insulation materials.

Only when a building achieves a 'low heat' standard does insulation's embodied energy become significant.

The durability of insulation affects its performance, e.g. settlement, physical degradation, vapour permeability and air movement.

Careful detailing is needed to avoid the risk of moisture ingress into the insulation.

Most insulation materials differ in their capacity to reduce heat flow. This means that different materials require different thicknesses to achieve the same effect. These differences need to be considered when planning wall cavity widths. It is therefore necessary to keep up to date with the Building Regulations.

Good insulation performance requires careful site supervision.

Insulation only provides reduction of heat loss through the building fabric. Equally important is the energy lost through ventilation and glazing.

TYPES OF INSULATION

There are numerous types of insulation available for the construction designer to choose from, the most common being polystyrene and mineral wool slabs

The many alternatives on the market have varying properties and can be divided into three groups:

- insulation derived from organic sources
- insulation derived from naturally occurring minerals
- insulation derived from fossilized vegetation.

It is important when selecting the type of insulation that consideration is given to the function it has to serve and the type of construction in which it has to serve. Very few insulation materials are capable of all functions.

Insulating materials obtained from organic sources should be chosen when sustainability is required. These include:

- wood fibreboard – recyclable, renewable, biodegradable and safe to use
- strawboard – recycled, renewable and biodegradable
- cellulose batts – recyclable if kept dry, biodegradable and safe to use.

Insulation materials from naturally occurring materials include:

- glass mineral and wool batts – recyclable and reusable
- foamed fibreglass – reclaimable
- exfoliated vermiculite – reclaimable and safe to install.

Insulation materials from fossilized vegetation include:

- expanded polystyrene – recyclable and reclaimable
- polyurethane board – reclaimable and very low conductivity
- extruded polystyrene board – reclaimable and recycled.

Functions of building materials

WATER

If water is allowed to penetrate the building a great deal of damage can be caused.

Once a building has become damp problems such as efflorescence, mould growth, corrosion of metals and damage to all finishes may occur

Most building materials admit water by capillarity. Many building materials contain voids (small air pockets) and water is drawn into these voids by capillarity.

Once water is inside a material it can cause damage to the material itself or to other materials when it tries to get out. When the water inside a facing brick freezes, it expands and can push off the face of the brick.

Timber has natural pores inside and can easily admit moisture if unprotected.

Timber that is allowed to get wet can suffer from wet rot. Wet rot is found in very damp conditions, especially cellars, basements and under ground floors. It occurs when the moisture content of the timber is very high. The affected wood breaks down into cubes.

Dry rot is found in damp conditions. It is more prevalent in older buildings and has a fusty smell. Dry rot is brown in colour and leaves the affected wood dry and friable.

If metal is allowed constantly to become wet and dry this can cause corrosion of metal. Some metals undergo destructive corrosion in the atmosphere. When this occurs, a new substance is formed that does not have the same properties as the original metal. Iron is a prime example of a metal that corrodes in the presence of oxygen and water, e.g. moist air. The process by which iron corrodes is known as rusting.

FROST

Frost can cause damage to certain building materials that are porous and allow moisture to penetrate.

Most bricks are frost resistant. Frost damage in bricks can cause the face of the brick to be pushed away. This is known as spalling.

Frost does not usually cause damage to old concrete, but it can affect new concrete if it is not protected.

Timber and metal are virtually unaffected by frost attack.

CHEMICALS

One of the most common attacks on building materials is sulphate attack. This occurs in cement mortars.

Sulphate attack will only occur in damp conditions and could cause severe damage to concrete or brickwork.

When using concrete and bricks in ground that contains salts it is essential to use sulphate-resistant cement.

HEAT AND FIRE

The effects of heat and fire on a building vary according to the type of building.

Brick and blocks have good fire resistance.

Timber is combustible and emits smoke and toxic gases. Most timbers have class 3 spread of flame. Surfaces char in fire at a rate of approximately 1 mm per minute. However, unburnt timber retains its strength and there is negligible expansion in length, so that typical members survive longer in fires than equivalent unprotected steel.

Thermoplastics soften at high temperatures and all plastics are combustible.

The spread of flame over the surfaces of some plastics is high, and burning plastics generally produce large volumes of smoke – the worst menace to life in a fire – and toxic gases, mainly carbon monoxide.

Metal can expand in very hot fires and can cause the building to move, causing further damage to the structure. Therefore, all steel beams and columns should be fire protected.

Smoke can cause damage to a building and water used by the fire brigade can also cause lasting damage.

DETERIORATION

Nearly all building materials can be treated against deterioration.

Most materials can simply be painted with a layer of moisture-resistant paint. This will need replacing periodically according to the type used.

To stop the corrosion of iron and steel, the metal surface must be protected from moist air, by coating or electrolysis:

- The surface of a metal can be protected by coating with paint, grease, bitumen and plastic.
- The surface of the metal can be coated electrolytically with a thin layer of another metal which does not corrode in the atmosphere, e.g. zinc, chromium or tin. A zinc coating will protect iron even if the surface becomes scratched. A galvanic cell is formed and the zinc corrodes, being higher in the reactive series.

Building sites: future design considerations

Construction sites are responsible for significant impacts, especially at a local level. These arise from disturbance, pollution and waste.

Impacts such as energy and water use are also significant, although minor in relation to the overall impacts of the building.

POLLUTION

Construction has the potential for major pollution, largely through pollution to air (through dust emission) and to water (via watercourses and ground water).

ECOLOGY

Wherever homes are constructed, there is a risk that however environmentally benign the building or development, it may present a threat to local ecology or areas of natural beauty.

Damage can be minimized either by selecting a site of low ecological value or by developing a site in a way that protects the most important ecological features.

House building need not reduce the ecological value of the site; it may enhance it in many cases. There will always be some temporary disturbance to the local ecology, but wildlife will return once construction is complete, providing an appropriate habitat is provided.

Whilst it may be an attractive option to build on and revitalize a previously derelict site, care must be exercised if it has been derelict for some time. The site may be inhabited by rare, protected or locally important species and, therefore have high, but hidden, ecological value.

PROTECTION OF TREES

There should be a physical barrier to prevent damage to existing trees. Trees should also be protected from direct impact and from damage to the roots.

Physical barriers should be provided to prevent damage to existing hedges and natural areas. If such areas are remote from site works or storage, construction activity should be prevented in their vicinity.

BUILDING FOOTPRINT

Land available for development is becoming increasingly expensive. The use of greenfield sites is already being limited and developers are likely to experience hostility from the local community.

To make best use of the available land and other resources, including materials and energy, it is important to ensure effective use of the building footprint by maximizing the usable space.

ENERGY

Energy management on site has been a key focus for the Construction Confederation, which has published specific guidance to optimize this.

Monitoring and reporting at site level are the key factors in raising awareness of the impacts of energy consumption.

One of the biggest energy losses on construction sites is through the wastage of materials. Therefore, it is important to encourage construction sites to be managed in an environmentally, socially considerate and accountable manner.

Internal energy efficiency

New buildings are being designed with energy-saving appliances, and the design of the building should improve insulation, ventilation and prevention of heat loss.

Energy production

As the world's resources of fossil fuels disappear there is a need to try other methods of energy production, such as:

- wind turbines
- small-scale hydroelectric
- fuel cells
- solar panels.

Multiple-choice questions

Self-assessment

This section of the book is designed to allow you to check your level of knowledge. The section consists of revision questions for this chapter. The questions are all multiple choice and have four possible answers. The answers are to be found at the end of the book.

The main type of multiple-choice question will be the four-option multiple-choice question. This will consist of a question or statement, known as the stem, followed by a choice of four different answers, called the responses. Only one of these responses is the correct answer; the others are incorrect and are known as distracters.

You should attempt to answer the questions by choosing either (a), (b), (c) or (d).

Example

The person employed by the local authority to ensure that the Building Regulations are observed is called the:

(a) clerk of works
(b) building control officer
(c) council inspector
(d) safety officer

The correct answer is the building control officer, and therefore (b) would be the correct response.

The construction industry

Question 1 Identify the type of building shown:

(a) industrial
(b) commercial
(c) domestic
(d) residential

Question 2 Identify the type of structure shown:

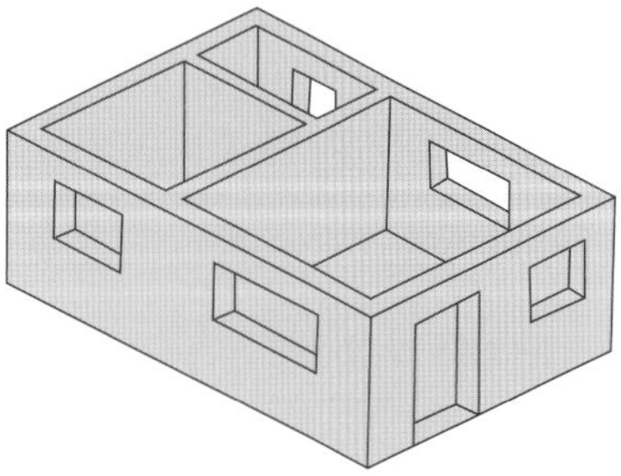

(a) frame
(b) cellular
(c) cross-wall
(d) hollow

Question 3 Identify the type of foundation shown:

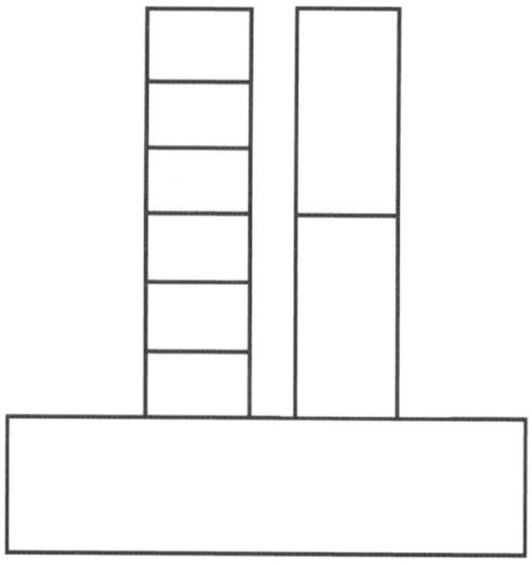

(a) pad
(b) strip
(c) pile and beam
(d) raft

Question 4 What type of wall is built on oversite concrete to support timber ground floors?

(a) supporting wall
(b) joist wall
(c) hollow wall
(d) sleeper wall

Question 5 Which of the following best describes sustainable building?

(a) building to last
(b) building with recycled materials
(c) building with reclaimed materials
(d) building with materials with low environmental impact

Question 6 Where is stone extracted from?

(a) quarries
(b) holes
(c) pits
(d) mines

Question 7 The approximate growing cycle for a tree is:

(a) 20 years
(b) 30 years
(c) 40 years
(d) 50 years

Question 8 Identify the type of floor shown:

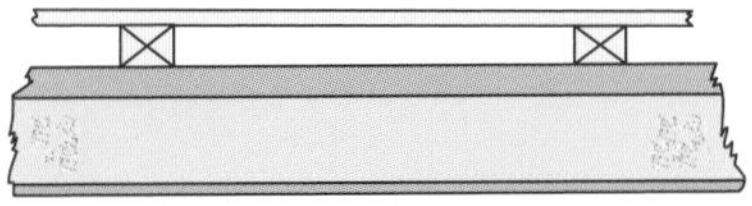

(a) solid floor
(b) hollow floor
(c) floating floor
(d) insulated floor

CHAPTER 2

Health and Safety in the Construction Industry

This chapter will cover the following NVQ and Diploma units:

- NVQ VR01
- CC 1001K

This chapter is about:

- Awareness of relevant current statutory requirements and official guidance
- Personal responsibilities relating to workplace safety, wearing appropriate personal protective equipment and compliance with warning/safety signs
- Personal behaviour in the workplace
- Security in the workplace
- Relationships

The following NVQ performance criteria will be covered:

- Identification of hazards
- Workplace safety
- Security arrangements
- Emergency procedures

The following Diploma outcomes will be covered:

- Know the health and safety regulations, roles and responsibilities
- Accident, first aid and emergency procedures
- Identify hazards
- Health and hygiene
- Safe handling of materials
- Working platforms
- Electricity
- Personal protective equipment
- Emergency procedures
- Signs and notices

Safety legislation

The construction industry is often involved in very difficult and often hazardous sites. It is therefore very important that the new recruit is aware of these dangers and that there are various regulations in place to control and reduce these possible hazards.

Prevention of hazards in the workplace

Hazards within a workplace can occur because of several circumstances. There may be faults in equipment, tools, stored substances, dangerously stacked materials, materials obstructing safe access, or simply a lack of site safety.

The health and safety of employees at their workplace and any other persons at risk through work activities are covered through various Acts of legislation and regulations.

Legislation

Legislation means that the following regulations have all been passed by parliament. Each of the following states the duties and responsibilities of both the employer and the employee. Ignoring the following could lead not only to injuries but also to a fine or imprisonment.

- The Health and Safety at Work Act 1974
- The Control of Substances Hazardous to Health Regulations 2002 (COSHH)
- The Noise at Work Regulations 2005
- Work at Height Regulations 2005
- Reporting of Injuries, Diseases and Dangerous Occurrences Regulations 1995 (RIDDOR)
- The Personal Protective Equipment at Work Regulations 1992
- The Fire Precautions (Workplace) Regulations 1997
- Provision of the Use of Work Equipment Regulations 1998 (PUWER)
- The Electricity at Work Regulations 1989.

The main health and safety legislation applicable to building sites and workshops is covered by the Health and Safety at Work Act 1974.

HEALTH AND SAFETY AT WORK ACT 1974

The HASAWA is applicable to all types of construction sites and includes all employers, employees, self-employed, subcontractors and suppliers.

The four main objectives of the HASAWA are:

- To secure the health, safety and welfare of all persons at work.
- To protect the general public from risks to health and safety arising from out of work activities.

- To control the use, handling, storage and transportation of explosives and highly flammable substances.
- To control the release of noxious or offensive substances into the atmosphere.

These objectives can only be achieved by involving everyone in health and safety matters.

Employers should have a health and safety policy if they employ five or more people; the policy should be in writing.

The self-employed should, so far as reasonably practicable, ensure their own health and safety and make sure that their work does not put other workers or members of the public at risk. Employees must co-operate with their employer on health and safety matters and not do anything that puts them or others at risk.

Employees should be trained and clearly instructed in their duties.

Outline of the Act

The Act itself is very complex and is an extensive document with numerous parts and sections.

Employers' and management duties:

1. Provide and maintain a safe working environment.
2. Ensure safe access to and from the workplace.
3. Provide and maintain safe machines, equipment and methods of work.
4. Ensure the safe handling, transport and storage of all machinery, equipment and materials.
5. Provide their employees with the necessary information, instruction, training and supervision to ensure safe working.
6. Prepare, issue to employees and update as required a written statement of the firm's safety policy.
7. Involve trade union safety representatives (where appointed) with all matters concerning the development, promotion and maintenance of health and safety requirements.

Employees' duties:

1. Take care at all times and ensure that they do not put themselves, their workmates or any other person at risk by their actions.
2. Co-operate with their employers to enable them to fulfil the employer's health and safety duties.
3. Use the equipment and safeguards provided by the employers.
4. Never misuse or interfere with anything provided for health and safety.

General legal requirements

All construction and demolition work is subject to the Health and Safety at Work Act and to certain provisions of the Factories Act.

The following information sets out some of the basic legal requirements that may apply to you.

CONTROL OF SUBSTANCES HAZARDOUS TO HEALTH (COSHH)

The Control of Substances Hazardous to Health 2002 (COSHH) must be consulted when dealing with the handling, moving, storage and finally disposal of potentially hazardous materials or products.

Alongside these regulations it is important to use any detailed codes of practice and manufacturer's advice which is often found on the packaging of the material.

A material or product that is hazardous to health could be anything that could affect your health.

There will normally be a sign stating the possible dangers (Figure 2.1).

What the regulations are about

Whether you are an employer, a contractor, a subcontractor or self-employed, COSHH requires you to protect people who may be exposed to health risks arising from hazardous substances with which you work.

The construction industry has always involved risks to the health and safety of its operatives, some of which are peculiarly its own, e.g. masons subject to silicosis from working with siliceous stones and painters being exposed to lead poisoning.

FIGURE 2.1
Asbestos warning

In more recent years, additional risks have been added by the use of new materials and processes being used, e.g. the use of epoxy resins.

The control of health hazards on construction sites is difficult owing to the continually changing working environments with the process of work in hand.

Risks common to all construction workers

Dangers to health can arise from:

- Ingestion of poisonous materials – care should be taken to avoid hazardous substances entering the mouth. Personal cleanliness is essential before food is eaten.
- Absorption through the skin by dangerous substances – continuous contact with the skin of certain materials can cause dermatitis and skin cancers.
- Exposure to certain physical conditions – without adequate protection, excessive noise can cause permanent damage to the ears.
- Inhalation of dangerous materials – breathing in any dangerous substance. Pollutants in the atmosphere that can be breathed into the body can be grouped into:
  - dust – small particles produced when emptying bags of powder such as cement and lime
  - fumes – paints and adhesives give off fumes and vapours
  - gases – these constitute a risk, particularly carbon monoxide.

The 'substances' covered by COSHH include:

- chemicals
- poisons
- solids
- timber
- liquids
- pesticides
- acids/alkalis
- dust – e.g. cement, gypsum, wood dust
- gas or fumes – e.g. welding fumes, hydrogen sulphide, carbon monoxide, nitrous oxide.

THE NOISE AT WORK REGULATIONS 2005

Building sites can be noisy places to work in. These should display signs as shown in Figure 2.2.

Excessive noise can be harmful, either by causing hearing damage or by creating nuisance, which may lead to stress.

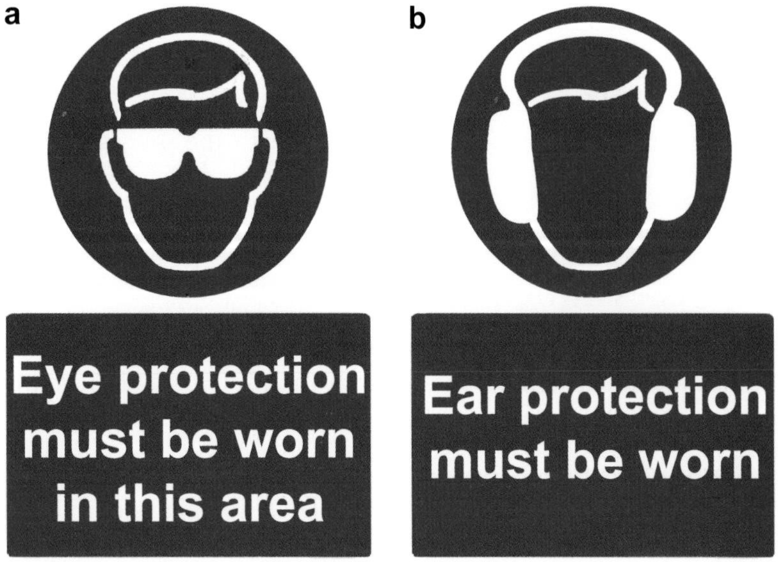

FIGURE 2.2
Eye and ear protection signs

Loud noise can cause a temporary partial loss of hearing, with recovery time varying from around 15 minutes after the noise stops to a few days, depending on the level of the noise. This temporary loss may be accompanied by ringing in the ears, or tinnitus, and this should be regarded as a warning: temporary partial hearing loss may become permanent with repeated exposure.

People usually first notice permanent hearing damage when ordinary conversation starts to become difficult to understand as they permanently lose part of their hearing range. This can gradually worsen if exposure to harmful noise continues.

Even a few minutes' exposure every day to very noisy machines in the construction industry can be enough to initiate permanent hearing damage.

Employers should do as much as possible to reduce noise. However, it may not be possible to quieten all machines enough to ensure that there is no hazard, and therefore proper ear protection should also be available.

WORK AT HEIGHTS REGULATIONS 2005

This regulation has been put in place to protect the workforce from injury or death from working at heights.

Your employer must provide the necessary equipment for working at heights, such as ladders, working platforms and scaffolding.

As an employee you must follow any instructions and training received while using any equipment provided and report any possible hazards to your supervisor.

More detailed information will be given in the section on working platforms.

REPORTING OF INJURIES, DISEASES AND DANGEROUS OCCURRENCES REGULATIONS 1995 (RIDDOR)

All employers have a duty under RIDDOR to report accidents, diseases and dangerous occurrences. Reporting accidents and ill-health is a legal requirement.

This information helps the Health and Safety Executive (HSE) and local authorities to identify where and how risks arise, and to investigate serious accidents. They can then help with providing advice on how to reduce injury and ill-health in the workplace.

PERSONAL PROTECTIVE EQUIPMENT AT WORK REGULATIONS 1992

Personal protective equipment (PPE) is defined as 'all equipment which is intended to be worn or held by the person at work and which protects him/her against one or more risks to health or safety'.

The main requirement of the regulation is that PPE is to be supplied and used at work whenever there are risks to health and safety that cannot be adequately controlled in other ways.

To allow the correct type of PPE to be chosen the employer must carefully consider the different hazards in the workplace. This will enable the employer to assess which types of PPE are suitable to protect the workers from the hazard and allow the work to be completed in a safe manner.

THE FIRE PRECAUTIONS (WORKPLACE) REGULATIONS 1997

The Fire Precautions (Workplace) Regulations 1997, as amended, cover places of work where one or more person is employed, e.g. commercial premises, universities, hospitals, shops, hotels and offices.

The regulations state that premises with five or more workers must have a written fire risk assessment detailing the appropriate fire safety work required, although some premises can be exempt.

Following the fire risk assessment the employer must, where necessary in order to safeguard the safety of employees in case of fire and to the extent that it is appropriate, provide:

- emergency exit routes and doors – the final emergency exit doors must open outwards and not be sliding or revolving
- emergency lighting to cover the exit routes, where necessary
- fire-fighting equipment, fire alarms and, where necessary, fire detectors.

Fire exit signs, fire alarms and fire-fighting equipment must be provided with pictograph signs, in accordance with the Health and Safety (Safety Signs and Signals) Regulations.

Employers must train employees in fire safety following the written risk assessment.

An emergency plan may have to be prepared and sufficient workers trained and equipped to carry out their functions within any such plan.

All equipment and facilities such as fire extinguishers, alarm systems and emergency doors should be regularly maintained and faults rectified as soon as possible. Defects and repairs must be recorded.

Employers must plan, organize, control, monitor and review the measures taken to protect employees from fire while at work. If there are five or more employees, then a record must be maintained.

Employers must appoint an adequate number of competent people to assist them to comply with their obligations.

PROVISION AND USE OF WORK EQUIPMENT REGULATIONS 1998

These regulations require risks to people's health and safety, from equipment that they use at work, to be prevented or controlled.

The regulations require that the equipment provided for use at work is:

- suitable for the job it has to do
- regularly maintained to ensure that it is safe to use.

They also require that:

- training is provided for those who use it.

In general, any equipment that is used by an employee at work is covered.

All electrical and mechanical tools, equipment and machinery are potentially hazardous if misused, worn or damaged.

During your career in the construction industry you will come into contact with a wide variety of mechanical equipment designed to be used by a variety of trades. Some machines are very specialized and only likely to be operated by trained and certificated workers.

Static plant should be sited on firm level ground, with brakes on and chocks in place (Figure 2.3).

MAIN TYPES OF PLANT AND EQUIPMENT

Mobile plant

Mobile plant is defined as plant that moves under its own power. It is used to transport materials to or around building sites, and in construction operations.

On a large site, mobile plant could be controlled by restricting speeds, confining vehicles to site roads and authorized (one-way) routes. Places may also be set aside for unloading, turning and manoeuvring.

One of the most common items of plant for moving materials around the site is the fork-lift truck (Figure 2.4).

Mobile plant should as far as possible be kept clear of structures, building activities, cranes, excavation and other hazards.

FIGURE 2.3
Small mixer ready for use

FIGURE 2.4
Mobile plant

Static plant

Compressors, generators, pumps and concrete mixers are examples of static plant (Figure 2.5).

This type of equipment works in a fixed location, although the location may change as work proceeds and from site to site.

If static plant is powered by electricity, care must be taken to site the cables where they cannot be damaged by vehicles, and to avoid laying them on rough or sharp projections.

Mechanical tools and equipment

These have exposed moving parts such as grinding wheels, sanders, drills, chipping hammers, portable saws, rotary wire brushes and air compressors. A typical tool and sign are shown in Figure 2.6.

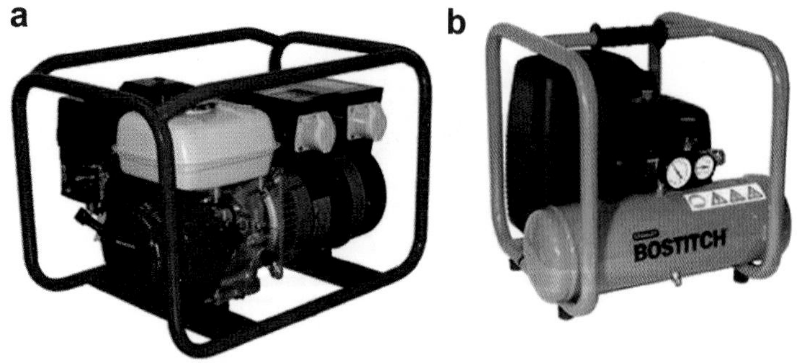

FIGURE 2.5
Static plant: (a) generator; (b) compressor

FIGURE 2.6
Mechanical tools and equipment

There are special regulations concerning grinding wheels and portable saws, and people must be authorized to use them.

- Never use any mechanical equipment that is unfamiliar.
- First read the manufacturer's instructions or seek advice.
- Never lay a tool down while it is still rotating.
- Never wear loose fitting clothes when using tools with fast moving parts.

Most accidents are caused by lack of knowledge, misuse, makeshift repairs or using faulty tools and equipment.

THE ELECTRICITY AT WORK REGULATIONS 1989

Any equipment that uses electricity is covered by the Electricity at Work Regulations. Appropriate signs should be positioned as shown in Figure 2.7.

Your employer has a duty to make sure that the whole construction site is a safe area and that there is no chance of coming into contact with a live electrical current.

Electricity

Electricity is something that you cannot see or hear. Strict rules are laid down for its use and *must* be rigidly obeyed by everyone.

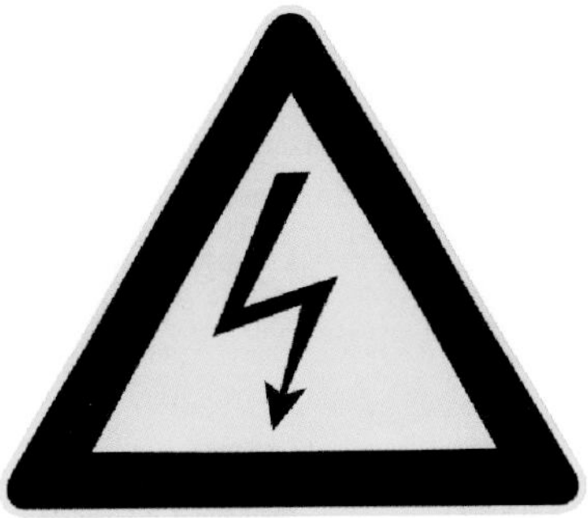

FIGURE 2.7
Caution sign for electricity

All tools should comply with British Standard 2769 and, except for 'double-insulated' tools (Figure 2.8), must be effectively earthed. Double-insulated tools, which have their own built-in safety system and bear the 'kite mark' and 'squares symbol', do not require an earth lead.

Any electrical tool or equipment must be operated at a reduced voltage of 110 V (volts), or lower if possible. Below 60 V the risk of death is greatly reduced.

Remember

As a non-skilled or certificated person, you may only carry out safety checks and MUST NOT attempt to repair any piece of electrical equipment.

If faulty electrical equipment is found you should return it to your supervisor who will see that the necessary repair work is carried out by a competent person.

Most electrical equipment has to be checked and tested annually.

Each tool or piece of equipment is made to be used at a certain voltage.

Check that the tool operating voltage and the supply voltage are the same before connecting the tool to the supply.

Transformers

A transformer (Figure 2.9) may be used if the supply voltage and the operating voltage of the tool or equipment are different.

Perhaps the most common transformer is the 110/240 V unit. This can be used to raise a 110 V supply to 240 V, or reduce a 240 V supply to 110 V.

Always check that the correct transformer is being used, and that it is of sufficient capacity to take the current flowing.

If you are using equipment operating at 230 V or higher a residual current device (RCD) can provide additional safety and rapidly switches off the current. An RCD plug is shown in Figure 2.10. If the RCD trips, this is a sign of a fault.

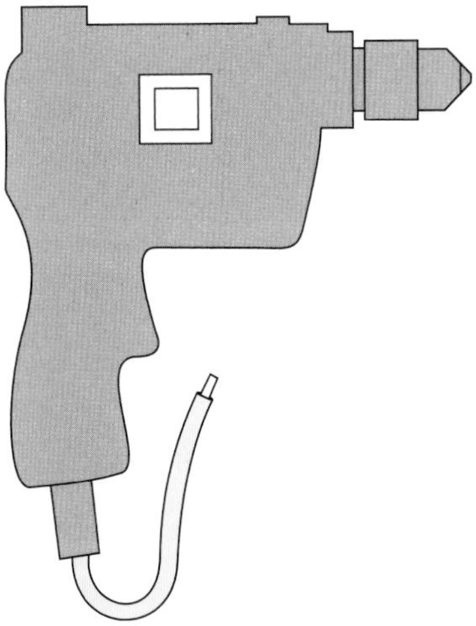

FIGURE 2.8
Double-insulated drill

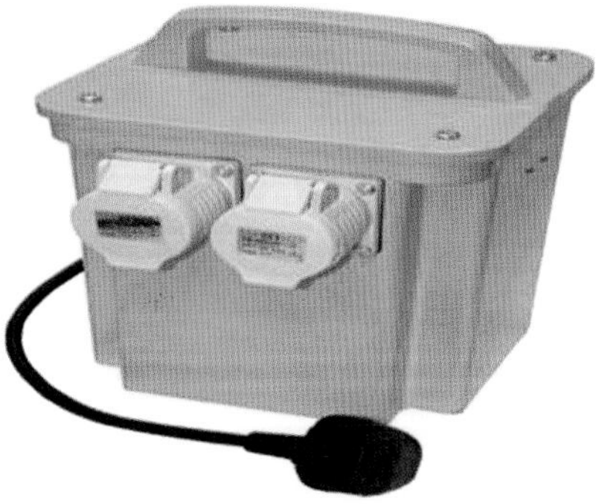

FIGURE 2.9
Transformer

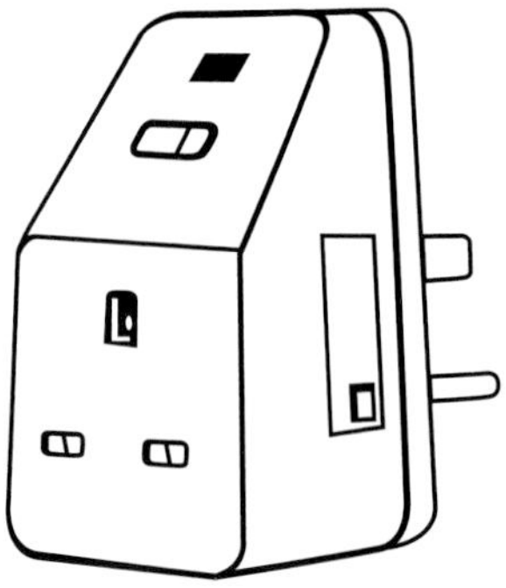

FIGURE 2.10
RCD

Connectors

The most common 110 V plugs and couplers are normally of the splash-proof type and are designed to make the connection to an incorrect voltage supply impossible.

Plug

The plug (Figure 2.11), connected to the end of the flex, fits into the electricity supply socket.

Electric fuses

A fuse is a safety device. It is a deliberate weak link in a convenient part of the circuit, usually in a form of a cartridge containing a wire which melts and therefore breaks the circuit when an excessive current flows.

Electric shock

- Cause: contact with low-voltage (240 V) supply.
- Action to be taken:
    1. Do not touch the casualty with hands.
    2. Break the electrical circuit by switching off or removing the plug from the socket.

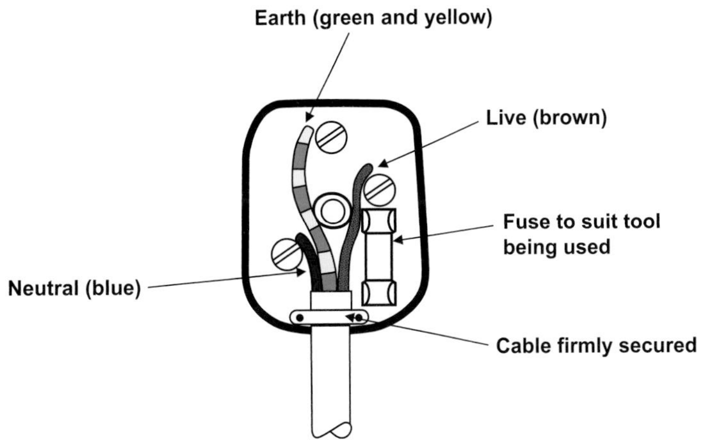

FIGURE 2.11
Wiring to a 240 V three-pin plug

3. If this is not possible, immediately break the casualty's contact with the supply by standing on dry insulating material and – using dry wood, a folded newspaper or a rubber object – push the casualty out of contact with the supply.
4. Treat burns.
5. Apply respiratory resuscitation if breathing has stopped.
6. Apply cardiac massage if the heart has stopped. Ensure that there is no heartbeat before commencing this procedure.

Emergency resuscitation

In the event of someone collapsing through injury, it may be necessary to attempt emergency resuscitation while a medical team or ambulance is being called.

Two methods can be used: the mouth-to-mouth with chest compressions method (Figures 2.12 and 2.13) and the Silvester method. The mouth-to-mouth method of resuscitation should be used if the casualty's mouth and face are not damaged. The Silvester method should be used if the casualty's mouth or face is damaged, but the chest is not damaged.

Workplace safety

Accidents

When joining the construction industry it is important to remember that you will be joining an industry with one of the highest injury and accident rating. It therefore it cannot be stressed enough that you could be at constant risk unless you start as you intend to continue, with a good safety attitude.

Any type of work carried out by the construction industry is often difficult and hazardous. Every site will be different, and therefore every site will bring possible new dangers.

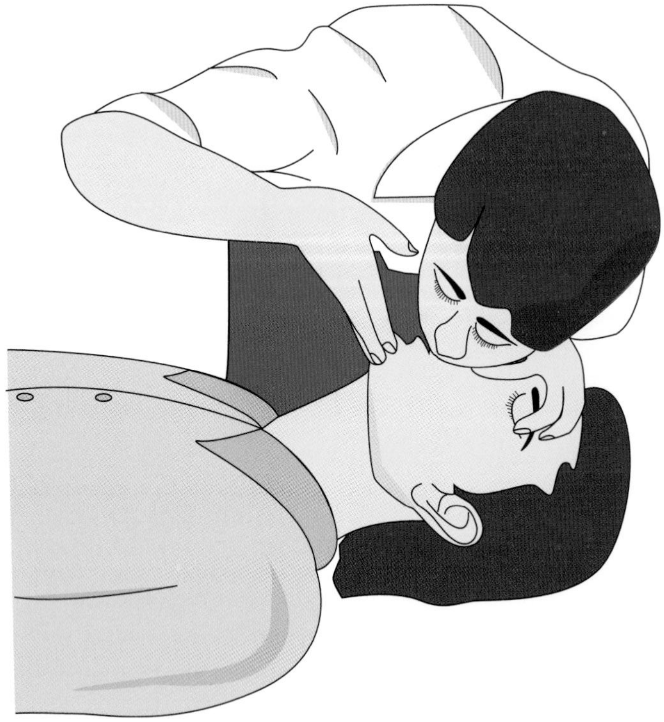

FIGURE 2.12
Mouth-to-mouth resuscitation

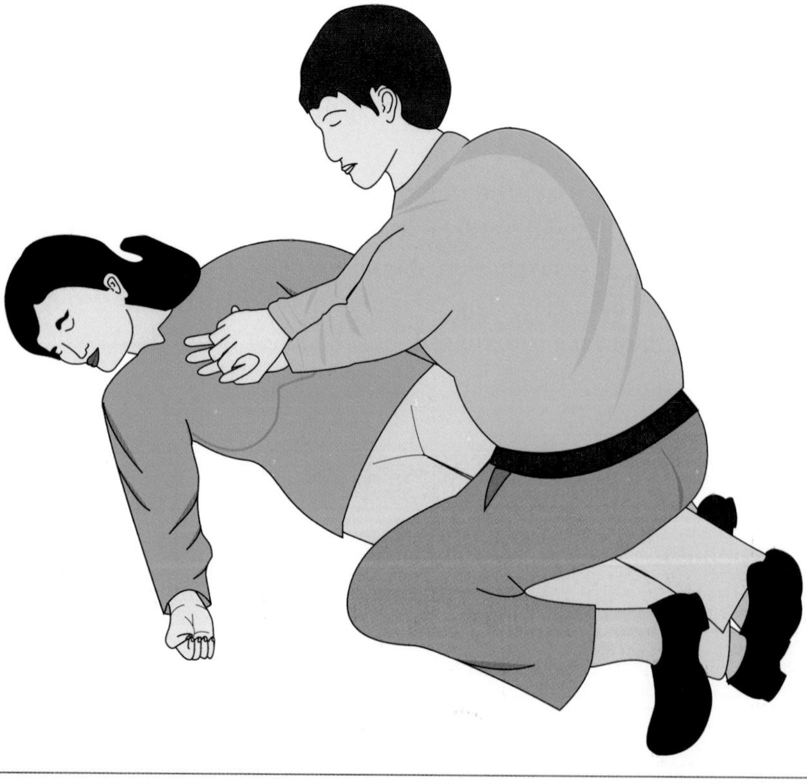

FIGURE 2.13
Chest compressions

An accident is an unexpected or unplanned happening which results in personal injury or damage, sometimes death.

Reported accidents are those which result in death, major injury and more than three days' absence from work or are caused by dangerous occurrences reported to the HSE.

Every day a large number of the industrial accidents that are reported involve construction workers.

ACCIDENTS DO NOT JUST HAPPEN, THEY ARE CAUSED.

Finding out what causes them is the first step towards preventing them. Usually an accident is the last link in a chain of events consisting of a series of dangerous conditions and dangerous actions.

CAUSE OF ACCIDENTS

Accidents are caused in various ways and may well be attributed to:

- trying to get the job done too quickly
- too little preparation before commencing
- taking short cuts
- distraction by others causing a lapse in concentration
- lack of concentration due to lack of interest in the job
- failure to observe the rules of safety; not wearing the correct PPE
- horseplay, that is acting irresponsibly, creating a danger and a hazard to yourself and others.

Remember

It is your responsibility to act in a safe manner.

Types of hazard

Everyone involved in the construction industry should be aware of the possible dangers and hazards in the construction site.

Site safety will be improved if everyone is safety conscious.

Types include the following:

- falling objects
- falls of operatives
- transportation of plant and materials
- electricity
- machinery and equipment
- fire and explosions.

HAZARD SPOTTING

Look at accidents that have occurred. The law requires that your company keeps records of all accidents to employees.

- When did you last look at your accident book?
- What are the most common accidents?
- What action has been taken to prevent similar accidents happening in the future?

Even in the best companies accidents still happen.

The next time there is an accident try looking at it more closely. An analysis procedure can help you to prevent accidents in the future. Start off by finding the facts. After analysing several accidents a pattern may emerge.

The causes of accidents may be similar, although these happen in different workshops or on different sites. Looking at the whole company may highlight accident prevention training needs.

How to spot hazards

Looking at any situation may be fun, but the action in spotting hazards in your own area is deadly serious. So too is training others to spot them.

As you walk around your work area use your senses: look, touch, hear and smell.

Practice in looking at hazards in more depth will help you to become more proficient at hazard spotting.

Writing them down will help you to analyse them and make your thinking clearer. Very soon you will find that you can dispense with the written method and carry out the procedure mentally.

Personal protective equipment

Depending on the type of workshop or site situation, the wearing of correct safety clothing and safe working practices are the best methods of avoiding accidents or injury. On some sites certain PPE is compulsory.

All construction operatives have a responsibility to safeguard themselves and others. Making provision to protect oneself often means wearing the correct protective clothing and safety equipment.

Your employer is obliged by law to provide the following:

- suitable protective clothing for working in the rain, snow, sleet, etc.
- eye protection or eye shields for dust, sparks or chippings
- respirators to avoid breathing dangerous dust and fumes
- shelter accommodation for use when sheltering from bad weather
- storage accommodation for protective clothing and equipment when not in use
- ear protectors where noise levels cannot be reduced below 90 dB(A) for 8 hours

- adequate protective clothing when exposed to high levels of lead, lead dust fumes or paint
- safety helmets for protection against falls of materials or protruding objects
- industrial gloves for handling rough abrasives, sharp and coarse materials.

Although our skin is not proof against knocks, bumps, cuts, acid, alkalis or boiling liquids, it is waterproof. Even so, we do have to cover up at times to protect ourselves.

Workers in the construction industry are liable to injury or even death if they are not protected. Because of this, protective clothing has been developed to help prevent injury.

Some items of PPE are shown in Figure 2.14.

Safety signs

As you go about your work on the building or construction site you will see various signs and notices. Your employer will give you instruction on what they mean and what you should do when you see one.

Safety signs fall into four separate categories, which can be recognized by their shape and colour. Sometimes they may be just a symbol; others may include letters or figures and provide extra information such as the clearance height of an obstacle or the safe working load of a crane.

The four categories are:

PROHIBITION SIGNS

- Shape – circular
- colour – red border and cross-bar; black symbol on white background
- meaning – shows what must not be done
- example – no smoking.

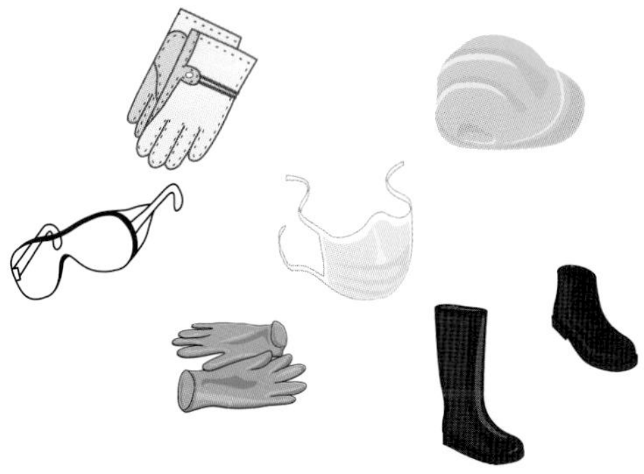

FIGURE 2.14
Selection of PPE

MANDATORY SIGNS

- Shape – circular
- colour – white symbol on blue background
- meaning – shows what must be done
- example – wear hand protection.

WARNING SIGNS

- Shape – triangular
- colour – yellow background with black border and symbol
- meaning – warns of hazard or danger
- example – caution, fork-lift truck working.

INFORMATION SIGNS

- Shape – square or oblong
- colour – white symbols on green background
- meaning – indicates or gives information of safety provision
- example – first aid point.

SIGNS WITH SUPPLEMENTARY TEXT

Any of the symbols shown above may also contain text. A few examples of these are shown in Figure 2.15.

Signs may be produced and erected according to the situation, as shown in Figure 2.16. For example, where there is a dangerous hazard a sign can be produced and placed in a prominent position to warn anyone approaching to take care.

FIGURE 2.15
Selection of signs: (a) prohibition sign; (b) mandatory sign; (c) warning sign; (d) information sign

FIGURE 2.16
Types of notice

Security arrangements

It is the responsibility of everyone on the work site to ensure that the security of that site is maintained. Security can take many forms and they are all equally important:

- alarms – positioned in an accessible place within view of the general public
- bars, mesh and locks – fitted to glass panelled doors and windows
- padlocks, padlock and chains – fitted to compound gates, pieces of plant and machinery (Figure 2.17).
- lighting – flood lights and movement-activated lights
- security firms.

If, despite the security measures taken, your site is breached, there are certain procedures you should follow. These should be given to you by your line manager, and may include:

- reporting the incident to the site supervisor
- reporting the incident to the police
- checking the inventory to find out what has been taken
- recording damage done to the premises and/or equipment.

Remember

Vandalism and theft do not always happen at night.

Always take care of tools and equipment you are using and report any suspicious people or events to your line manager.

Emergency procedures

Responding to emergencies

To save life and reduce the risk of injury occurring, all new people at work must be told of the current safety and emergency procedures.

From day one you should be aware of what to do in the event of a fire or an accident. Should an emergency happen you should be able to:

- Know what to do – acting quickly and calmly, carry out the correct procedure

FIGURE 2.17
Site security

- Follow the fire procedure – take the correct action in the event of discovering a fire:
    1. Select and use the correct type of fire extinguisher. (Only if the fire is small enough for you to put it out.)
    2. Call for help, sound the alarm.
    3. Telephone the Fire Service: 999. Give the correct address of the building.
    4. Leave the building by the nearest exit.
    5. Go directly to the assembly area. Await the roll call.

Accidents

An accident involving injury to a person can happen at any time. It may be a workmate who has fallen off a ladder or someone with a burn or a cut, or who has fainted. To help them when they most need it you should know what to do!

IMMEDIATE ACTION

1. Unless you are a fully trained first aider – *do not attempt to treat the injured person.* (Only move an injured person if their life is in danger, e.g. danger from fire.)
2. GET HELP. Report the accident to a person in charge or:
3. Telephone the Emergency Services: 999.
   When your call is answered you should have the following information at hand:
    - type or types of services required – fire, ambulance, police
    - type of accident
    - location/address at which it has happened
    - telephone number you are calling from
    - your name

Remember

An accident is an event causing injury or damage that could have been avoided by following correct methods and procedures.

An accident is an unplanned event or occurrence which is outside the planned or expected procedure of work.

Although every worker should do their best to prevent accidents happening, there is still that unfortunate time when the worst happens.

FIRST AID

If you are a qualified first aider you will know what to do. If you are not qualified in first aid, what can you do?

Do you know where the first aid box is kept?

According to the Health and Safety (First Aid) Regulations 1981 you employer must provide first aid equipment and facilities and appoint a qualified first aider or, on a smaller site where only a small number of people are working, should arrange for someone to take charge in the event of an injury.

On very large sites, there may be a first aid room with a qualified first aider in charge.

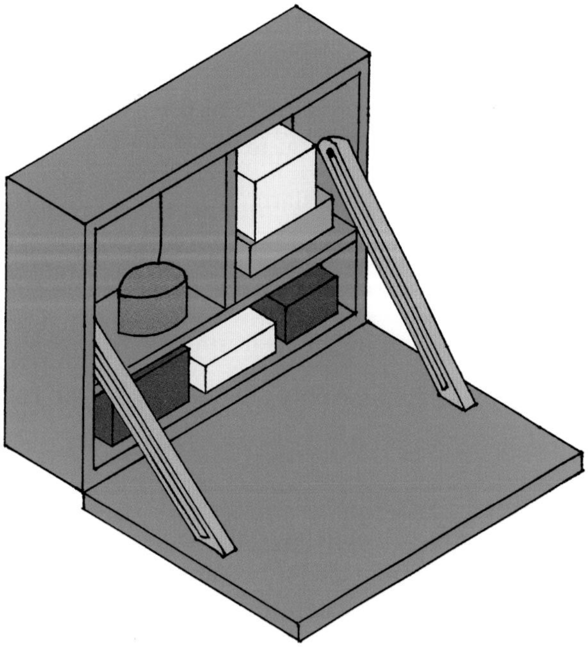

FIGURE 2.18
First aid box

First aid boxes should be situated at various points around the site and clearly marked with a white cross on a green background.

A small portable first aid kit (Figure 2.18) should be carried by an employee when working alone or in a small group well away from the main site.

If you have to use the first aid kit, report it and state what you have used so that it can be replaced.

ACCIDENT REPORTING

Every accident should be reported – an accident report book should be on every site or workshop, usually with the site supervisor, or whoever is in charge of the site or workshop. A typical page from an accident book is shown in Figure 2.19.

Make sure that you report any accident in which you are involved as soon as possible.

Obviously some accidents are more serious than others. Accidents that result in death, major injury or more than three days' absence from work are called 'reported accidents'. Any such accident should be reported to the HSE.

Accidents where people require hospital treatment must be recorded at the place of work, even if no treatment was given there.

Fire and emergency procedures

All fires must be taken seriously and action taken immediately to prevent harm to people and damage to property.

1 About the person who had the accident	**2** About the person filling in this page If you did not have the accident, complete the following
Full Name	Full Name
Address	Address
Postcode	Postcode
Occupation	Occupation

3 Please sign and date	
Signature:	Date:
The person who has had the accident should sign and date if they have not filled in the book (as confirmation that they agree the accident recorded is a true and accurate record	
Signature:	Date:

4 About the accident When and where it happened	
Date:	Time :
In what room or place did the accident happen?	

5 About the accident - What happened?
How did the accident happen?
Materials used in the treatment

6 Reporting of injuries, diseases and dangerous occurrences
For the employer only – complete the box provided if the accident is reportable under RIDDOR
How reported
Date reported:
Employer's name and initials:

FIGURE 2.19
Typical page from an accident book

If a small fire cannot be controlled quickly, the building must be evacuated.

Fires require three factors to start. The removal of any one of these factors will extinguish the fire. Check out these three factors in Figure 2.20.

Fires, if they occur, need to be extinguished. Only tackle a fire if it is safe to do so or it is small enough to extinguish with a fire extinguisher.

The fighting of a fire should only be carried out by people who are fully trained. In some situations inexperienced site personnel are involved with the early stages of fighting a fire.

The fire-fighting should not continue if:

- the fire becomes too dangerous
- there is the possibility that the escape route might be cut off

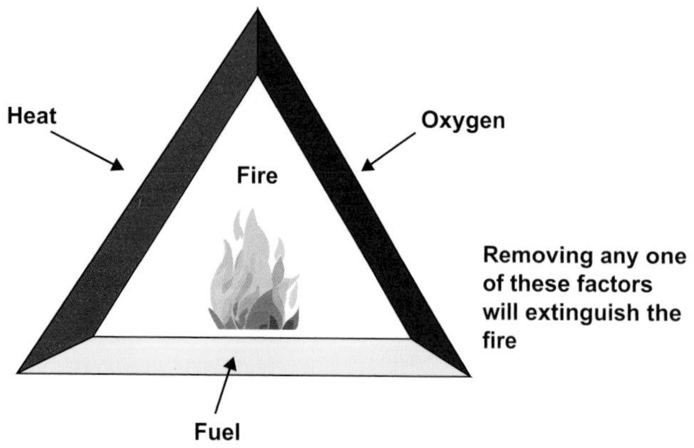

FIGURE 2.20
Factors in a fire

- the fire continues to spread
- there are gas cylinders or explosives in the immediate area that cannot be removed or soaked continually with water.

Exits and oxygen

Never work in a position where a fire may block the exit.

Since all fires need an ample supply of oxygen, if a fire does break out try to close all windows and doors.

FIRE-FIGHTING EQUIPMENT

To prevent fires the workplace should be kept clean and tidy. Rubbish should not be allowed to accumulate in rooms and storage areas. Fire prevention is therefore mainly good housekeeping.

When using flammable materials a suitable fire extinguisher must always be kept handy and ready for use. Before starting work on any job, make sure that the extinguisher operating instructions are fully understood.

Make sure that the extinguisher is the right type for the fire that may occur. Use of the wrong type can be disastrous. Fires can be classified according to the type of material involved:

- class A – wood, cloth, paper, etc.
- class B – flammable liquids, petrol, oil, etc.
- class C – flammable gases, liquefied petroleum gas, propane, etc.
- class D – metal, metal powder, etc.
- class E – electrical equipment.

The following types of fire-fighting equipment are available:

- Fire blankets (Figure 2.21) are useful for wrapping around a person whose clothing is on fire. They may also be used to smother a small, isolated fire.

FIGURE 2.21
Sand bucket, fire blanket and hose

- Sand (Figure 2.21) is also useful for dealing with small isolated fires, such as burning paint droppings.
- Water hoses (Figure 2.21) are fitted in some buildings.
- Fire extinguishers – pressurized extinguishers can be filled with various substances to put out a fire. Fire extinguishers are now all the same colour (red), but they have a colour band which identifies the substance inside:
  - Water: the colour band is *red* and the extinguisher can be used on class A fires.
  - Foam: the colour band is *cream* and the extinguisher can be used on class A fires. A foam extinguisher can also be used on class B fires if the liquid is not flowing and on class C fires if the gas is in liquid form.
  - Carbon dioxide: the colour band is *black* and the extinguisher can be used on class A, B, C and E fires.
  - Dry powder: the colour band is *blue* and the extinguisher can be used on *all classes* of fire.

A range of extinguishers is shown in Figure 2.22.

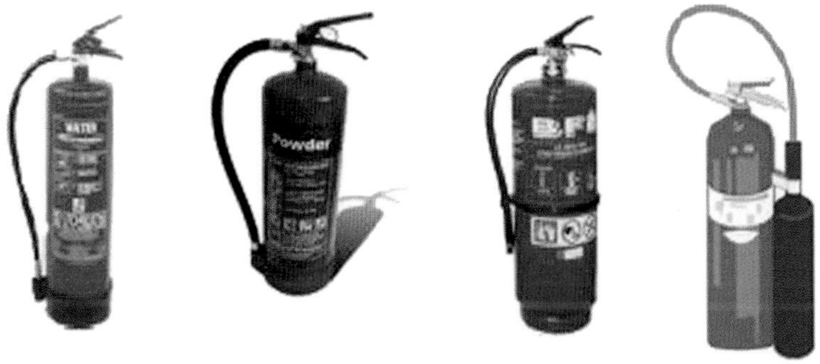

FIGURE 2.22
A range of pressurized extinguishers

Risk assessments

The employer has a duty to protect the workforce as far as is reasonably practicable. Risk assessment is a very important part of protecting all site operatives.

Risk assessment is simply a careful examination of what could cause harm to the workforce. The workforce has a right to be protected from harm caused by a failure to take reasonable control measures.

In the construction industry risk assessments are carried out by experienced people who have been taught to identify what risks are possible when carrying out tasks.

There are five main steps to risk assessments:

- Step 1 – Identify the hazard.
- Step 2 – Decide who might be at risk and how.
- Step 3 – Evaluate the risks and decide on the best precautions.
- Step 4 – Record your findings and implement them.
- Step 5 – Review your assessment and update if necessary.

Remember

Do not overcomplicate the process.

If you run a small business and you are confident you understand what is involved, you could do the assessment yourself. You do not have to be a health and safety expert.

Most of the risks come from tripping, slipping and moving heavy loads.

If you are an employer of a large company you should ask a health and safety advisor to help out.

Health and hygiene

Certain precautions must be taken to ensure that the health of employees in construction firms is protected against hazards, as mentioned in the previous section. As far as is practicable, their health must also be protected

Remember

A hazard is anything that can cause harm, such as chemicals, electricity or working from ladders.

The risk is the chance, high or low, that an employee could be harmed by these and other hazards.

Vulnerable parts of the body

The health of the site operatives can be divided into the following areas of the human body:

- Skin – One of the most common problems with the skin is dermatitis. To reduce the problem barrier creams could be used or appropriate gloves worn. Most construction sites now provide numerous types of gloves for every situation.
- Eyes – Protection of the eyes has been mentioned previously, but as they are the only ones you have it is important to take extra care and use the appropriate glasses or goggles for the job.
- Ears – Again, most sites provide a selection of ear protectors or plugs to be used when working with or close to noises.
- Lungs – Protective breathing apparatus or simple disposable masks should be available.

Other areas to consider

COLD

This is most damaging to health when it is associated with damp weather. In adverse conditions the body cannot maintain normal body temperature.

Construction workers who get cold and wet frequently and for substantial periods, and who wear inadequate clothing, will have an increased risk of illnesses such as bronchitis and arthritis.

Workers are particularly at risk from cold when the temperature around them is below 10°C. It is important to realize that the wind-chill factor could lower this temperature considerably.

Remember

You are not being the tough guy in trainers, jeans and lightweight jackets. You are being stupid.

Note

If you are very cold and your skin turns blue and you are suffering from shivering or dizziness or become stupefied, then you may be in the early stages of *hypothermia*.

HEAT

Heat exhaustion could occur, which is dangerous if working on ladders, trestles, etc. Sunburn and sunstroke are also possible.

Sweating excessively causes a loss of body fluids and salt, creating severe muscular cramps.

PERSONAL HYGIENE

Always keep yourself clean and tidy; just because you are in one of the dirtiest occupations, there is no need to look untidy.

If you are working in a client's home you need to present yourself correctly.

Always wash regularly and have your work clothes washed regularly.

Wash your hands after going to the toilet and before eating and drinking.

GENERAL

Most construction sites undertake regular blood checks for drugs and alcohol.

It important that you arrive at your place of work in a fit state to carry out your duties in a manner safe to yourself and your workmates.

All the major sites have permanent health and safety officers, and one of their roles is to give 'toolbox talks' on various areas of health and safety.

All personnel new to the site should have a health and safety induction before they are allowed to work on the site. This will include the rules and regulations involved and the procedures to be followed if any are broken. This could result in someone being banned from entering the site.

Construction Skills Certification Scheme

As part of the certification process towards NVQ and Technical Certificates each candidate has to pass the Construction Health and Safety test. This compulsory test consists of a number of multiple-choice questions.

New entrants to construction will also have to apply for a Construction Skills Certification Scheme (CSCS) card. Most large sites now require proof of certification before a worker is allowed to enter.

A trainee will be given a *red* card if he or she is registered for an NVQ or a Technical Certificate but has not yet achieved Level 2 or 3. The cards are valid for three years.

The aim of the scheme is to raise the standards of health and safety. It also provides a record of all workers in the construction industry who have achieved a recognized level of competence, and provides a means of identification.

Handling materials and components

The trainee should be able to select and use appropriate safety equipment and protective clothing when handling different materials, and to select and use appropriate equipment and aids to carry materials.

The trainee should also be able to demonstrate safe manual handling techniques.

Lifting gear

Numerous items of small lifting equipment are available to assist with handling materials on site and in the workshop. Only use these if you are qualified to do so.

They range from small brick lifts, slings, barrows and dumpers to mechanical fork-lift trucks. A selection is shown in Figure 2.23.

- Barrows are the most common form of equipment for moving materials on site.

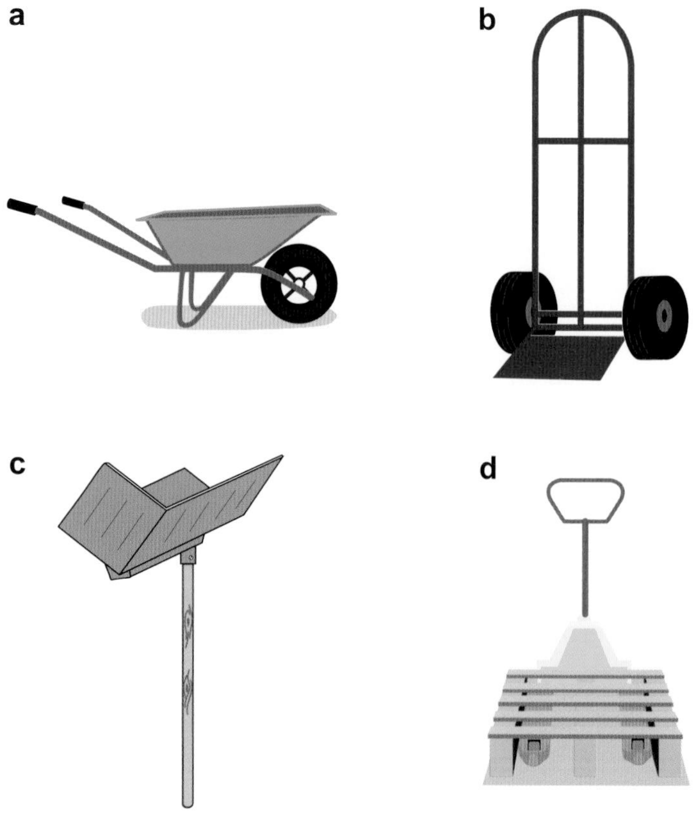

FIGURE 2.23
Moving equipment: (a) barrow; (b) sack truck; (c) hod; (d) pallet truck

Note

Always use lifting gear if it is available.

Remember

The sequence for lifting heavy and awkward loads:

1. Plan the task.
2. Bend your knees.
3. Get a good grip.
4. Lift, with your legs taking the strain.
5. Place the load down.

- A sack truck can be used for moving bagged materials and paving slabs.
- A hod can be used for moving bricks on to higher levels such as scaffolds.
- A pallet truck can be used on hard areas for moving heavy loads.

Many materials are delivered to the site on lorries equipped with mechanical off-loaders. Once the material has been off-loaded it is the builder's responsibility to move the materials to a secure place until required for use.

Working platforms

The Work at Height Regulations 2005 applies to all work at height where there is a risk of a fall liable to cause personal injury.

There are on average 65 fatal accidents and over 4000 major injuries in the construction industry each year. They remain the single largest cause of workplace deaths and one of the main causes of injury.

These regulations place a duty on all employers to provide a risk-free environment in which to work.

- Duty holders must avoid working at height where possible.
- Where working at height cannot be avoided, equipment or measures to prevent falls must be in place.
- If there is a risk of a fall, equipment to minimize the distance of the fall must be used.

Remember

Before any work is carried out at height a risk assessment must be produced.

Before work starts

No scaffold should be erected, substantially added to, altered or dismantled except under the immediate supervision of a *competent person* and, as far as possible, by *competent workers* possessing adequate experience of such work.

The competent person should also be given sufficient and sound materials for the job. It is false economy and highly dangerous to skimp on materials, and if faulty materials are provided the dangers may be hidden from those who use them.

Levels 1 and 2 dealt with hop-ups, trestle scaffold and towers. This book, Level 3, will deal with tubular scaffolding.

Types of scaffold

There are four main types of scaffold:

- independent
- dependent – putlog – bricklayer's
- unit or frame

INDEPENDENT SCAFFOLD

This is a scaffold which, apart from the necessary ties, stands completely free of the building (Figure 2.24).

The main applications for this scaffold are:

- access to stonework on masonry buildings
- access to solid or reinforced concrete structures
- maintenance and repair work
- cladding work.

The scaffold consists of two rows of standards joined together by ledgers which, in turn, are joined by transoms. Transoms may extend beyond the inner standard to support one or two boards between inner standards and the wall.

It is essential to include cross-bracing at every lift to ensure rigidity.

Although it is called an independent scaffold it does include normal ties for stability.

DEPENDENT SCAFFOLD

This is more commonly known as a 'putlog' or bricklayer's scaffold (Figure 2.25).

It is similar to the independent access scaffold, but has only one row of standards, with the inner row replaced by the brickwork. This means that the inside ledger and ledger bracing are not required. The remaining scaffold functions in the same way as the independent scaffold.

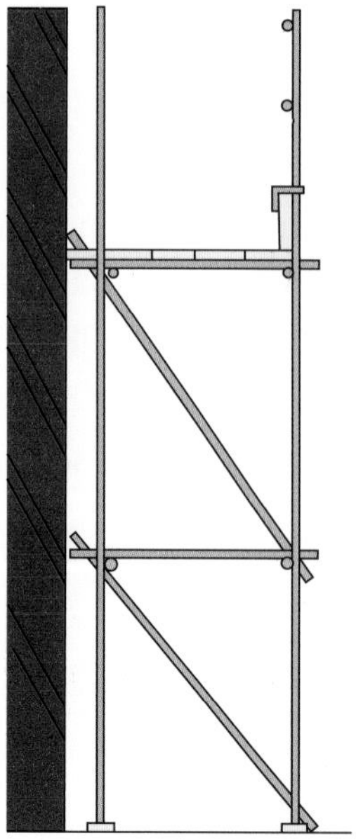

FIGURE 2.24
Independent access scaffold

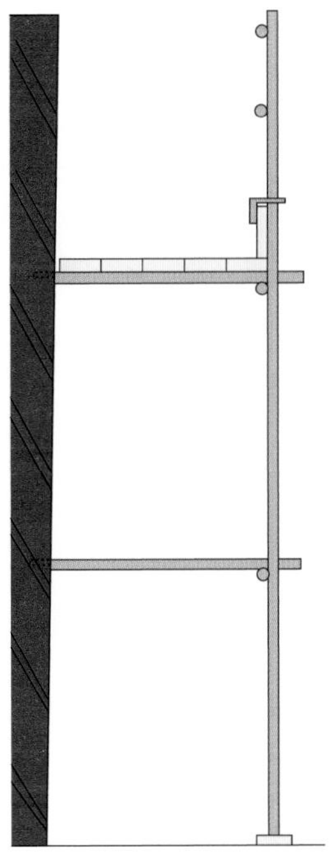

FIGURE 2.25
Dependent access scaffold

The scaffold can be erected against existing brickwork, but is usually erected along with new building work.

The platform in this type of scaffold is supported by putlogs and not transoms. The putlog is supported by the new brickwork by allowing the spade end of the putlog to rest flat on the brickwork. Putlogs should never be removed.

If this scaffold is being used for existing walls putlog holes will have to be chopped out in the existing brick joints with a special plugging chisel.

The putlog spacings are exactly the same as for independent scaffolds, but as only one scaffold clip is used it must be a right-angled coupler.

Although the putlog is fixed into the wall it does not act as a tie for the scaffold. It should be tied to the building at least every 4 m vertically and 6 m horizontally.

UNIT OR PROPRIETARY SCAFFOLDS

Proprietary scaffold

A proprietary system of scaffolding consists of special components, each with their own fittings. There are no loose fittings to lose or maintain.

All the components are of reasonable length and weight to make them easier to handle and less susceptible to damage.

Pressed steel battens are also available with non-skid, durable surfaces.

UNIT SCAFFOLD

This system comprises preformed 'H'-shaped units that are slotted into one another to form the structure (Figure 2.26).

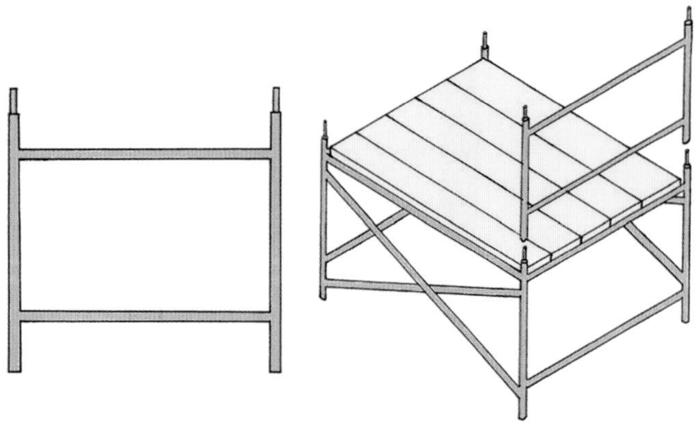

FIGURE 2.26
Unit access scaffold

These are quick and easy to erect but are often very expensive to purchase. They are not as versatile as other systems.

Scaffold fittings

Scaffolding components should all be examined by a competent person.

Tube ends should be square with the tube axis and cut cleanly. Tubes should be free from distortion, corrosion, splits, laminations, surface flaws and undue rust.

All couplers and fittings should be free from rust and distortion, worn threads and damaged bolts, and should be maintained in an oiled condition.

These are the main points to observe and should cover most of the faults found in tubes, fittings and boards.

HOWEVER: IF IN DOUBT NDASH; LEAVE IT OUT!

There are numerous scaffold fittings (Figure 2.27); each one should only be used for the purpose it has been designed for.

Scaffold boards

All scaffold boards should be made to British Standard specifications or they should be 'specials'

To prevent boards from splitting, the ends should be bound with a galvanized metal band (Figure 2.28).

Boards for use in platforms can be 38, 50 or 63 mm thick. The width is normally 225 mm and the length can be up to 3900 mm.

The most common sized scaffold board is 22 × 38 × 3900 mm.

Scaffold boards must be:

- made from straight-grained timber
- free from knots and shakes

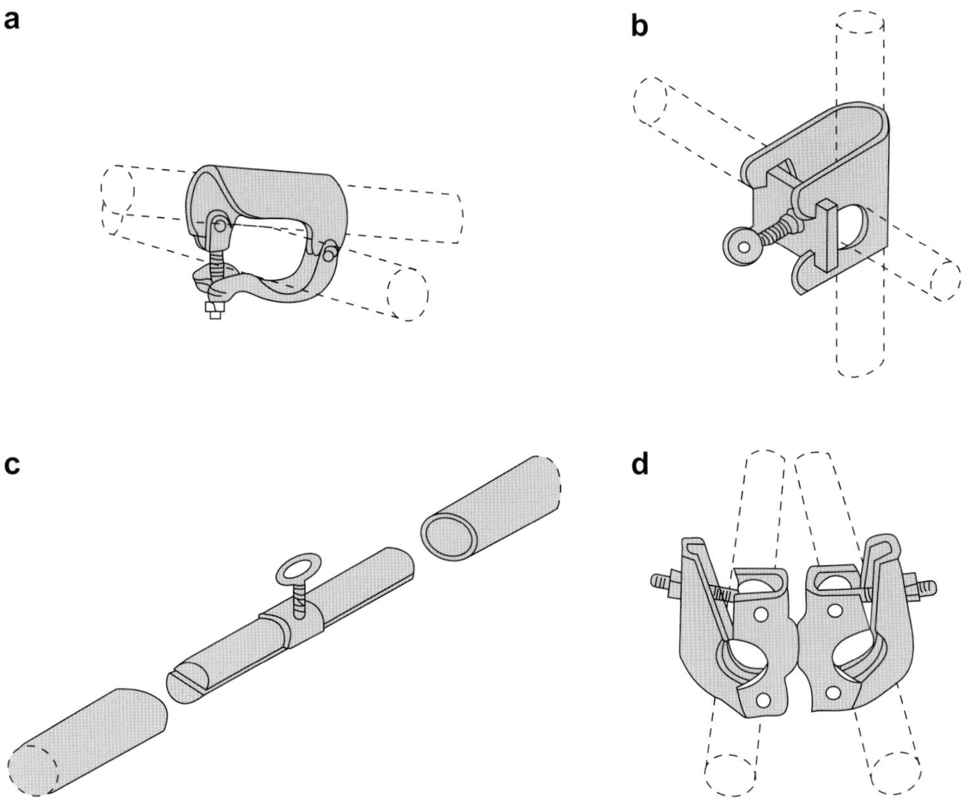

FIGURE 2.27
Selection of scaffold fittings: (a) putlog coupler; (b) right-angled coupler; (c) jointing coupler; (d) swivel coupler

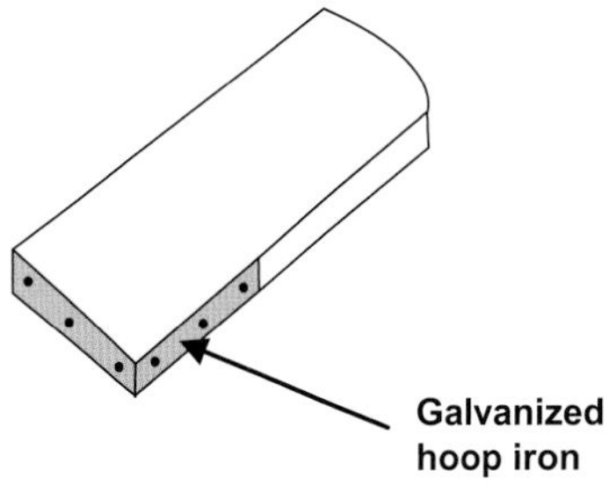

FIGURE 2.28
Boards prevented from splitting

- free from decay
- clean and free from grease and thick paint, etc.

Scaffold boards must not be twisted or warped or have split ends.

The distance between the supports governs the thickness of the board used:

- 1.2 m for graded boards
- 1.5 m for BS boards.

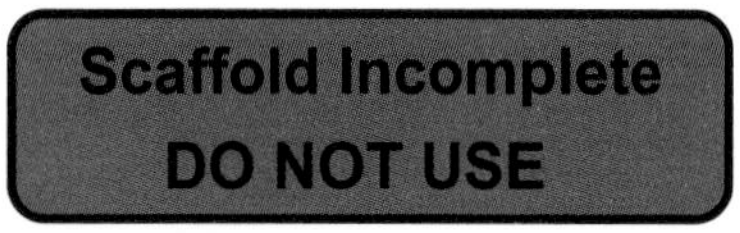
Scaffold Incomplete
DO NOT USE

FIGURE 2.29
'Do not use' notice

OVERHANG

No board should overhang its supports by more than four times the board thickness, or less than 50 mm.

GUARDRAILS

Access platforms more than 2 m high must have guardrails and toeboards.

The risk of falling materials causing injury should be kept to a minimum by keeping working platforms free from loose materials and debris.

In addition, materials or other objects must be prevented from rolling, or being kicked, off the edges of working platforms. This can be achieved by fixing toeboards, solid barriers, brick guards or similar at open edges.

Guardrails should be fixed at a minimum height of 910 mm. Any gap between the top rail and any intermediate rail should not exceed 470 mm. Toeboards should be suitable and sufficient and adequately secure.

Note

Two scaffold boards must not be used one on top of the other.

Access

It is now good practice to access a bricklayer's scaffold from a fixed ladder, usually inside the scaffolding, and it must be positioned so that this can be done easily and safely. If the height is more than 9 m then a safe landing should be provided.

Relevant barriers and notices

A scaffold which is partly erected or dismantled should have its access blocked off and have a notice displayed saying that the scaffold should not be used (Figure 2.29).

Remember

The Work at Height Regulations do not ban the use of ladders.

They should only be used when a risk assessment has shown that the use of other suitable equipment is not appropriate.

The use of a ladder should be restricted to short durations only.

Multiple-choice questions

Self-assessment

This section of the book is designed to allow you to check your level of knowledge. The section consists of revision questions for this chapter. The questions are all multiple choice and have four possible answers. The answers are to be found at the end of the book.

The main type of multiple-choice question will be the four-option multiple-choice question. This will consist of a question or statement, known as the stem, followed by a choice of four different answers, called the responses. Only one of these responses is the correct answer; the others are incorrect and are known as distracters.

You should attempt to answer the questions by choosing either (a), (b), (c) or (d).

Example

The person employed by the local authority to ensure that the Building Regulations are observed is called the:

(a) clerk of works
(b) building control officer
(c) council inspector
(d) safety officer

The correct answer is the building control officer, and therefore (b) would be the correct response.

Health and safety in the construction industry

Question 1 Which safety sign has a red circular border with a red cross-bar?

(a) mandatory
(b) warning
(c) information
(d) prohibition

Question 2 Out of the current legislation listed below, which one covers hard hats?

(a) Work at Heights Regulations
(b) Control of Substances Hazardous to Health
(c) Personal Protective Equipment at Work Regulations
(d) Fire Precautions Regulations

Question 3 Which of the following is the official body that enforces Health and Safety controls?

(a) the local authority

(b) the public health authority

(c) the Health and Safety Executive

(d) the employer

Question 4 What is the item of equipment shown called?

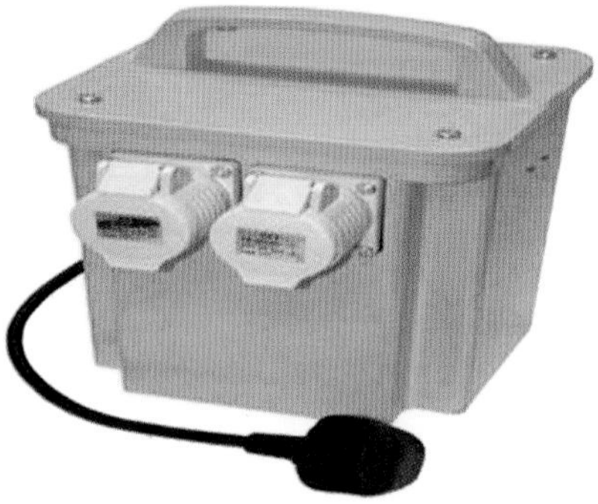

(a) transformer

(b) RCD

(c) generator

(d) compressor

Question 5 Which of the following could be prosecuted for an offence under the Health and Safety at Work Act?

(a) employer

(b) visitor

(c) general public

(d) spectator

Question 6 A person who is allowed to erect a scaffold should be:

(a) a bricklayer

(b) a scaffolder

(c) a site supervisor

(d) fully trained and competent

Question 7 A reportable accident is one:

(a) where attendance at the hospital was necessary
(b) where more than three consecutive days are lost from work
(c) where the site supervisor has been informed
(d) where an injury to a fellow operative has been caused.

Question 8 What is contained in a fire extinguisher which has a blue label?

(a) dry powder
(b) water
(c) carbon dioxide
(d) foam

CHAPTER 3

Programming and Resources for Work

This chapter will cover the following NVQ and Diploma units:

- NVQ VR209
- CC 3002K

This chapter is about:

- Identifying the work activities involved
- Adopting safe and healthy working practices
- Identifying resources to carry out the work
- Confirmation of a work programme/schedule for the occupational area of work being carried out

The following NVQ performance criteria will be covered:

- Programme and resources
- Clarification and advice on the resources
- Project requirements and external factors
- Work activities
- Alterations to the work programme

The following Diploma outcomes will be covered:

- Know about producing drawn information
- Know how to estimate quantities and price work
- Know how to ensure good working relationships

Programme and resources

The purposes of planning

Planning is the process where procedures are laid down so that certain objectives may be achieved within a specific time at an optimum cost.

Construction work is rarely simple and planning the project commences long before the first piece of turf is removed from the site.

The objective of planning is to complete a contract within the contract period to the satisfaction of the client at the least cost to the contractor.

Without planning the completion of a construction contract by a specified date would be pure chance.

Planning, to be fully effective, must be:

- based on realistic standards of performance
- aimed at using all available resources to the full
- balanced throughout
- flexible enough to meet with changing circumstances.

Planning is concerned with:

- men
- materials
- machines
- method
- money.

These are referred to as the 5 Ms of management. Without any one of these the project would be difficult to organize.

Planning provides the answers to:

- What is required?
- Where is it required?
- When is it required?

Advice on resources

Time and cost are the influencing factors when planning decisions are made.

Time

The majority of construction contracts have to be completed within a certain period, otherwise financial penalties may be incurred.

This contract period must influence planning decisions.

It is pointless completing a contract weeks ahead of time if the production costs are going to be greater and there is no financial gain to the contractor; the aim must be to complete the work as quickly as possible but at the optimum cost.

Cost

The costs of operations must be considered in relation to the time that is available for the operations to be performed.

Often the time available will indicate the method and cost.

External factors

Problems vary according to the type of site and contract.

No two projects are the same and each site brings with it new problems, such as:

- material hold-ups
- labour shortages
- plant breakdowns
- accidents
- non-appearance of subcontractors
- lack of information
- hold-ups in supplies of specialist equipment and machinery
- interference from the weather.

Planning will provide information on:

- the dates when labour, plant and materials are required
- the methods that will be adopted
- the dates when subcontractors are required
- the dates when plans, details and information must be available.

The degree of detailed planning will obviously bear some relation to the type and magnitude of the contract, but even planning in its broadest sense will highlight problems and provide a way to overcome them.

Planning the programme

Purpose

- To provide the broad guidelines for completing the contract within the time allowed.
- To provide for continuity of work for all site labour and plant.

- To consider possible difficulties, and resolve such difficulties so that maximum co-ordination and progress are maintained.
- To highlight requirements of:
  - labour
  - plant and equipment
  - materials
  - subcontractors
  - schedules
  - drawings, details and information, with relative dates.
- To illustrate the planning in a suitable form for easy reference.
- To provide a means for recording progress.

General rules of planning

- Decide what needs to be done.
- Ascertain the time required to carry out the work.
- Break the contract down into operations.
- Decide on the sequence and phasing of the operations.
- The target times used for the operations must be realistic and based on fact.
- Organize all resources to ensure that they are on site as and when required.
- Set down the planning in a suitable form for use on site.

Planning and programming

Do not confuse planning with programming.

Planning is the thinking which predetermines how the work shall be carried out to a successful completion.

Programming illustrates the planning in a graphical form as a means to achieve greater efficiency on site.

Programmes are usually in the form of bar charts. A bar chart consists of a horizontal timescale in hours, days, weeks or months, against which the estimated duration of the various construction operations are plotted.

Another graphical form of programming used in the industry is known as network analysis (critical path method), which uses arrow diagrams to decide the sequence of construction operations and to highlight the key critical operations that must be carefully controlled.

Method statements

The method statement shown in Figure 3.1 illustrates the labour, plant and equipment required for each operation.

This is prepared with the view of setting down in a logical sequence the best approach to the execution of the work, at the minimum cost, using resources effectively and giving a suitable production flow.

Details may include:

- key operations for the contract
- method and sequence of working
- plant requirements
- labour requirements
- contractor's labour and subcontractors' labour
- any additional notes for guidance.

Calculation sheets

When producing a programme a duration is given to each operation. This is calculated from previous experience and knowledge and should give acceptable times.

A typical calculation sheet is shown in Figure 3.2.

Bar charts

The bar chart is simple in concept and equally simple to understand. It is hardly surprising therefore that most site supervisors prefer this method.

Prior information is required before the bar chart can be completed:

- method statements
- calculation sheets.

The information from these two sheets is transferred to the bar chart (Figure 3.3).

Client: Contract: Ref No:		METHOD STATEMENT	Sheet No: Prepared by: Date:		
No	Operation	Method /Sequence	Plant	Labour	Notes
1	Excavate foundation trench	Excess material removed from site by lorry	Exc 360 2 lorries	1 lab	Timber if required
2	Concrete foundations	Begin immediately after trench is completed. Ready-mix concrete.	Dumper	4 lab	

FIGURE 3.1
Method statement

Client: Contract: Ref No:		CALCULATION SHEET				Sheet No: Prepared by: Date:	
No	Operation	Quantity	Rate Hours	Total Hours	Plant Gang size	Duration Hours	Duration Days
1	Excavate foundation trench	160 m^3	10 m^3 per hour	16 hours	Exc 360 1 lab	16	2
2	Concrete foundations	32 m^3	1 m^3 per hour	32 hours	Dumper 4 lab	8	1

FIGURE 3.2
Calculation sheet

The activities are listed down the left-hand side in the sequence they will take place on site. The timescale is drawn horizontally and the bars represent the time when the work will proceed on the activities.

Bars may be suitably shaded or coloured to distinguish individual trades.

It is usual for only half of the bar to be used for programming. The remainder is completed as progress is achieved. Alternatively, progress can be recorded immediately above or below the bar or in a separate chart specially provided for that purpose.

A good bar chart should:

- be easy to produce
- be simple to understand
- be simple to update due to changing circumstances
- be well illustrated
- be easily adapted to monitor progress
- use simple symbols as a warning to supervisors to chase information
- provide, as far as possible, for all resources to be used continuously while on site.

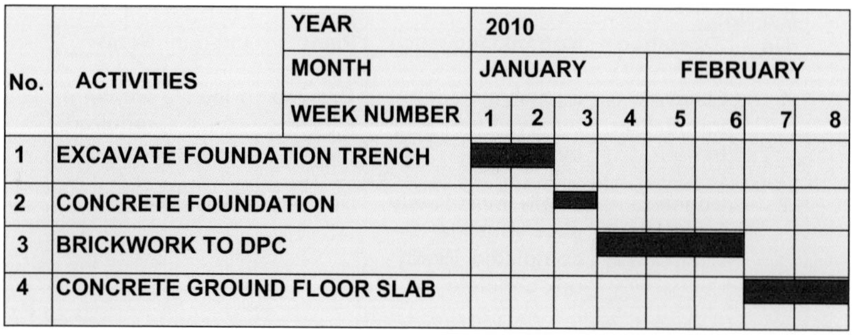
No.
ACTIVITIES
YEAR
2010
MONTH
JANUARY
FEBRUARY
WEEK NUMBER
1 2 3 4 5 6 7 8
1 EXCAVATE FOUNDATION TRENCH
2 CONCRETE FOUNDATION
3 BRICKWORK TO DPC
4 CONCRETE GROUND FLOOR SLAB

FIGURE 3.3
Typical bar chart

The person drawing the bar chart could decide on the layout, or it may be company policy to adopt a standard format.

The minimum information on the bar chart will comprise the activities and their duration, but plant, labour and material deliveries could also be included.

The example in Figure 3.3 shows the activities in their correct sequence. Against each activity is a bar showing the anticipated duration. Each bar should produce a flow through the various activities until final completion of the project.

Sequencing the work

The construction sequence is the order in which activities should be carried out.

The correct sequence is found by asking three questions of each activity:

- What activity must precede it?
- What activity can be done at the same time?
- What activity must follow?

The answer to the above questions will create the correct sequence to add to the programme irrespective of which method is used.

A typical sequence for a small detached bungalow may be as follows:

1. Set up site, including setting out the building.
2. Excavate foundations.
3. Concrete foundations.
4. Brickwork up to DPC.
5. Hardcore to ground floor, drainage.
6. Concrete to ground floor.
7. Brickwork to wall plate, including scaffolding and windows.
8. Roof structure.
9. Roof tiling, watertight.
10. First fix, joiner, plumber and electrician.
11. Plasterer.
12. Second fix, joiner, plumber and electrician.
13. Painter, internal and external.
14. External works.
15. Clean out.
16. Hand over.

There will often be differences of opinion over the sequences depending on the workings of each individual construction firm.

Certain operations can be concurrent, but others definitely cannot be concurrent.

Remember

Even though bar charts are simple, they will be more useful if planned and drawn with care.

The draughting process can be made easier by providing good-quality standard bar charts.

Progress charts

The programme represents an anticipated work flow which, if adhered to, will allow the project to run smoothly and be completed on time without any financial penalty. It is therefore necessary to mark progress on the chart to show the site manager when any areas are falling behind so arrangements can be made to bring the project back on programme.

Such action is usually the outcome of an enquiry as to the reasons for insufficient progress. This enquiry will lead to a reallocation of existing resources or the input of increased labour or plant.

It is most unlikely that any construction project will run entirely in accordance with the programme. Some activities will be ahead of and some behind schedule.

The moment of progress recording should always be shown on the document. In the case of bar charts a date cursor should be placed on the actual date. This is usually a ruler or piece of string with drawing pins from the top to the bottom of the programme (Figure 3.4).

Critical path

The critical path method (CPM) is another form of programming.

This planning method identifies the critical operations within the programme and shows the site agent which operations to monitor closely if the contract is to be completed on time.

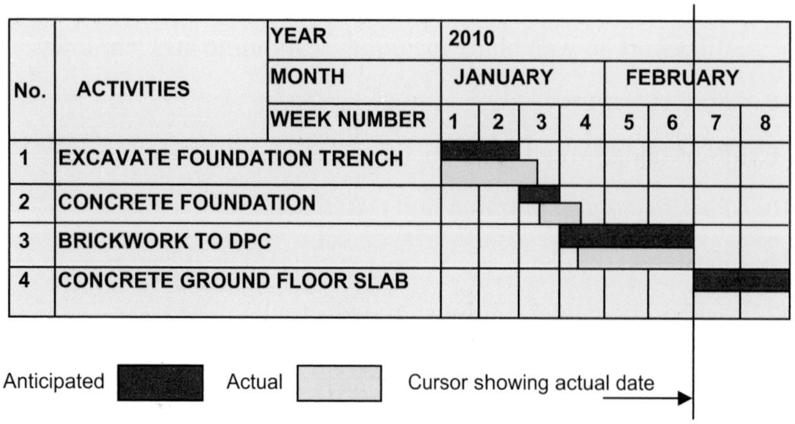

FIGURE 3.4
Marking progress

The calculation sheet and method statements are used similarly to preparing the bar chart, but the method of presentation and programming are completely different.

As described with bar charts, a construction project can be considered as being made up of a set of operations. In CPM these operations are called activities. They are:

> *THINGS WHICH HAVE TO BE DONE, WHICH TAKE TIME AND RESOURCES.*

On most construction projects a time factor is involved; the aim of construction planning is to complete the project in the shortest time at the least cost.

To complete a contract on time not only must the various activities all be completed, but they must be completed in a certain sequence.

We must ask ourselves the same questions as with bar charts:

- What activity precedes?
- What activities can be done at the same time?
- What activities follow?

The basis of a CPM is the logic network. This is a diagrammatic plan drawn up using three basic symbols to show in logical form the interconnected relationships of the activities involved in carrying out a construction project.

The three basic symbols are:

- activity
- event
- dummy activity.

An *activity* is represented by an arrow.

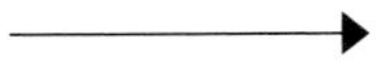

This is an activity which takes time and resources. The arrow is not drawn to scale. The head of the arrow indicates where the activity ends.

The activity name is written along the top of the arrow and the duration is written underneath the arrow.

Always keep arrows horizontal to aid with inserting the activity name.

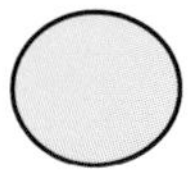

An *event* is represented by a small circle, called a node.

An event is a point in time, or intersection, at which activities can start or finish.

The event circle is identified by an event number inside the node.

A *dummy activity*, shown as a dotted or dashed arrow, indicates a relationship between activities, but does not have any time or resources allocated to it.

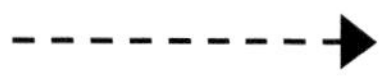

A dummy is used only to make clear the logical relationships between activities or to ensure that no two activities have the same event numbers at both head and tail.

The event number at the head of the arrow must be greater than the event number at the start.

The node can be completed in several ways according to the decisions of the company.

A typical example could be as follows:

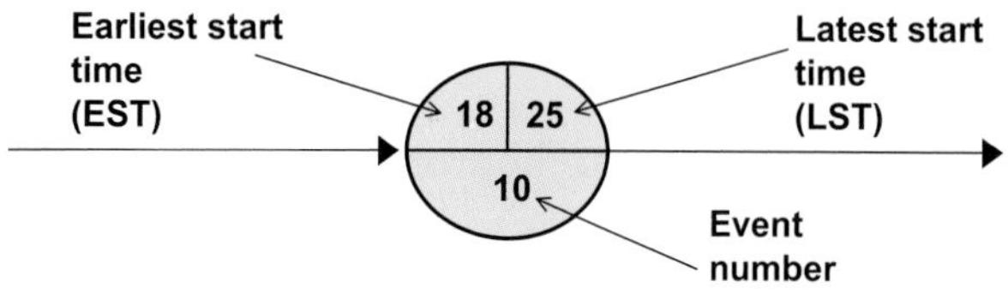

BASIC ARROW DIAGRAMS

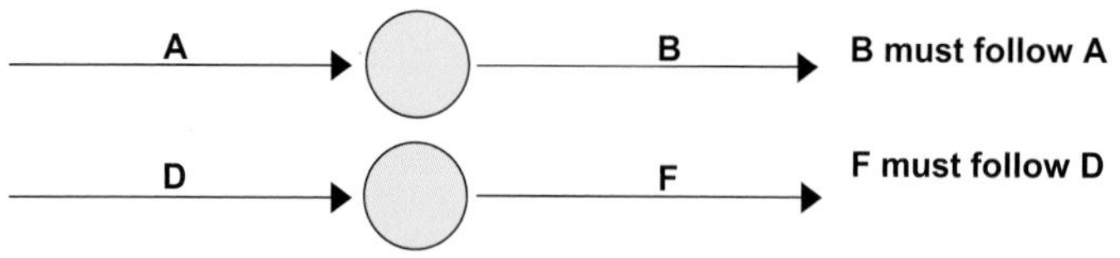

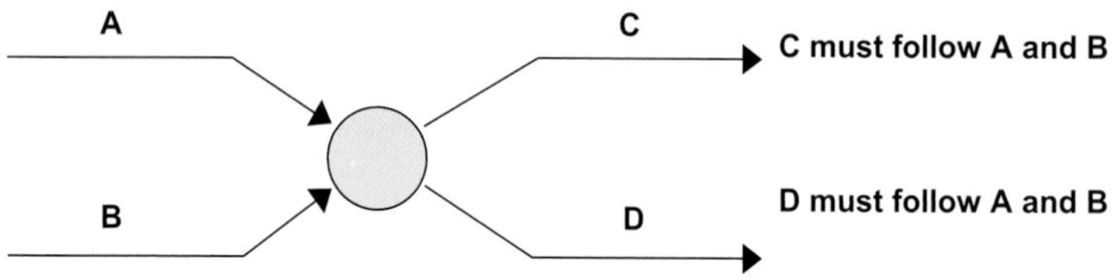

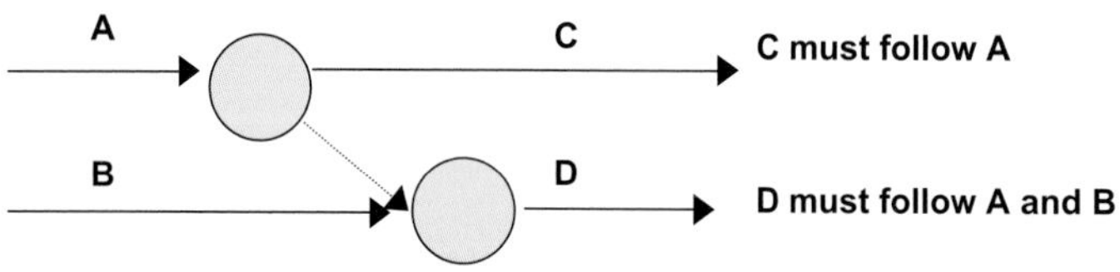

An activity cannot start until the event preceding it has occurred. An event cannot occur until all activities leading to it are completed.

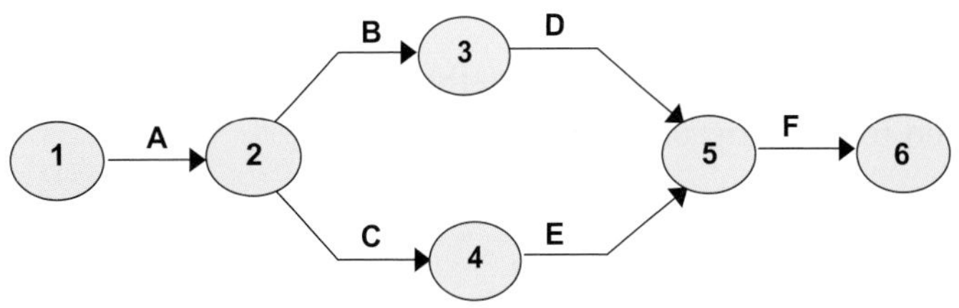

In the above example, B and C cannot start until event 2 has occurred, i.e. A has been completed. F cannot start until event 5 has been completed, i.e. both D and E have been completed.

If a dummy arrow was inserted from event 3 to event 4 what would be the result?

D can be started as soon as B is completed but E cannot be started until both B and C are completed.

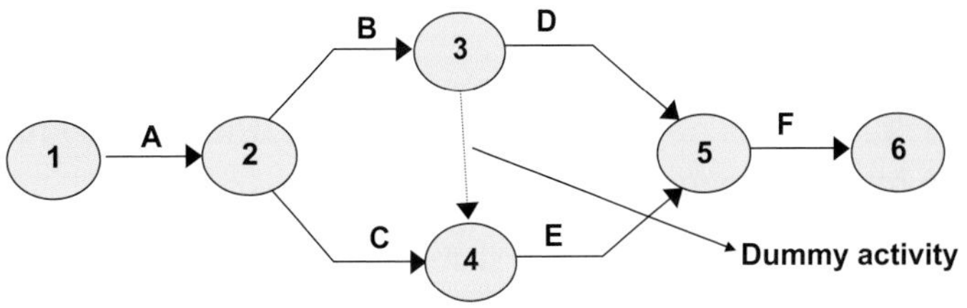

Event 4 cannot now occur until event 3 has occurred, i.e. B has been completed.

If a new activity G is introduced, which can start as soon as event 5 has occurred, and must finish before event 6, it must not be drawn as the example shown below:

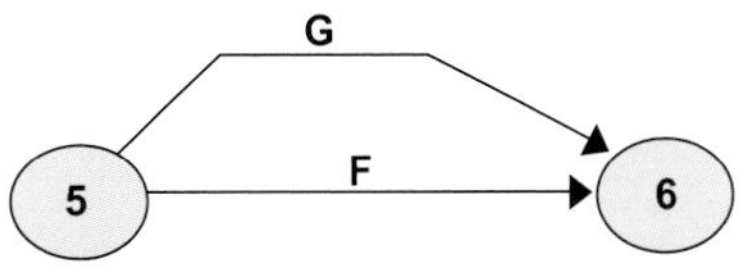

This would give both F and G the same event numbers (both 5 and 6), which would confuse the following analysis.

A dummy activity and a new event node must therefore be inserted and renumbered.

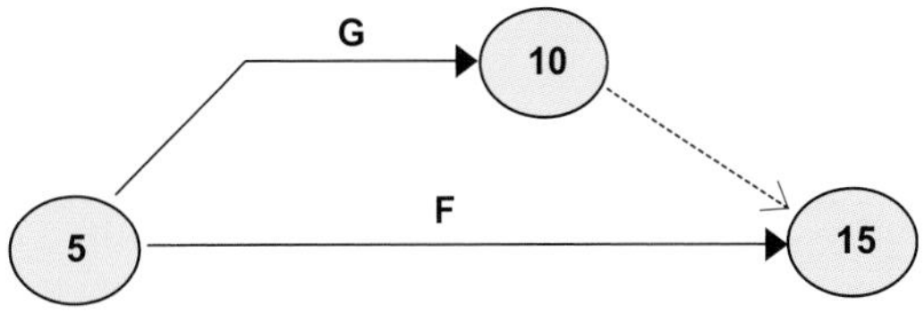

This highlights another problem with numbering event nodes. It is beneficial to number in 5s, therefore allowing for the insertion of a node without renumbering.

BASIC TECHNIQUES

The information received from the method statement and calculation sheet can be used to build up the critical path.

Put the list of activities into their correct sequence and determine their priority, i.e. for each activity:

- What activity must it follow?
- What activities must it precede?
- What activities can be done at the same time?

Express the activities in an arrow diagram, remembering to allocate only one activity per arrow.

The arrows should always run from left to right.

At this stage the arrow diagram indicates only the sequence of operations and the interrelationships between them. The duration of the activities have not been considered.

Because the diagram is not drawn on a time scale, this allows full attention to be given to the sequencing of the operations.

Using information from calculation sheets and method statements we now have to assign duration to each activity, and indicate these times on the arrow diagram.

This should be done underneath the activity arrow.

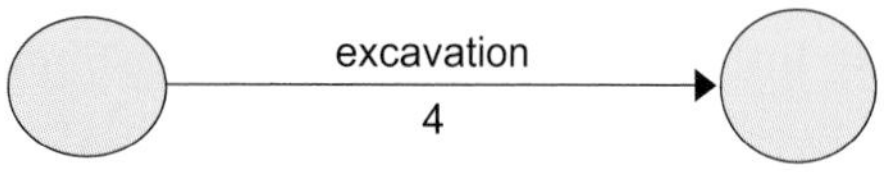

Example 1

Consider the following six activities as part of a construction project:

- Site access: 2 days
- Site clearance: 4 days
- Temporary services: 2 days
- Site accommodation: 1 day
- Security fencing: 3 days
- Set out site: 4 days

The following points must be taken into consideration:

- Site access has to be completed before any other work.
- Site accommodation cannot commence until the site has been cleared.

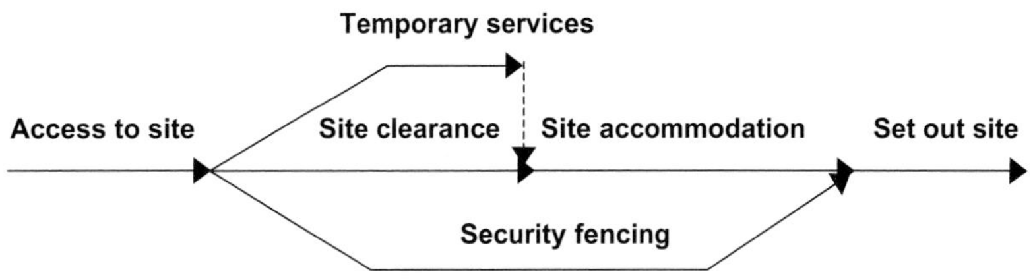

FIGURE 3.5
Network outline

- The site cannot be set out until the security fence has been completed.
- The following network shows the interrelationships of the activities (Figure 3.5).

Procedure

1. Draw up the network. Once the framework has been completed the nodes have to be completed (Figure 3.6).
2. The second task is to add the durations to the network, remembering to place them underneath each activity arrow.
3. The third task is to number the nodes, remembering to number in 5s (Figure 3.7).
4. The fourth task is to calculate the earliest event times.

THE FORWARD PASS

The earliest event time for each event is the earliest time when all preceding activities have been completed.

Working from left to right on the network, record under each event the earliest time an activity can start and finish.

The earliest start time is entered in the left quarter of each node.

The first node is always 0 (Figure 3.8).

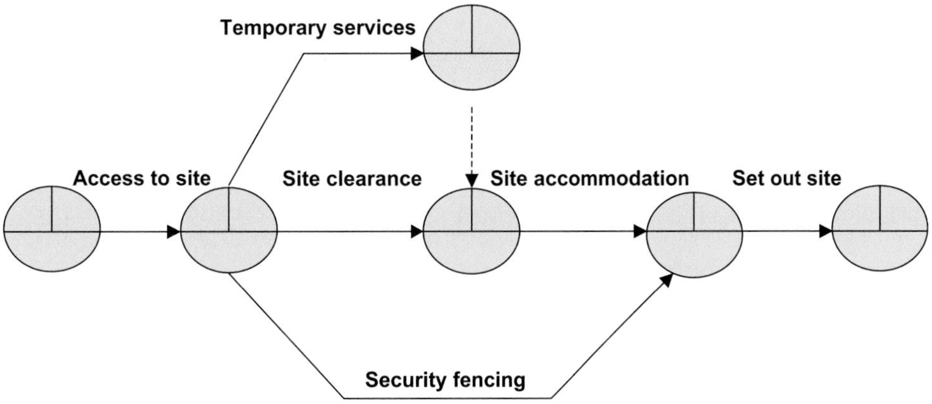

FIGURE 3.6
Network with nodes

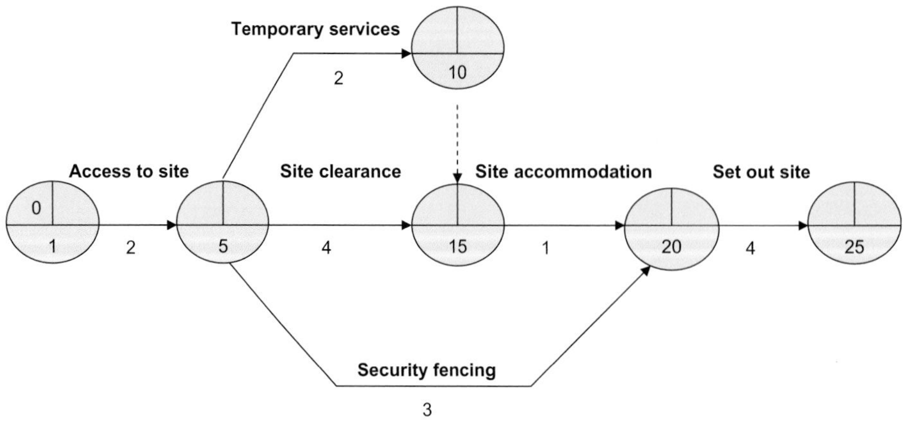

FIGURE 3.7
Network with duration and node numbers

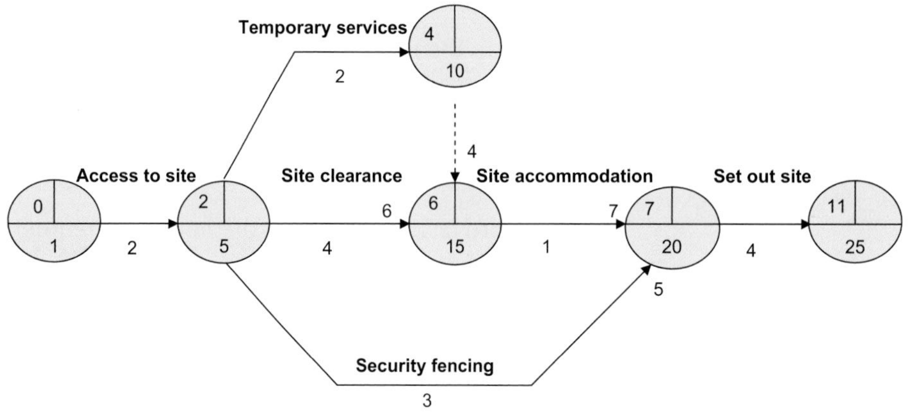

FIGURE 3.8
Network with earliest event times

By adding the duration of the next activity to this start time you can calculate the earliest finish time:

$$0 + 2 = 2$$

This is inserted in the left-hand quarter of the next node.

This should be continued until all earliest event times are completed.

Remember: the latest finish for one activity is the earliest start for the next.

When two or more pathways merge the calculated times should be placed outside the node and the highest number placed inside the node.

5. The fifth task is to calculate the latest event times.

The latest event time is the latest time which allows all the following events to be completed within the overall project time.

THE BACKWARD PASS

The procedure is the reverse of establishing the earliest event times, i.e. by working from right to left.

The first number to enter in the right-hand quarter of the last node is the same number as was placed in the left-hand quarter of the last node.

The calculated time is entered in the right-hand quarter of all the nodes.

This time, when more two or more pathways merge the lowest number is placed inside the node (Figure 3.9).

6. The final task is to calculate and mark in *red* the critical path.

This is the route through the network for all activities that must be started without delay, following the completion of all preceding activities, and which need all the available time until the next event.

It passes through all events having equal latest and earliest times, i.e. critical activities (Figure 3.10).

FLOAT

In the example it will be noticed that activities that are critical have no spare time. All the other activities have spare time, and this is called float. Float

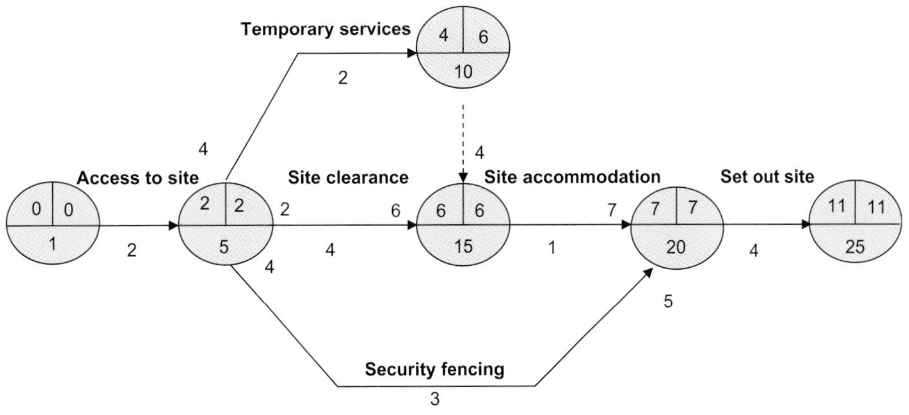

FIGURE 3.9
Network with earliest and latest event times

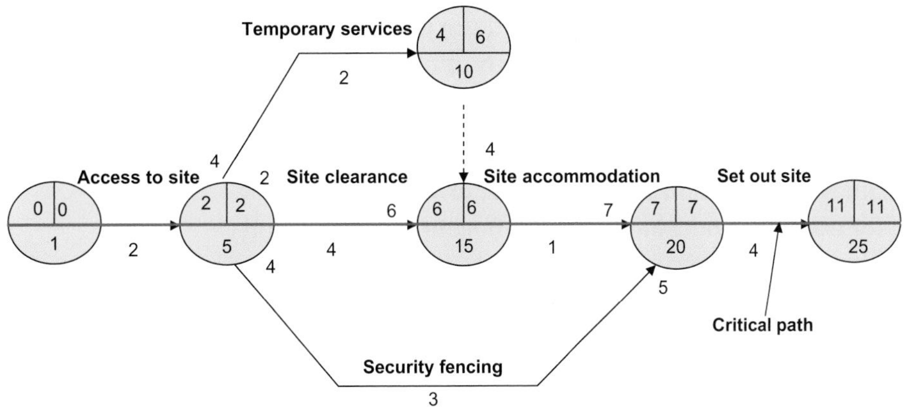

FIGURE 3.10
Critical path

represents time that can be used without affecting subsequent activities, providing the activity starts at its earliest time.

Alterations to the work programme

Causes of change

It is an unfortunate fact of building life that contracts often run behind schedule. There are contractual obligations placed on the contractor with regard to contract duration. If it becomes apparent that the programme is behind schedule the contractor should notify the architect in writing, stating the reason for the delay.

Not only is this good management, but the success of the project depends both financially and contractually on conformity with the initial programme.

There may be many reasons for a delay. Although extra payment may be received for some, others can cause a financial loss.

Possible reasons for extending the programme duration include:

- changes in design
- architect's instructions.

DESIGN CHANGES

There may be many reasons for the design to be changed, either by the client or by the architect.

ARCHITECT'S INSTRUCTIONS

However carefully the design team has prepared the production information, almost invariably, there will be a need to issue further instructions, drawings and schedules after the contract has been signed. These additional detailed requirements and items of information are known as architect's instructions.

For example, the clerk of works or building control officer may require extra depth to the foundations at the inspection stage, and the main contractor has to comply with that instruction. This extra depth of excavation must be confirmed with an architect's instruction.

Architect's instructions have a defined place under the contract. They should always indicate what alteration is being made, or what provisional sum is being spent, for example, and what the total effect of the instruction is on the contract sum.

They can include the following:

- compliance with statutory requirements
- discrepancies in documents
- levelling and setting out the works

- variations
- making good faults
- removal of work or materials not in accordance with the contract
- expenditure of provisional sums
- subcontractors' work.

Variations

Variation orders, issued during the course of the contract, alter the originally specified work.

They are brought about for one of the following reasons:

- alterations due to statutory requirements
- alterations, additions and omissions, due to either the employer or the architect changing their minds
- errors or omissions in the bill of quantities.

The architect can only issue a variation within the terms of the contract. He or she cannot issue a variation that would radically change the whole contract; for example, to turn an office block into a factory unit.

Variation orders generally mean a variation in the amount of the contract. This is achieved by using either prices for a similar item in the bill of quantities or daywork sheets.

A variation is an architect's instruction, but not every architect's instruction is a variation, as some merely amplify previously given information.

External conditions

There are also numerous external causes of changes in the duration of the contract. The main reason is the weather. During the planning stage the planners will have taken the time of year into consideration, but despite well-laid plans, the weather can cause havoc.

It is important to monitor the weather each day of the contract and the site diary should be filled in accordingly, usually morning and afternoon.

This information can be used as evidence when putting in claims for extra contract time.

Lack of resources

A lack of resources may result from a delay in receiving information from the architect, or the material supplier may be having problems with delivery.

If the lack of resources is due to the architect's delay or the material supplier is nominated by the architect then these are reasonable claims for an extension of time. If the material supplier has been chosen by the contractors then they have no basis for a claim.

The resource could also be either plant or labour. The responsibility for these lies with the contractors. It is essential that labour and plant requirements are constantly checked and if there are any discrepancies in the programme extra labour or plant should be brought in immediately.

A further resource is the financing of the contract. A budget should have been produced before the start of the contract. This should be checked weekly to analyse whether the contract is on schedule.

Financial problems

If the contract goes over time there will be financial losses. They could be either:

- liquidated damages against the contractor
- extra costs for the contractor for wages.

EFFECTS ON CONTRACTUAL OBLIGATIONS

There are two key dates in the contract for the contractor:

- date for possession
- date of completion.

The architect may insist on liquidated and ascertained damages forming part of the contract conditions. The contractor therefore has to complete the contract on time unless there are legal extensions to take into consideration.

It is vital that the contracts manager keeps meticulous details of all architect's instructions, variations and weather details.

Despite these contractual procedures the main contractor must still complete the contract by the date of completion, if at all possible. Therefore, good planning, organization and monitoring of progress are essential.

Improved efficiency

The contractor has an obligation to complete the contract on time, even with all the problems thrown at him. This may be achieved through extra efficiency from the site staff.

Project requirements and external factors

When planning a site it is important to check the following because each one could affect the final programme and cost of the contract:

- occupiers
- near neighbours
- public access
- existing utilities

- operational area conditions
- operational area transport routes
- application of codes of practice
- health, safety and welfare
- manufacturers' instructions
- operating instructions
- waste disposal requirements.

OCCUPIERS

It is important when you are working on a property that is occupied that the occupier in not put under any stress while work takes place. The client should be taken into consideration with all planning and development work.

Remember: a satisfied client may bring in other work by word of mouth.

NEAR NEIGHBOURS

It is only polite to take into consideration the neighbours near to the property you are working on.

Any noise and dust could cause discomfort and should be taken into consideration during planning.

PUBLIC ACCESS

Most greenfield sites are difficult to close off completely to protect the public from harm. Small sites can easily be cordoned off with hoarding or fences to prevent the public from straying onto the site and causing damage to the site and themselves.

It may be necessary to provide a temporary footpath while work is being carried out, but special permission has to be granted from the local authority.

EXISTING UTILITIES

Any existing utilities must be located before any construction work takes place as damage could be caused when excavating on the site.

The various authorities will assist in the operation to locate them.

OPERATIONAL AREA CONDITIONS

The site must be visited before programming to take into consideration any particular problems, for example:

- access and egress, temporary roads, the siting of offices, cabins and materials
- the type of ground, boundaries, etc.
- any existing services or buildings to be removed.

All these items could affect the programming and the eventual contract price.

OPERATIONAL AREA TRANSPORT ROUTES

The situation of the site could affect transport costs, not only for your own staff but also for the cost of material deliveries.

APPLICATION OF CODES OF PRACTICE

All sites have to be organized taking into consideration the various appropriate British Standards, codes of practice and Building Regulations.

HEALTH, SAFETY AND WELFARE

All sites come under the Health and Safety at Work Act. The welfare provisions for the workforce on site are of particular importance when costing out the contract.

MANUFACTURERS' INSTRUCTIONS

When using any plant or equipment that is new or different it is important that the manufacturer's instructions are adhered to. If not, it could affect any insurance claim.

OPERATING INSTRUCTIONS

As before, any insurance claim could be refused if the operating instructions were not adhered to.

WASTE DISPOSAL REQUIREMENTS

It is important that the nearest tipping facilities are located, and the distance and cost of tipping taken into consideration when costing.

Suppliers

It is essential to liaise with suppliers at all times during the contract. One of the best methods is to hold site meetings.

Organizational procedures

There will be numerous occasions when corrective action is required.

Remember that a programme is what should happen. In realistic terms it is a rare case when the contract runs exactly to programme.

All activities must be recorded for future reference. The most common method is the use of a site diary.

The building site diary

This has often proved a most useful document and if well kept, with the right type of information recorded in it, its value in matters of dispute later in the project cannot be overemphasized. Even long after project completion, the diary has often been of utmost importance in cases that have gone to court.

A typical site diary and report sheet is shown in Figure 3.11.

The main objective of the site diary is to record events and information that do not warrant special records being kept.

Different supervisors will use the diary in different ways, placing more importance on certain items than others. As a general rule, items that may be recorded could be:

- telephone promises from subcontractors, suppliers, etc.
- verbal instructions from the architect
- visits to the site by the client, architect, quantity surveyor, factory inspector, etc.
- details of weather conditions, especially in winter months
- delays in programme due to late delivery of materials or late start of subcontractors
- verbal instructions from head office or the contracts manager

SITE DIARY AND DAILY REPORT		
Contract	Contract No:	Week ending: ....
Weather	Temperature AM	Temperature PM ....
Labour on site: Own staff / Subcontractors		Plant on site: Own plant / Hired plant
Job stoppages	Man hours lost	Reason
Drawings received	Information received	Telephone calls
Visitors to site	Accidents	Reported accidents
Brief report of work and progress on site		
Signed : Site manager	Contracts manager	

FIGURE 3.11
Site diary and report sheet

- any matters of unusual occurrence, decisions or actions it is felt should be recorded
- working drawings.

Working drawings

The working drawings can be classified as:

- location drawings
- component drawings
- assembly drawings.

LOCATION DRAWINGS

These are further classified into:

- block plans
- site plans
- location plans

Block plans

These are used to identify the site in relation to the surrounding area (Figure 3.12).

The scale is usually 1:2500 or 1:250 and is too small to allow much more than an outline of the site and boundaries, road layouts and the other buildings in the near vicinity.

The orientation of the site is always shown with a suitable logo depicting north. The actual site should be outlined in red.

It is unlikely that dimensions would be added to these drawings.

Plans of this sort are usually based on the Ordnance Survey sheet for the area; however, if such a source is used, permission should be obtained for its reproduction.

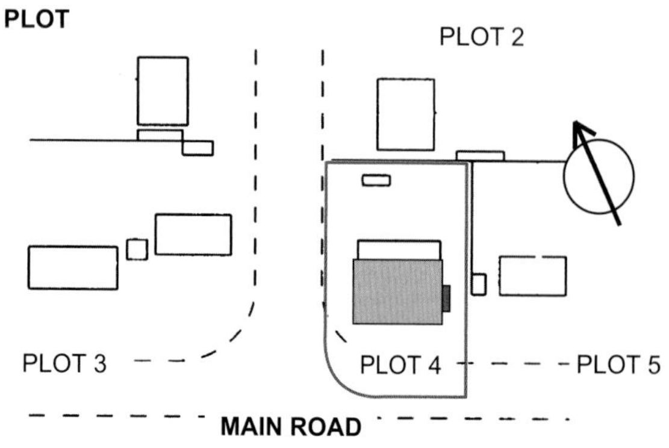

FIGURE 3.12
Block plan

These drawings will show the estimator where the site is and whether travelling is involved.

Site plans

These are used to show the position of the proposed building on the site, together with information on proposed road, drainage and service layouts, and other site information such as levels. A typical site plan is shown in Figure 3.13.

Again, the orientation of the site should be shown.

Scales of 1:500 and 1:200 are often used.

The information on the drawings is used by both the design team and the contractors. These drawings will show the estimator the amount of excavation and drainage required.

Location plans

These are used to show the size and position of the various rooms within the buildings and to position the principal elements and components (Figure 3.14).

Location plans are usually drawn to a scale of 1:200, 1:100 or 1:50.

These scales are sufficiently large to allow dimensions to be added.

Further information from the drawing includes the wall construction and position and type of doors and windows.

DRAWINGS

The design team will be required to produce working drawings for the builder to use on the site.

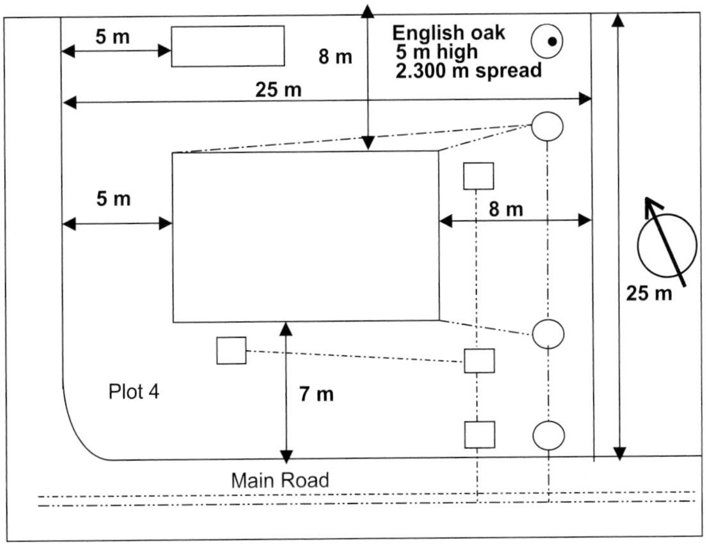

FIGURE 3.13
Site plan

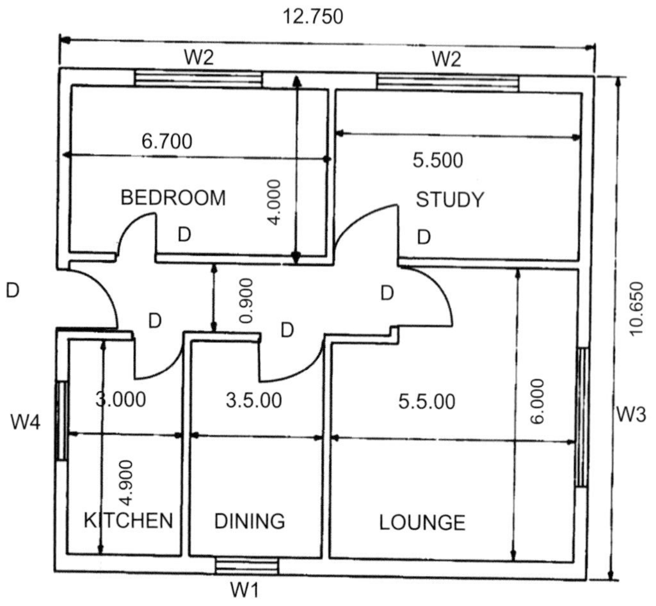

FIGURE 3.14
Ground floor plan

Drawings, schedules and specifications will have to be prepared, explaining how the design team requires the building to be constructed. To be able to read these drawings it is essential that the trainee is able to understand them.

Drawings should be produced according to the current British Standard. These recommendations apply to the sizes of drawings, the thickness of lines, dimension of lettering, scales, various projections, graphical symbols, etc.

The person carrying out a task should be able to read drawings and extract the required information.

Information concerning a project is normally given on drawings and written on printed sheets.

Drawings should only contain information that is appropriate to the reader; other information should be produced on schedules, specifications or information sheets.

DETAIL DRAWINGS

These are used to show the information necessary for the manufacture of the various components, i.e. doors, windows, concrete units, cupboard units, etc. Figure 3.15 shows a typical detail drawing of a coping stone.

The information from these details is only for the manufacturer – the architect will produce these details either in full size, i.e. 1:1, or at 1:5, 1:10, etc.

They should include every tiny detail required for manufacture.

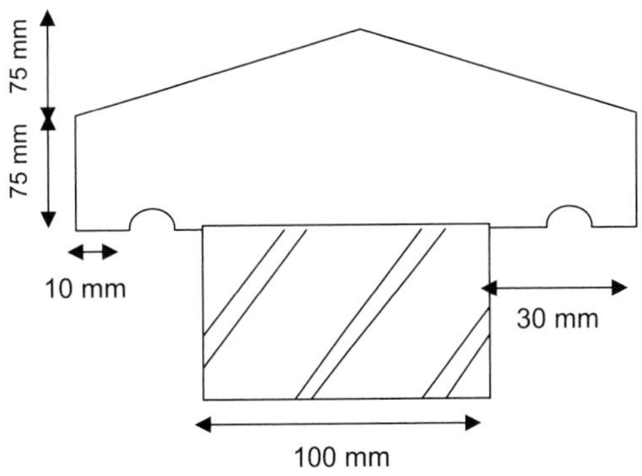

FIGURE 3.15
Typical detail

ASSEMBLY DRAWINGS

These are used by the architect to show in detail the junction between the various elements and components of the buildings. For example, Figure 3.16 shows a section through the eaves of a building.

These details are necessary for the builder to know exactly how the architect requires the construction to be completed.

The scales are usually 1:20, 1:10 or 1:5 and should be fully dimensioned and annotated.

PROJECTIONS

Drawings should aim to give as much information as possible. They usually show several views on a single drawing.

Orthographic projection is mainly used for illustrating three-dimensional objects. The basic layout for a drawing in orthographic projection is shown in Figure 3.17.

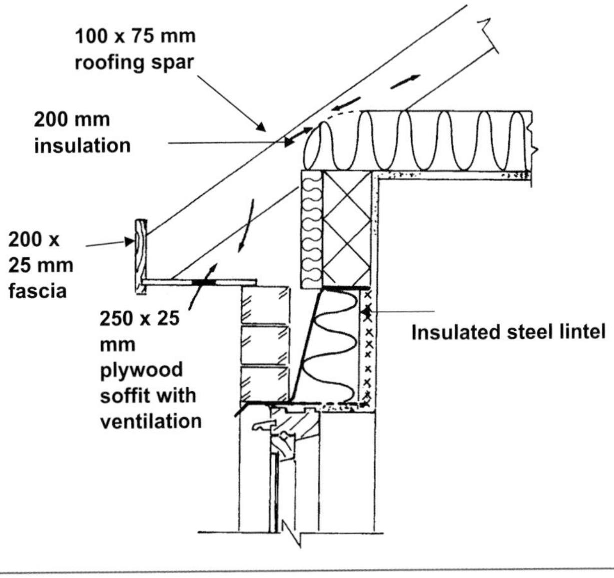

FIGURE 3.16
Assembly drawing

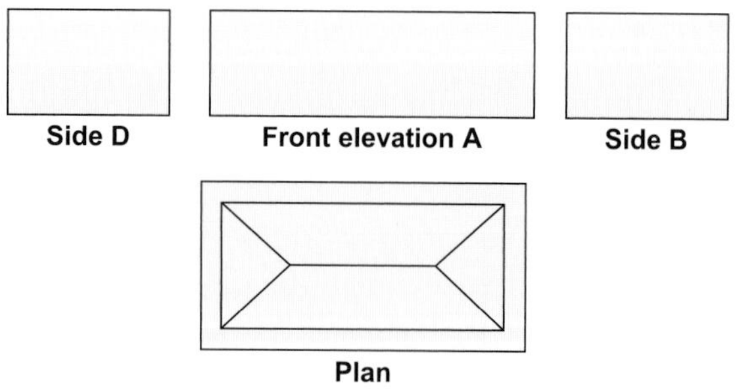

FIGURE 3.17
Orthographic projection

There are occasions when it is an advantage to be able to represent more than one side of an object in the drawing. This can be achieved using either isometric or oblique projections, when two sides and the top or bottom of the object are shown.

Isometric projection

This is a method of presenting the detail in pictorial format (Figure 3.18).

The definition of isometric projection is that all vertical lines remain the same but the horizontal lines are all drawn at 30 degrees to the horizontal.

Oblique projection

This is another method of presenting the detail in pictorial format.

There are two main types of oblique projection, cavalier and cabinet (Figure 3.19). Both are drawn with the front face drawn to true shape and size, similar to a front elevation.

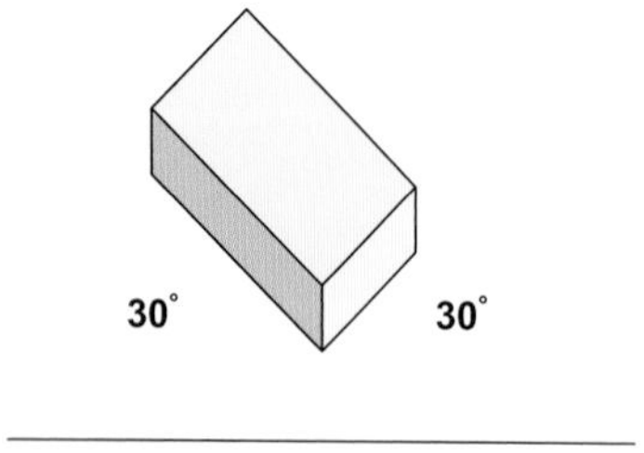

FIGURE 3.18
Isometric projection

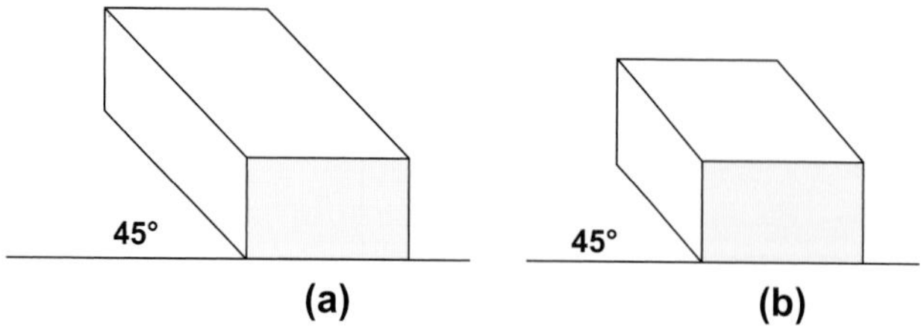

FIGURE 3.19
Oblique projection: (a) cavalier; (b) cabinet

The faces that recede are drawn at an angle of 45 degrees, and the difference between the two methods is the scale of these receding lines. The cavalier method uses the full scale, whereas with the cabinet, half full size scale is used, which creates a more normal appearance.

Computer-aided design

Computer-aided design (CAD) gives an accuracy which is greatly superior to manual drawing. The scale of the drawing can be enlarged on screen to allow the user to design more accurately. The drawing can be moved around the screen as required and can be either two-dimensional or three-dimensional. The designer can zoom in to give greater detail and for ease of working.

There are common libraries of wall, floor and roof types. Many manufacturers illustrate their products on a CD-ROM for use by designers.

The most important benefit of CAD is the option to alter and amend as required at the touch of a button, compared with the use of an eraser in traditional drawings. Once the CAD designers are proficient the speed of drawing is much faster than manual drawing.

The main downside is the initial cost of the computer hardware and programs. In addition, the use of any computer equipment can be detrimental to health. It is recommended not to spend too long at the screen in order to avoid glare and eye strain.

With either method the drawings are only as good as the person drawing them.

Estimating

Bills of quantities

The bills of quantities (Figure 3.20) are produced by the quantity surveyor working for the architect.

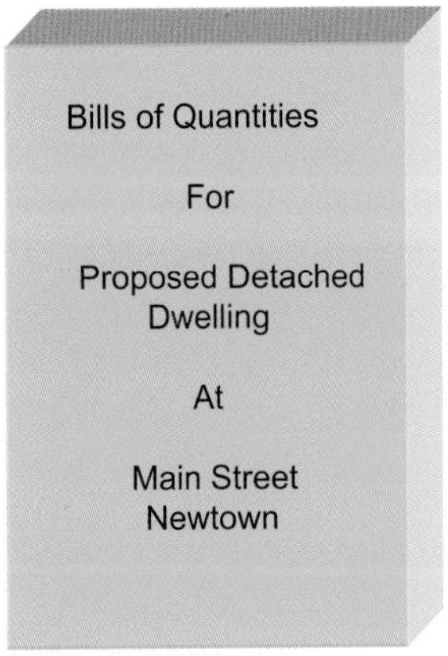

FIGURE 3.20
Bills of quantities

These documents give a complete description and measure of the quantities of labour, material and other items required to carry out the works, based on the working drawings, specifications and schedules.

The estimator should also be able to take off the quantities of materials required for the contract directly from the working drawings.

Specifications

Except in the case of very small building works the drawings cannot contain all the information required by the contractor, particularly the required standard of materials and workmanship.

For this purpose, the architect will prepare a document known as the specification (Figure 3.21) to supplement the working drawings

The specification is a precise description of all the essential information and job requirements that will affect the price of the work but cannot be shown on the drawings.

Typical items included in the specification are:

- restrictions to working hours
- limited access
- availability of services
- description of materials, quality, size and tolerance
- description of workmanship, quality, fixing and jointing
- other requirements, e.g. site clearance, making good on completion, nominated suppliers and subcontractors
- inspection of the work.

When drawings are received on site they should be carefully studied so that the work to be done is fully understood.

Drawings showing the work to be carried out are drawn to scale, in one of the general scales used in the building industry. These can be numerous

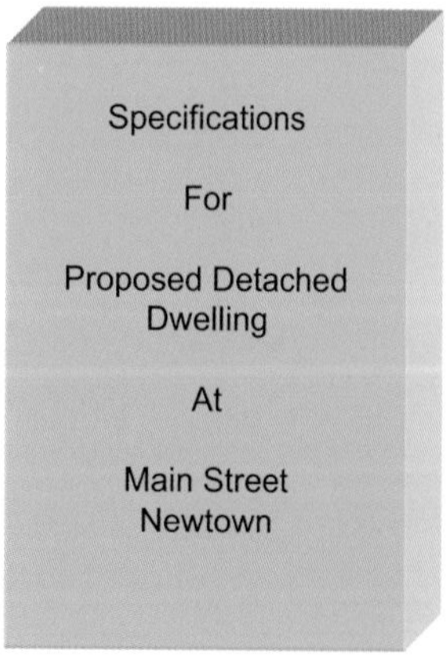

FIGURE 3.21
Typical specifications

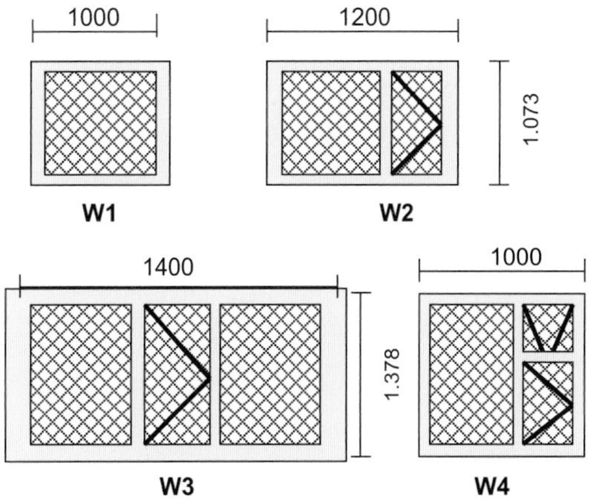

FIGURE 3.22
Window drawings

on large site, so only those relating to the quote required from the supplier are sent.

The example in Figure 3.22 could be used for a quotation for windows. The drawings of windows can be obtained from any manufacturers' literature with their references, i.e. W1, W2, W3, etc. This reference is all that is required by the architect when completing the working drawings as the builder is provided with a window schedule for cross-referencing.

Schedules

These are used to record repetitive design information about a range of similar components, such as:

- doors
- windows
- ironmongery
- sanitary ware
- radiators
- finishes
- floor and wall tiling.

The information contained in schedules is essential when preparing estimates and tenders.

Schedules are also extremely useful when measuring quantities, locating work and checking deliveries of materials and components. A window schedule is shown in Figure 3.23.

Estimating quantities and price

Once the resources have been identified, such as labour, materials and plant, the next important step is to produce an estimate for the work.

WINDOW SCHEDULE					
Job Number		Site Address			
Window Reference	Location	TYPE			
		W1	W2	W3	W4
External Double glazed	Dining	✓			
External Bow window Double glazed	Lounge		✓		
External Double glazed	Kitchen			✓	
External Double glazed	Study				✓

FIGURE 3.23
Window schedule

The estimator's role is to calculate a cost for carrying out a building contract. He or she will break down each item of work found in the bill of quantities into labour, materials and plant, and apply a rate to each. These rates are usually costs involved in previous contracts and the application of years of experience.

Added to the total cost of all the items will be a percentage for overheads and profit. This is usually 10–15 per cent according to the firm's policy.

The first stage is the pricing of the bill by the application of costing rates.

The cost of the material itself comes from expected market prices.

The cost of labour comes from the application, to the total quantities, of labour rates per unit quantity, e.g. £10.50 per cubic metre for excavating, £8.60 per cubic metre for clearing away, and so on through the trades, or by the estimated costs of complete operations such as those used for bonus targets.

Plant and other expenses may be allowed for in the labour rates or, preferably, estimated separately.

All this budgeting, and the tender figure when obtained, comprise a budget of cost, including profit.

Plant and equipment

With the modern trends in building techniques, coupled with the high cost of labour, mechanization in the form of mechanical handling and excavation is having to be accepted on contracts.

A contracting company will have to decide whether to form their own plant organization or to obtain plant from recognized hire companies. To assess the problem, one should first investigate the economics of hiring versus purchase.

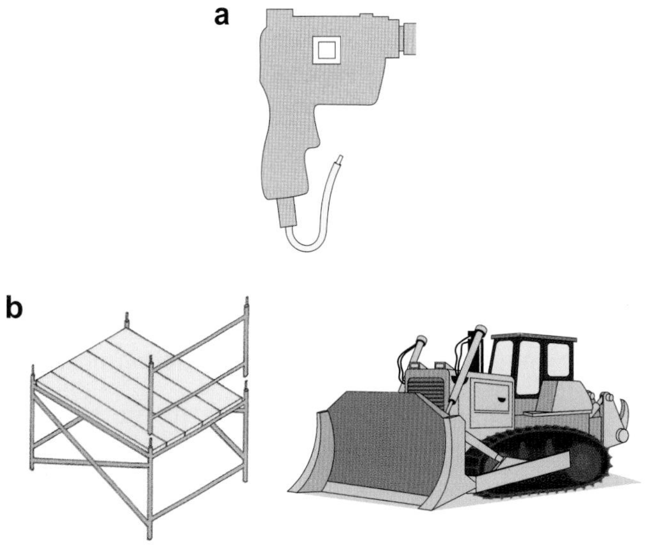

FIGURE 3.24
Items of plant and equipment: (a) small plant and tools; (b) large mechanical plant and scaffolding

Mechanical plant operated by the builder falls into one of two categories, shown in Figure 3.24.

PURCHASE OR HIRE

While either category can be owned by the builder or hired from a plant hire company, small plant and tools such as hand-held power drills and saws are normally purchased by the builder.

The builder will cover the costs of such tools by a percentage addition to the contract.

In pricing an item in the bill of quantities which includes plant items, the estimator will use either the hire charges quoted by a plant hire company or a rate which is normally quoted as a rate per hour, either including or excluding the operator.

Example 1

If it assumed that an excavator with a 1 m^3 bucket can carry out 30 operations per hour and costs £45.00 per hour, then the excavation costs with this machine will be £1.50/m^3.

There are a number of factors associated with this example, such as how the £45.00 per hour was derived, the number of operations per hour, the disposal of the excavated material and the balancing of the excavator with the transporting plant.

Faced with the alternatives of plant purchase or plant hire, a contractor must decide which practice is more advantageous for them to adopt. In doing so, he will consider the capital outlay involved, the length of time for which he requires the machine and its employability.

Plant hire

Under a hire agreement the company can obtain plant at a fixed rental which covers all maintenance, insurance and, if necessary, the provision of a trained operator.

Transport and other overheads are considerably reduced and depreciation can be ignored.

The large hire companies stock their fleets with the latest and most up-to-date machines, in order to provide an efficient service. Contractors can therefore obtain the right machine for their purpose and, at the same time, can test new equipment before deciding whether or not to buy.

One may think that hire would appeal more to the smaller company, faced with limited capital resources. This is not altogether true, for although a large company may have financial strength, its commitments are also likely to be proportionally greater. It is often the large contractor who, to meet peak demands, hires to supplement their own large plant fleet.

In general, it is cheaper to buy a machine if adequate capital is available and if the machine will be continually employed over long periods. For short- or medium-term use, hire is a better proposition. It would be unsound to buy new plant that may be idle for long periods.

THE PLANT RATE

Running costs

The plant hire company or the firm has to calculate the cost of plant, which is normally on a rate per hour basis.

The plant rate is built up from the following factors:

- initial cost and finance
- depreciation of the item of plant
- life of the plant
- hours worked per year
- repairs and renewals
- insurance and licences
- fuel, oil and grease
- inflation.

Initial cost

There are two ways to consider the purchase of plant: by purchasing from own resources or borrowing to purchase.

Depreciation

Several methods are used for calculating depreciation, but the straight-line method is usually preferred for the purposes of calculating a plant hire rate.

The straight-line method

The method simply involves the estimation of two factors:

- the economic life of the plant
- the scrap value of the plant at the end of its economic life.

If the item of plant costs £100 000.00 and has a scrap value after seven years of £10 000.00, then the amount of depreciation per year is calculated as follows:

Cost of plant	£100 000.00
Scrap value	– £10 000.00
	= £90 000.00
Years	÷ 7
Depreciation per year	= £12 857.00

Estimated working life

This is very difficult to assess, especially with the way in which operatives use each item. A guide for the most common plant is as follows:

- excavators: 7 years
- mobile cranes: 7 years
- compressors: 7 years
- tower cranes: 5 years
- pumps: 3 years
- fork-lift truck: 5 years

The anticipated life of these items of plant has been arrived at through experience.

Hours worked per year

Plant is never used for every working hour of every working day of the year.

Certain types of plant, such as concrete mixers, wheeled tractors and excavators, are likely to be in use for the majority of the year. In contrast, specialist types of plant, such as motorized scrapers and mobile concrete pumps, will be restricted in their use by the weather and demand.

The number of working hours per year for a site is approximately 1900.

Example 2: Worked hours

During the on-hire period certain factors will reduce the efficiency of the item of plant.

For an excavator the following time may be lost:

Total days per year	365
Weekends × 52	– 104
	= 261
8 public holidays	– 8
	= 253
3 weeks annual holidays	– 15
	= 238

This makes a possible 238 days per year or 1904 hours (using an eight-hour day)

But a further percentage could be lost due to:

Weather	10%
Manoeuvring	10%
Breakdowns	5%
Operator efficiency	5%
General waiting	10%
Total	40%

This extra 40 per cent could reduce the above hours from 1904 to 1143 hours.

Such factors need to be taken into account when calculating the effective output of plant.

Repairs and renewals

All mechanical plant suffers from breakdown due to failure of components. This has two cost effects, the first being the cost of the repair and the second being time lost.

The cost of repairs and renewals will vary for different types of plant, but a general figure would be 10 per cent of the purchase price per year.

Insurance and licences

It is normal for builders to have a contractor's all-risks insurance policy. This will cover all risks associated with the construction of a building.

For items of plant that will travel on the highway it is necessary to have a vehicle-specific insurance and also a current road fund licence.

Fuel, oil and grease

Fuel consumption is generally given in terms of litres per hour. Diesel is available as derv for use on public roads, and as gas oil for use on site. The latter is less expensive owing to the tax concessions.

A useful rule of thumb is that a petrol engine will consume approximately 0.275 litres of fuel per brake horsepower (bhp) hour, and a diesel engine will consume approximately 0.2 litres of fuel per bhp hour.

Example 3

A hydraulic excavator with a maximum power output of 130 bhp will be working for 5 seconds on full power during a 20 second cycle.

For the remaining 15 seconds it is likely to be operating at about half this rate.

The diesel fuel is therefore calculated as:

$$(0.25 \times 130 \times 0.2) = 6.5$$

$$+ (0.75 \times 65 \times 0.2) = 9.75$$

$$= 16.25 \text{ litres per hour}$$

The cost of oil and grease can be taken as 10 per cent of the fuel cost.

Tracks and tyres

The cost of maintenance on the tracks and the repairing or renewal of tyres has to be taken into consideration.

The tyre life for normal operating conditions is as follows:

Loaders: normal 2500 hours; harsh 1500 hours

Trucks: normal 3000 hours; harsh 2000 hours.

Inflation

A plant rate is calculated at the time that the plant is purchased. If no account were to be taken of inflation, then during the last year of the life of the plant the rate would not be competitive.

Rates are therefore reviewed on an annual basis and increases for the current rate of inflation are added.

PLANT SCHEDULING

Plant requirements should be ordered as soon as possible to allow the plant manager time to ensure availability from their own stock or by hiring from a reputable plant hire company.

It is normal practice for the planning officer within a construction firm, in conjunction with the estimator, contracts manager and plant manager, to fix levels and types of plant required for key operations. The site manager has therefore no alternative but to follow instructions regarding types of plant to be used.

There are times when the general foreman is expected to use his or her own initiative and to select, order and use plant, the suitability of which would depend entirely on the site manager's past experience.

A typical plant and equipment schedule is shown in Figure 3.25.

NEW CONTRACTORS LTD
PLANT AND EQUIPMENT SCHEDULE

Contract: .... No: ....
Job No.: .... Date: ....
Prepared by: ....

Plant Type and Equipment	Date required		Plant hire firm of supplier	Remarks
	Date on site	Date off site		
Concrete mixer	Oct 20th	Feb 10th	Own plant	Call off when required
Excavator	Nov 10th	Dec 12th	Hire firm	Maintained by hire firm
Dumper	Nov 10th	Dec 12th	Own plant	Call off when required

FIGURE 3.25
Plant requisition sheet

Materials

The cost of materials used in the construction industry exceeds 50 per cent of the overall costs.

The site manager has little or no control over the initial price of materials; every contractor has to pay approximately the same price. However, by employing strict methods of control in handling and using materials, the most efficient contractors will involve themselves with minimizing the costs in their utilization. In this respect the general supervisor plays a major role.

Effective planning and control procedures instigated by the site manager and management will result in considerable economies.

COST OF MATERIALS

It is very difficult to state the cost of each material. This information should be obtained directly from the manufacturer before ordering.

The cost of materials can be affected by:

- the amount required – bulk buying can be cheaper
- the distance from the works – cost rises the farther the delivery distance.

QUANTITIES OF MATERIALS

Taking off quantities of materials from drawings

Taking off is the first stage in which the site manager may be involved in material control; company policy will determine who is responsible for taking off.

In a large- to medium-sized contractors' organization, taking off will be regarded as the responsibility of a specialist department, usually under the control of a buyer. In smaller organizations the responsibility could be either the contract surveyor's or site manager's.

Whenever possible, taking off should be done directly from the main contract drawings, with cross-reference to bills of quantities and specifications.

Any discrepancies, especially those of major consequence, should be reported to the architect as his or her instructions are required.

Preparation and checking of material delivery schedules

This follows taking off. The items on the buying list, i.e. the materials required for the contract, the delivery dates, the rate and sequence of delivery, and suppliers' names and order numbers from the materials delivery schedule, should be checked.

A copy of the delivery schedule is held on site with the copy orders. This gives the site and forward delivery programme, which enables site staff to see when materials are due for delivery, and allows them to prepare for receipt and to adjust deliveries to suit the progress of work.

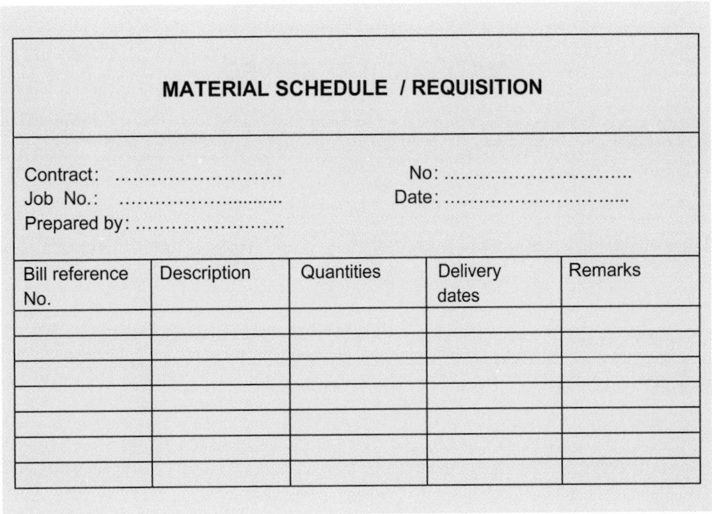

MATERIAL SCHEDULE / REQUISITION

Contract: .... No: ....
Job No.: .... Date: ....
Prepared by: ....

Bill reference No.	Description	Quantities	Delivery dates	Remarks

FIGURE 3.26
Material schedule/requisition sheet

On receipt of materials it is essential that careful checks are made on quantity and quality, and the materials are not damaged. Any discrepancies should be advised in writing to the suppliers and the delivery ticket marked accordingly.

A material schedule and requisition is shown in Figure 3.26. This sheet will have been completed before commencement of the site, and gives the site manager the information required for controlling the materials on site. If materials arrive on site very early they are prone to damage or pilfering.

MATERIAL DELIVERIES

The site manager should receive prior notification of any deliveries. This may take the form of an advice note when the order was made originally. The driver delivering the materials will present the site manager with a delivery note to be signed as proof of delivery and quality of the goods.

Delivery notes should not be signed until the recipient is satisfied of the quality and quantity of the goods.

Each individual delivery ticket should be sent to the head office and a record of the delivery kept on site (Figure 3.27).

MATERIAL TRANSFERS

When materials need to be transferred between sites, correct documentation should be completed. Figure 3.28 shows a typical example of such a sheet.

CALCULATING MATERIAL REQUIREMENTS

When calculating the quantities of materials required while scheduling, a number of important details must be observed, which may save the site supervisor many hours of wasted time and energy.

MATERIALS RECEIVED

Contract: .... No: ....
Job No.: .... Date: ....
Prepared by: ....

Date	Delivery Note No	Supplier	Materials And quantity	Rate	Value	Remarks

FIGURE 3.27
Materials received sheet

TRANSFER OF MATERIALS

Sheet No: .... **Date** ....

From site:	To site:
Description:	Quantity:

Issued by:	Received by:	Driver's signature

FIGURE 3.28
Materials transferred sheet

EXCAVATIONS

In calculating volumes of earth to be removed from foundation trenches the following are considered:

- the length of the trench
- the width of the trench
- the depth of the trench.

The calculation is straightforward for a single straight trench, but slightly more complicated for the rectangular building shown in Figure 3.29.

The volume of this trench is calculated as follows:

$$\text{Length} = A + B + C + D$$

This gives the total length of the outside of the trench.

It does not give a true answer as it would include extra measurements at the corners, shown in Figure 3.30.

To reach the centre line you have to deduct two times/half the thickness of the wall at each corner (Figure 3.30).

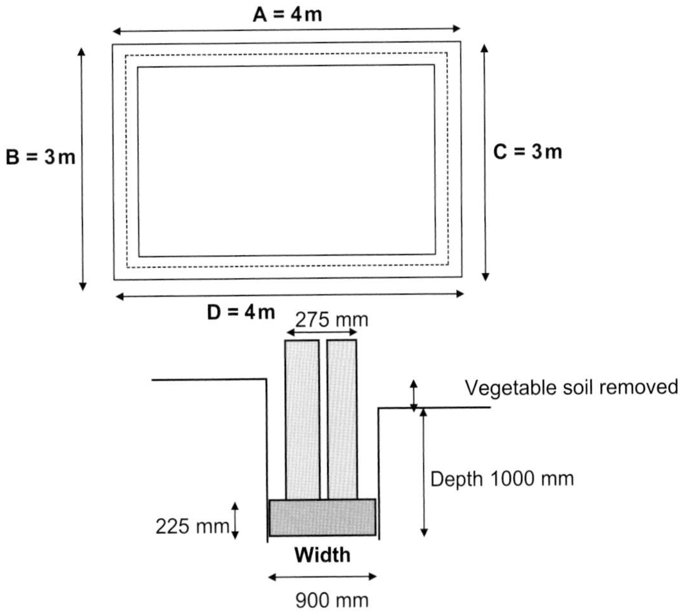

FIGURE 3.29
Foundation excavations

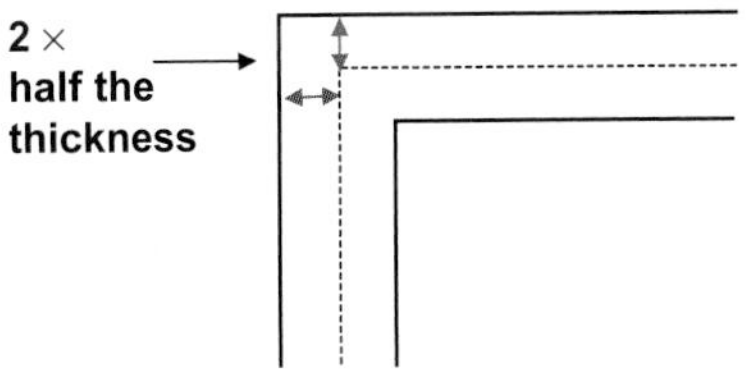

FIGURE 3.30
Centre line

The total length of the centre line is therefore:

$$(\text{Length of } A + B + C + D) - (4 \times \text{Width of trench})$$

$$\text{Length of centre line} \times \text{Width} \times \text{Depth}$$

Using the dimensions on the previous drawing:

A = 4 m

B = 3 m

C = 3 m

D = 4 m

$$\text{Width of trench} = 0.9\ \text{m}$$

$$\text{Centre line} = (4+3+4+3) - (4 \times 0.9)$$

$$14 - 3.6\ \text{m} = 10.4\ \text{m}$$

This centre line measurement can be used for several calculations.

Soil excavations

Soil expands when excavated, by the following amounts – this is called bulking:

- clay bulks: 33%
- sand bulks: 25%
- gravel bulks: 20%

This bulking has to be added to the volume of earth to give the true amount of earth to be carted away in lorries.

Example 4

Calculate the amount of earth to be removed from the trench.

$$\text{Centre line} = 10.4\text{ m}$$

$$\text{Volume of trench} = \text{Centre line} \times \text{Width} \times \text{Depth}$$

$$10.4 \times 0.9 \times 1\text{ m}^3$$

$$= 9.36\text{ m}^3$$

$$\text{Allow for bulking} = \text{say } 25\% = 2.34\text{ m}^3$$

$$\text{Total earth to be removed to the tip} = 9.36 + 2.34 = 11.7\text{ m}^3$$

CONCRETE FOUNDATIONS

Example 5

$$\text{Volume of concrete per m}^3 = \text{Centre line} \times \text{Width} \times \text{Thickness}$$

$$= 10.4 \times 0.9 \times 0.225\text{ m}^3$$

$$= 2.106\text{ m}^3$$

Ready-mix concrete

Before ordering ready-mixed concrete, a waste allowance of 2.5 per cent is added to the volume of concrete measured from the drawings or extracted from the bill of quantities.

Mixed on site

A percentage is added to both aggregates and cement to allow for a decrease in volume when the materials are mixed, due to the fine aggregate filling the voids between the coarse aggregate particles.

There may also be losses when transporting the mixed concrete to its place of deposit.

Allowances are usually made of 40 per cent for the decrease in volume and 2.5 per cent for wastage.

Concrete statistics

- Bags of cement – either 50 kg or (safer to handle) 25 kg
- Cement – approximately 1500 kg/m^3
- Aggregates – approximately 1600 kg/m^3

Example 6

If the ratio of the mix is 1:2:4, it contains seven parts. The amounts required of each material are as follows.

First, value the requirements for one part:

$$\frac{2.106\ m^3}{7\ \text{parts}} = 0.3\ m^3$$

Therefore,

Cement is 1 part of 0.3 m^3	= 0.3 m^3
Fine aggregate is 2 parts of 0.3 m^3	= 0.6 m^3
Coarse aggregate is 4 parts of 0.3 m^3	= 1.2 m^3
Check	= 2.1 m^3

(a) Cement quantity

Since cement is approximately 1500 kg/m^3

1500 kg × 0.3 m^3	= 450 kg of cement
Add for decrease in volume and waste: 42.5 %	= 191.25 kg cement
Total cement	= 641.25 kg

Since bags of cement weigh 50 kg,

$$\text{Total} = \frac{641.25\ \text{kg}}{50\ \text{kg}}$$

$$= 13\ \text{bags of cement}$$

(b) Fine aggregate quantity

Fine aggregate required	= 0.6 m^3
Add 50% for decrease in volume and waste	= 0.3 m^3
Total material	= 0.9 m^3

Hence

$$\frac{0.9\ m^3 \times 1600}{1000} = 1.440\ \text{tonnes}$$

(c) Coarse aggregate quantity

Coarse aggregate required	= 1.2 m^3
Add 50% for decrease in volume and waste	= 0.6 m^3
Total material	= 1.6 m^3

Hence,

$$\frac{1.6\ \text{m}^3 \times 1600}{1000} = 2.56\ \text{tonnes}$$

BRICKS

The centre line of the outer skin of facing bricks can now be calculated.

This time, because we are moving the centre line outwards we have to add on four times the cavity thickness (62.5 mm) and then four times the wall thickness (102.5 mm).

This will give the centre line of the facing bricks (Figure 3.31).

Note

The above calculations should be used as a guide for small jobs.

On large contracts the concrete will have been mixed according to the design.

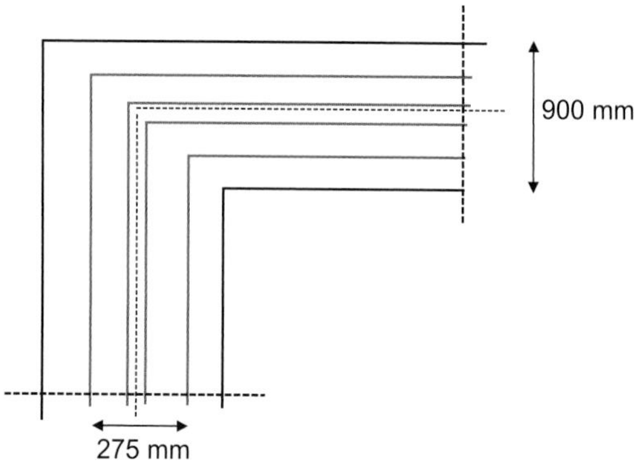

FIGURE 3.31
Centre line of bricks

Example 7

Centre line of facing brickwork $= 10.4 + (4 \times 62.5 + 4 \times 102.5)$
(cavity + outer skin)
$= 10.4 + 0.660$
$= 11.06$ m

Assume the building is 3 m high, then $11.06 \times 3 = 33.18$ m^2 of facing bricks.

This calculated figure is then multiplied by the number of facing bricks per square metre.

There are approximately 60 bricks per m^2 in ordinary stretcher bond.

Therefore $33.18 \times 60 = 1990$ facing bricks.

Other brick statistics

- Half-brick walls in English bond = 90 bricks/m^2
- Half-brick wall in Flemish bond = 80 bricks/m^2
- Half-brick wall in English garden wall = 75 bricks/m^2

An allowance of 5 per cent is normally made for waste on facing bricks and 2.5 per cent for waste on commons.

Mortar

Mortar materials are assessed in the same way as for concrete, using ratios as before.

However, one can assess the amount of mortar per m^2 of brickwork as:

0.03 m^3 for half-brick wall in stretcher bond

0.07 m^3 for one-brick walls.

Allow 10 per cent for waste on mortars.

Building blocks

Allow 10 blocks per m^2 and 0.005 m^3 of mortar per m^2 of blockwork.

Then 10 per cent is added for waste.

Hardcore

To find the volume of hardcore, the area to be covered is multiplied by the thickness to produce a volume in m^3.

Allow 20 per cent extra for consolidation.

Sand blinding is calculated in the same way, but 50 per cent is added for waste and consolidation.

Labour

The cost of labour represents about 40–60 per cent of the overall cost of building.

Objectives

To secure maximum production by the complete integration and co-ordination of all site operations to ensure a continuous, even flow of work for all concerned.

The site manager's responsibilities in relation to labour utilization are to ensure that the total labour force on site is of the right size.

The site manager should also ensure that working gangs are of the optimum size in relation to the type of work and the associated plant.

The site manager will have had the initial responsibility for making any necessary arrangements for site labour and, therefore, must be capable of forecasting labour requirements in advance.

Plant and labour balance

Assuming that the necessary materials will be available, it is important that the correct balance is achieved between *labour* and *plant*. The correct balance is essential if low operating costs are to be achieved.

The following points must be carefully considered:

- the appointment of craftsmen and labourers, depending on the type and situation of the work
- the appointment of labour to different types and size of plant
- the effective plant output
- the materials that the plant can handle.

The site manager should aim to ensure that:

- the plant is operating to full capacity
- the labour is kept fully occupied with a fairly distributed workload throughout the gang.

Overmanning operations

Very often construction operations are overmanned, this is often evident in relation to labourers and craftspeople in clearing up operations, unloading of materials, hand excavation and concreting operations.

Calculating labour requirements

The cost of labour is calculated as an hourly figure as the all-in hourly rate. This rate covers the actual wages paid to the operative plus a number of payments that a builder must make and are a direct result of employing operatives.

These payments are not always made to the operatives or always on their behalf. Most of these payments are outlined in the Working Rule Agreements, and fall into the following categories:

A. Payments made as a result of negotiation between unions and employers, including hourly wage rates, holidays with pay and overtime rates.

B. Expenses such as fares, meals and lodgings.

C. Statutory payments such as national health insurance, employer's liability insurance, sick pay and severance pay.

D. Payment made as a matter of good business management, including third party liability insurance and adequate supervision.

The rates of pay and conditions for building operatives are negotiated by the National Joint Council for the Building Industry (NJCBI).

Factors affecting the cost of labour are:

- earnings:
  - rate of pay
  - guaranteed weekly earnings
  - guaranteed minimum bonus
  - overtime

- distance to site:
  - travelling time
  - reimbursement of fares
  - provision of transport
  - lodging allowances
- extra payments:
  - continuous extra skill or responsibility
  - intermittent responsibility
  - bonus, incentive, productivity schemes
- holidays:
  - holidays with pay scheme
  - public holidays
- injury and sickness:
  - injury payments
  - sick pay
  - retirement and death benefit scheme
  - employer's contribution to National Insurance
  - redundancy funding
- other:
  - payment for difficult conditions
  - tool and clothing allowance
  - employer's liability insurance
  - Construction Industry Training Board (CITB) levy
  - supervision
  - tea breaks.

There are other items set out in the Working Rules under the heading of employer/employee legislation, but it is difficult to place these under the heading of labour costs.

Time sheets

Accurate timekeeping is the basis of ascertaining labour costs, and production can only be fully operational if each operative's time is accounted for.

Time sheets are normally of two types: daily or weekly.

Daily time sheets are completed where jobbing, daywork or component work for varying contracts is being done.

Weekly time sheets are suitable for repetitive work or site work on contract. The weekly sheet usually operates with the first day being that after wages are made up, e.g. if wages are paid on Friday, the week would end on a Wednesday or Thursday.

Time sheet preparation can be something of a tradition, which follows a set method or policy of the firm.

All time sheets should include reference to time and materials for work, relating to variations or extras to contract. All time sheets for jobbing or day-work should include provision for materials used.

If essential for labour costing, subdivisions of the time sheets could be made for craftspeople, apprentices and labourers. An example of a time sheet is shown in Figure 3.32.

Daywork

The Royal Institute of British Architects (RIBA) Form of Contract states:

> *'where work cannot be properly measured and valued, the Contractor shall be allowed day work rates on the prices prevailing when such work is carried out'.*

The types of job justifying daywork include:

- work that the quantity surveyor cannot estimate accurately, such as pumping water, underpinning, filling of craters and possible site clearance
- provision of openings, chasings or channels (for doors, windows, service pipes or drains), in walls or floors after completion

NAME				WEEK ENDING				
LOCATION				JOB NUMBER				
Description of work	Hours Mon	Hours Tues	Hours Wed	Hours Thurs	Hours Fri	Hours Sat	Hours Sun	Total
TOTAL								
Employees Signature : .... Date :								

FIGURE 3.32
Typical time sheet

- any variation to the contract that may be required by the client or architect.
- attendance on other trades, i.e. extra work which is done in attendance on other trades or for subcontractors, or specialists other than the normal type of service allowed for under the terms of 'waiting on other trades' in the contract.

RECORDING OF DAYWORK

Records must be kept on a weekly basis because of the weekly vouchers required by the architect.

A check should be made on the firm's method of submitting the vouchers – normally the craft supervisor submits them to the site agent or the head office by a certain day of the week.

The supervisor must record the number of hours worked, the materials used, and the use of any plant or plant hire, for the duration of the job.

The supervisor must ensure, as previously emphasized, that the recording of daywork on operatives' time sheets is accurately done. For easy recording or reference, the supervisor may consider separate daily or weekly time sheets especially for daywork.

Unless the supervisor has a good system of recording the materials used, which may be difficult on a large site, the tradesperson should be instructed to record an accurate statement of materials used in making out their time sheets.

Accurate recording of materials is essential, as this item, more than hours worked, will be closely checked by the quantity surveyor. A typical example is shown in Figure 3.33.

Reduction of waste

The wastage of materials for similar operations varies considerably on different construction sites. This difference is without doubt caused by the difference in control exercised by the site manager.

Expensive materials that are wasted, and in particular those in short supply, can make the difference between profit or loss.

Site management must, therefore, pay particular attention to the prevention of waste by instituting regular inspection and careful assessments of material requirements.

With regard to timber and reinforcement, the preparation of cutting lists before actual ordering takes place, combined with correct storage, can avoid unnecessary wastage.

DAYWORK SHEET
No. ....

Contract ....
Week ending ....

Description of work
....
....

Name	Trade	Hours M	T	W	T	F	S	Total Hours	Rate	£.	p.	£.	p.
MATERIALS								Add	% £ £				

Description	Rate				
	Add % £ £				
DIRECT CHARGES (fares, lorry, machines , etc)	£ Add % £				
	TOTAL COST £				

Signed .... Supervisor

FIGURE 3.33
Typical daywork sheet

Waste occurs on site for a number of reasons, most of which can be prevented. Some of the more obvious ones are:

- misinterpretation of drawings
- overestimating the quantity required
- faulty workmanship
- careless handling of materials
- uneconomic cutting of materials.

Multiple-choice questions

Self-assessment

This section of the book is designed to allow you to check your level of knowledge. The section consists of revision questions for this chapter. The questions are all multiple choice and have four possible answers. The answers are to be found at the end of the book.

The main type of multiple-choice question will be the four-option multiple-choice question. This will consist of a question or statement, known as the stem, followed by a choice of four different answers, called the responses. Only one of these responses is the correct answer; the others are incorrect and are known as distracters.

You should attempt to answer the questions by choosing either (a), (b), (c) or (d).

Example

The person employed by the local authority to ensure that the Building Regulations are observed is called the:

(a) clerk of works
(b) building control officer
(c) council inspector
(d) safety officer

The correct answer is the building control officer, and therefore (b) would be the correct response.

Programming and resources for work

Question 1 The method of planning shown is known as:

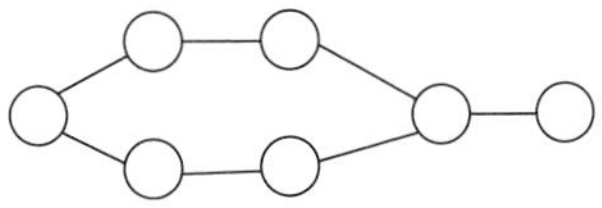

(a) critical path
(b) bar chart
(c) link chart
(d) method statement

Question 2 When a list of operations is produced it is known as a:

(a) sequence
(b) programme
(c) method
(d) statement

Question 3 Manufacturers' instructions for an item of plant are found in:

(a) the contract programme

(b) the site diary kept by the site manager

(c) the information pack which came with the item of plant

(d) the operator's log book

Question 4 The following drawing is known as:

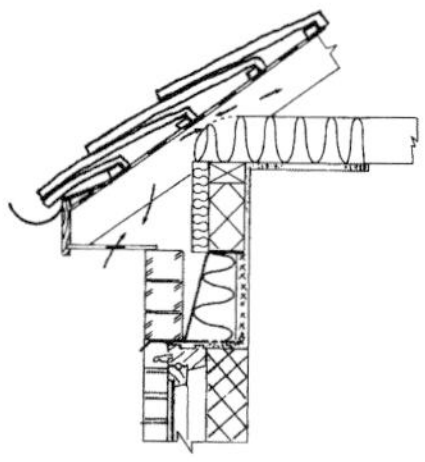

(a) site plan

(b) block plan

(c) location plan

(d) assembly details

Question 5 The document which contains a complete description and measure of the proposed contract is known as the:

(a) specification

(b) method statement

(c) bill of quantities

(d) contract programme

Question 6 What method is used in recording repetitive design information?

(a) specification

(b) schedule

(c) statement

(d) scheme

Question 7 The document which an operative completes each week before payment is made is known as the:

(a) time sheet

(b) bonus sheet

(c) site diary

(d) daywork sheet

Question 8 Taking measurements directly from the drawings to help to price the work is known as:

(a) estimating

(b) taking off

(c) pricing

(d) costing

CHAPTER 4

Working Relationships

This chapter will cover the following NVQ and Diploma units:

- NVQ VR 210
- CC 3002K

This chapter is about:

- Interpreting information
- Adopting safe and healthy working practices
- Working with people, informing and supporting people
- Developing and maintaining good occupational working relationships

The following NVQ performance criteria will be covered:

- Working relationships
- Informing people
- Offering advice
- Deal with alternative proposals
- Resolve conflicts

The following Diploma outcome will be covered:

- Ensure good working relationships

Working relationships

The building industry is rapidly becoming a very complex area of work.

A typical small site is shown in Figure 4.1.

New technologies together with new materials and construction methods have brought about new ideas in supervision.

Speedier building with a smaller labour force has demanded a more scientific approach to the building team and its form of management, resulting in an endeavour to select and train its supervisory staff and workforce.

Today, with better general working conditions and a good apprenticeship and training scheme, the trade offers better prospects and a good promotion ladder for the keen and energetic new entrant.

Although much has been done through the spread of further education and the various courses and examinations resulting from this, there is still a national shortage of properly selected and trained supervisors and site managers in the construction industry.

Team spirit

It is important that the site manager understands about human nature in order to encourage constructive working relationships and team spirit across the whole team.

It has long been understood that many workers do not like being supervised and have little regard for their fellow colleagues or hierarchy. However, investigations have shown that workers are products of circumstance and that changes in conditions around the workforce have remarkable results.

The site manager (Figure 4.2) should aim to have a better understanding of the entire workforce on the site.

Team spirit in the construction industry depends on the ability of the site manager to lead in a correct and fair manner. The workforce has a right to expect efficient leadership, and in return they should give their best.

FIGURE 4.1
Small building site

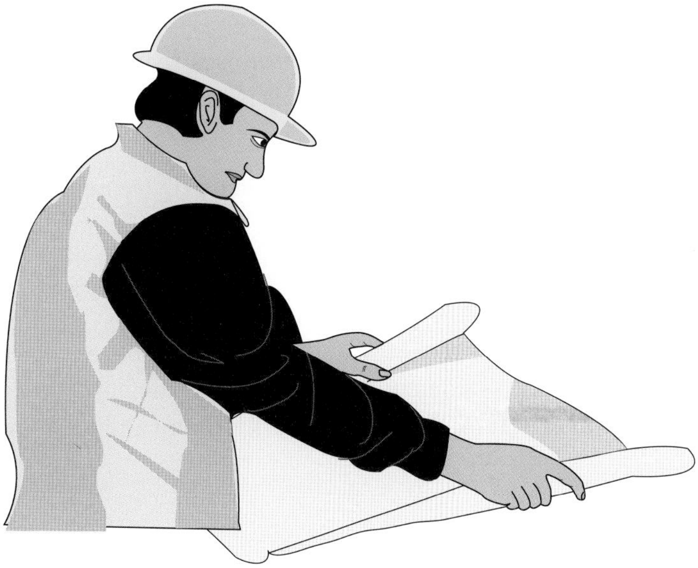

FIGURE 4.2
Site manager

Creating team spirit on site is difficult because humans are influenced by:

- feelings
- ambitions
- personal likes and dislikes.

People work for various reasons, but mainly to earn a living. They wish to receive a fair day's pay for a fair day's work.

Others work for the possibility of promotion or pride in their work.

To achieve good team spirit the firm's policy must be explained to the workforce and the entire workforce must be properly trained.

Creation of team spirit by the site manager is essential for the smooth running of any contract. All site managers are under pressure to complete on time and within budget, but this should not interfere with mixing with the workers on site and maintaining good working relationships.

ADVANTAGES OF TEAM SPIRIT

The workforce understands what is required of them and of the management team. Everyone knows that credit will be given where it is due. No one will be under threat and the whole site should run as clockwork.

DISADVANTAGES WHEN TEAM SPIRIT IS LACKING

The site will be an unhappy place to work and the workforce will also be unhappy. Targets will not be met, either in time or in cost.

Goodwill and trust

The employer has a duty to keep to the promises they have made to the client, employees and other undertakings. It is important to create honest and constructive relationships.

It is essential to create this goodwill to the clients by trying to keep to the proposed timetable and deliver the quality specified. To create this goodwill it is essential to have good relations with your employees, material suppliers, subcontractors, etc.

Never set standards that are too difficult to achieve, and never cut corners or use poor materials and components. Always provide the necessary personal protective equipment appropriate for the work in hand and treat employees fairly.

Organizational structure

Construction firms operate with a view to making a profit and to these ends it is essential, if success is to be achieved and maintained, that they are organized correctly and efficiently.

This is carried out in many ways but all have one thing in common: the breaking down of the whole into groups or sections, each of which will have certain given tasks or objectives. They should operate according to recognized rules, laid down in the objectives and policy of the company, whereby everyone is left in no doubt as to whom and for what they are responsible.

Is also usual for a company either to give a contract to each new employee to show where he or she fits within an organization, or to produce handbooks outlining each and everyone's place within the firm.

Organizations are not a new creation, as organizations were required to construct structures such as the Pyramids and Stonehenge.

CLASSIFICATION

Organizations are classified as either:

- formal
- informal.

It is very unlikely that an organization would be totally formal or totally informal.

Informal organizations

Informal organizations have a loosely defined structure and very rarely have any written rules of conduct. They have less defined objectives, are very flexible and can react spontaneously to new situations.

Formal organizations

The main characteristics of formal organizations are well-defined structures with precisely identified beginnings, lifespan and membership. A formal organization may contain a number of informal organizations with overlapping membership.

Other smaller formal organizations can also exist, such as a trade union or a board of directors.

There are various relationships within a formal organization, as follows.

Line or direct relationships

Certain individuals within a formal organization have the authority to take action and make decisions.

A line usually starts at the top with the most important or powerful person and stretches down to the least important subordinate. This is usually known as the chain of command.

Lateral or equal-level relationships

These relationships exist between people on a similar level within an organization, e.g. between one site manager and another site manager. They have equal status, responsibility and accountability.

Functional relationships

Some individuals are responsible for a particular section but act in an advisory capacity to other individuals at a more senior level.

The safety officer could advise the top management on safety. The marketing manager, personnel manager and security manager all have similar roles.

Staff relationships

Staff relationships exist between people who act as assistants to a superior and the superior.

An individual may have no authority of his or her own, but operates on the authority of his or her superior. For example, a site manager could have an assistant site manager, and a manager could have a personal secretary.

SPAN OF CONTROL

This is the recognized number of operatives under a person's control.

The maximum span of control should be no more than eight, but in practice much larger spans have been identified. Too wide a span leads to lack of communication and control, whereas too narrow a span leads to increased costs.

Normally each person in a small firm has a clear understanding of their and others' responsibilities. This is not always clear in larger firms.

Diagrams are used to help to describe organizational structures and can be divided into two main groups: head office (Figure 4.3) and site level (Figure 4.4).

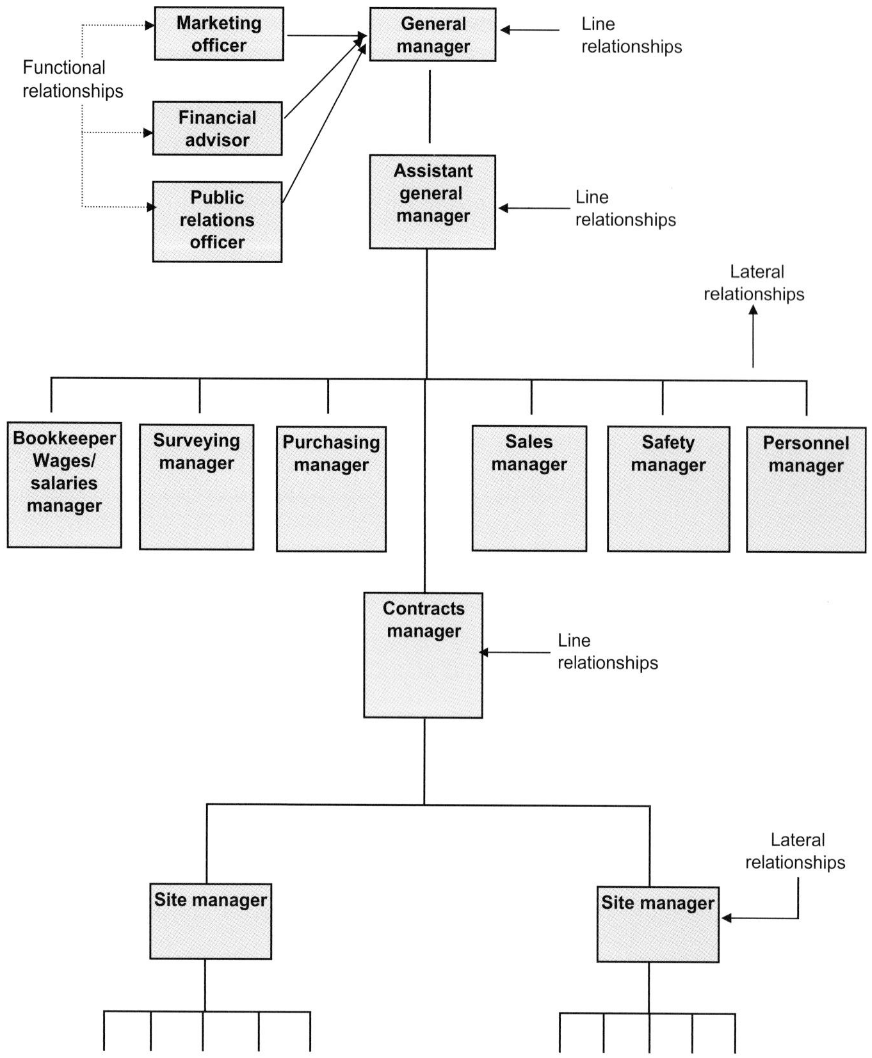

FIGURE 4.3
Typical head office organizational chart

CO-OPERATION BETWEEN CRAFTS

Co-operation in the workforce is essential if full team spirit is to be maintained and the whole workforce united into an effective working team. This co-operation between trades belonging to the main contractor is important and it must also flow through to the various subcontractors on site.

Almost all sites require different trades to work together to help one another complete on time. For example, the bricklayer and labourer will assist when holes are required through walls for the plumber or electrician, and make good afterwards.

Good co-operation should achieve better discipline, responsibility and response to authority. There should also be less wastage of time and resources and good use of labour. Economical use of tools and equipment can also be achieved.

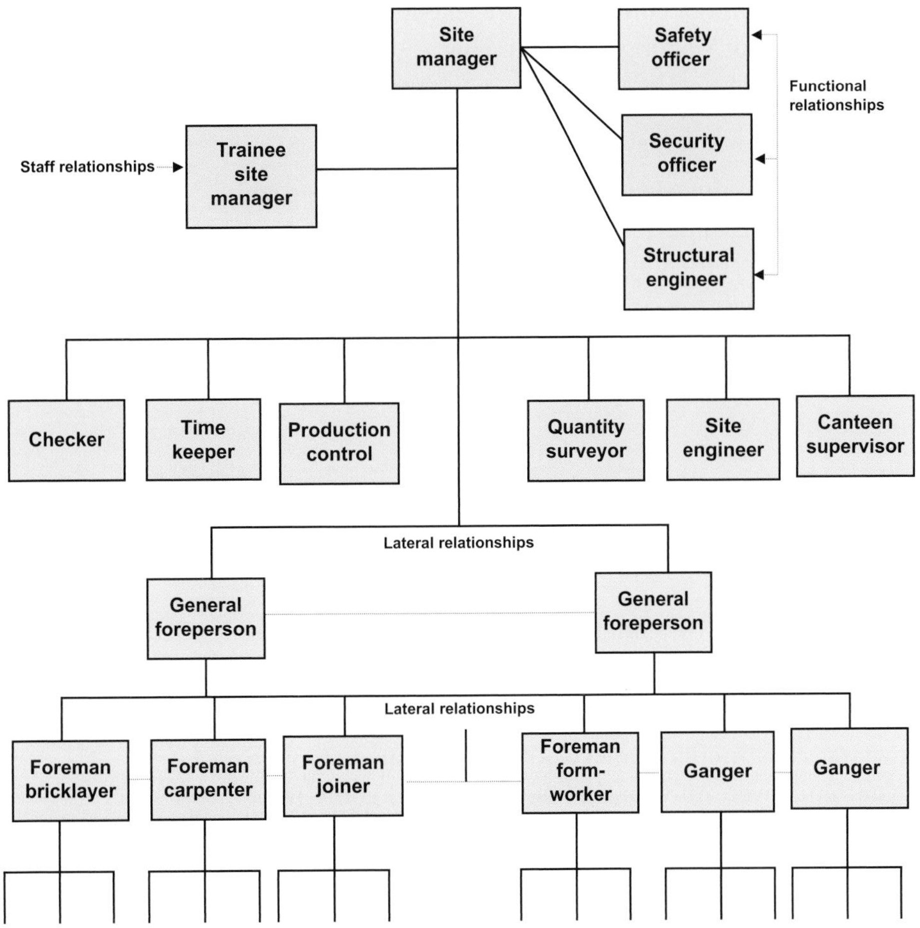

FIGURE 4.4
Typical site level organizational chart

Informing people

There are numerous people working on a construction site who need to be informed of work activities in an appropriate level of detail.

The various work activities that have to be communicated are:

- work progress
- achievement and results
- occupational problems
- health and safety requirements.

Humans have developed communication skills over many centuries, and even when communicating by satellite or computer we have to remember the basics of communication.

To be able to relay information to the appropriate people, good communication skills are required. Such skills are important in every aspect of our lives; in the workplace they are essential.

You may at times be the first point of contact with the client or customer, whether you are self-employed or working for a company.

Meeting and talking to people outside your organization can be one of the more interesting aspects of any job, but it is also a challenge.

You may be talking to people on the telephone or face to face, but in each case you are acting as the representative of your company and therefore you should present a positive and friendly image.

The people who you have to contact may include: employer, site manager, client or customer, colleagues, subcontractor, suppliers of materials and services, and any other visitors to the site.

Employer contact

The amount, if any, of contact with your employer will depend on the type of construction firm.

When employed in a small firm there is often a great deal of contact. Your employer may even be working with you on site, and could well be a colleague with whom you are working.

When employed by a large construction firm it would be very rare for you to have contact with your employer. All your instructions would come from the site manager.

Site manager

During the organization of the site there is a constant need to communicate with various people.

The site manager as an administrator must have some form of office provision, but this depends on the size of the job and the number of operatives under his or her control.

SITE DIARY

The site diary is one of the most important reports that a site manager has to produce during a contract.

It will include the following:

- the weather, morning and afternoon
- any visitors on site
- delays by subcontractors

- instructions from the clerk of works
- materials delivered
- number of staff on the site.

The site manager's main method of communication is through site meetings, at which the appropriate information will be disseminated.

The site manager will also have to deal with individual operatives when there are problems on site, such as discipline.

It is essential to communicate with suppliers and call off materials when required to keep the site on programme.

Contacting subcontractors and dealing with their requirements on a daily basis is also a very important role.

The site manager may also communicate with the client and the building control officer or clerk of works, depending on the type of client.

Client or customer

There are five important points to remember when dealing with a customer.

- Whether you are self-employed or you work for a company, you must remember that it is the customer who pays your wages, either directly or indirectly. Therefore, they should be treated in a polite and helpful manner at all times.
- Deal with any problems or questions they may have as quickly as possible. If you are confronted with a problem or complaint that you are unable to deal with, pass it on to someone of higher authority in the company. Remember, you are representing your company. If you project a professional image, it may result in your company gaining further business.
- Disruption to the customer should be kept to a minimum. Always treat the customer's property with care and respect. Particular care should be taken when moving furniture and electrical appliances so as to avoid breakages.
- Try to keep the area you are working in as clean as possible. This makes the area safer to work in and once again projects a professional image. Dust sheets should be used to protect carpets and furnishings when working internally, and pavements and surrounding areas externally. Clean up frequently and make a special effort when the contract is completed. If any problems arise or damage occurs, contact your supervisor.
- It is important to ensure a good standard of personal hygiene, especially when working in occupied premises, be they offices or someone's home. A person who presents him or herself in an unwashed manner with dirty overalls and boots will give the customer the impression that the work will be poor. This may result in your being

withdrawn from the contract and/or further work being given to another company.

It is therefore important:

- to wash frequently
- to use a deodorant if you have a smelly perspiration problem
- to wash overalls at least once a week
- not to wear muddy boots (particularly when working indoors).

Colleagues

It is essential that the workforce works together as a team.

There can be many different trades on a large site and it is important that they all know their roles and their responsibilities towards one another.

Many of the workforce could be subcontractors and again it is essential that they work alongside the main site operatives to achieve successful completion of the contract.

The site manager will organize the work and who is involved. There are numerous occasions on a long contract when site operatives have to work together or complete a section of work ready for the next person.

For example, the scaffolder has to erect and check the scaffold before any other operative can use it, and ground workers have to excavate trenches and ensure that they are timbered and safe before the drain layers or concreters can carry out their work in them.

Operatives who follow other operatives have to be able to trust in their work.

COMMUNICATING WITH COLLEAGUES

For a company to function properly and make a profit, it is important that the workforce is as happy and relaxed at work as it can be.

It is essential therefore to maintain good working relationships within the workforce. To achieve this, various sections and members of the workforce must communicate with one another as fully as possible. This will then create good team spirit and hopefully motivate the staff. In general, people work harder and more efficiently in a relaxed and friendly atmosphere.

Good working conditions are also very important, e.g. pay, holidays, status, security, future opportunities and a pleasant, safe working environment.

Obviously it is important to get on with colleagues, particularly those who may not be so friendly. When you join a company you are often the odd one out and therefore you should try hard to co-operate with your new work colleagues. This will mean being polite, acting on requests as quickly as possible and building a good working relationship with workmates. All of these things will make your working life much better and more rewarding.

Meeting and talking to people outside your organization can be one of the more interesting aspects of any job, but it is also a challenge.

You may be talking to people on the telephone or face to face, but in each case you must try to be both positive and friendly in order to give a good account of yourself as well as to promote good effective work relationships.

In the working environment, day-to-day methods of communication may be supplemented by written information that your firm may require, such as

- request forms
- job sheets
- materials orders
- time sheets
- day sheets.

TIME SHEETS

Time sheets are important as your company needs to know how much time has been spent on a job in order to charge its customers for work done.

The layout of time sheets will vary from company to company, but they are usually of three basic types.

Most firms require the following information on a time sheet.

- your name and address
- your job number and title or customer's name
- time started and finished work
- total hours worked on job
- travelling time and expenses
- report on work done
- materials used on the job.

Time sheets are also very important for the worker:

- You should always complete them clearly so that the office staff can understand them.
- The information must be accurate.
- The form should be handed in at the appropriate time.

If you fail to follow these guidelines you may find your paypacket is not ready on time.

A typical time sheet is shown in Figure 3.32 in Chapter 3.

Subcontractors

The main contractor will very rarely employ all the trades and specialists required to complete a particular project. The subcontractor is employed to carry out these other functions.

Subcontractors mainly fall into specialist areas, such as:

- scaffolders
- flat roofers
- heating and ventilation
- plastering
- suspended ceiling fixers
- insulation fixers.

The main method of communicating with subcontractors is the site meeting.

SITE MEETINGS

To ensure the contract is running to schedule, frequent meetings will have to be held during the contract.

The frequency of meetings will depend on the type of contract. For small, private jobs there will be a minimum of meetings, possibly monthly, but on the larger contracts the meetings could be weekly. When contracts are behind or going through difficulties, meetings may take place daily.

The contractor's site meeting will be an important co-ordination and monitoring meeting.

Organizing the meeting

For all meetings, adequate notice must be given to those who are to attend; the meetings of registered bodies will normally be considered invalid if every member has not been notified.

Notice may be oral or written. For informal meetings oral notice is often sufficient, but for formal meetings written notice should be given.

The notice must state the date, time and place of the meeting. It is essential that it be given early enough for the recipient to have a reasonable chance of attending, and keeping the occasion free of other engagements.

Agenda

An agenda is simply a list of items to be considered during a meeting.

The purpose of the agenda is to guide the chairperson through the meeting and give prior indication of what is to be discussed to those attending. This will allow them time to do their homework and prepare any necessary data.

The agenda should be circulated with or form part of the notice.

Items usually appearing on an agenda for a formal private meeting are as follows:

1. Apologies for absence.
2. Minutes of the last meeting.
3. Matters arising.

4. Contractor's report.
5. Clerk of work's report.
6. Quantity surveyor's report.
7. Any other business.
8. Date and time of next meeting.

The agenda will be distributed along with the minutes of the last meeting to the members of the team, who may include:

- contracts manager
- site supervisor
- clerk of works
- architect
- quantity surveyor
- structural engineer
- subcontractors.

During the meeting the site manager will report on progress. The bar chart will be shown and the manager will explain the position.

If the contract is on schedule there will not normally be any problems. If the contract is behind schedule then the site manager will give the reasons for this. These may be beyond his control, such as the weather or architect's alterations.

Meetings can also take place on site between the foreman or site supervisor and the contracts manager to discuss the weekly cost statement. This is necessary to ensure that the contract is financially sound.

The cost and bonus surveyor will normally produce information for this meeting, which will normally be a week behind. Therefore, action must be taken immediately if there are losses in the contract.

Suppliers

All building contracts, no matter how small, require suppliers. These organizations specialize in supplying building contracts with their requirements.

Materials are the main requirements on all sites. These include materials for the construction of the main structure of the building. Other suppliers deal with ready-mixed mortar and concrete.

The building will require internal supplies such as decorative and finishing materials. The completed building may also require furniture.

All these materials are very often designed, manufactured, imported and delivered by suppliers as and when they are required, to ensure that the site work flows smoothly.

Many of the materials will have been tested to ensure that there is no risk to the user and adequate information should be provided about its use, handling and storage to ensure site safety.

Site visitors

Consideration must also be given to appearance and conduct when greeting visitors to a site, such as the client, agents, general public, local authority and inspectors.

It is important to greet visitors promptly and politely.

Visitors should complete the site visitor's book and be aware of the safety requirements when visiting the site.

Communication methods

It may be necessary to communicate with customers, visitors and potential employers in a number of different ways, for example:

- orally – face to face
- in writing
- by telephone
- through drawings.

ORALLY – FACE TO FACE

When dealing with a customer or visitor face to face, always appear helpful and polite. Smile but do not appear to be over-friendly as this can make some people feel uncomfortable.

Never act in an aggressive or immature manner.

Remember, first impressions count. If your attitude comes across as 'couldn't care less', potential customers may think that this is your attitude towards your standard of work. Obviously this can frighten off potential customers.

Always offer as much advice and information as possible. If you cannot answer their questions or deal with their requests, find someone who can.

WRITTEN

Written communication can take many forms and one or more of the following will commonly be used:

- letters
- memos

Remember

Ask rather than tell.

Always be polite.

Keep instructions short and to the point.

Be friendly.

- reports
- records
- site diaries
- handbooks and regulations.

Letter writing

Writing a letter is an effective form of communication. We need to write letters on a number of occasions to:

- give information
- ask for details
- confirm things
- make arrangements.

Letters provide a permanent record of communication between organizations and individuals.

They can be handwritten, but formal business letters give a better impression of the organization if they are typed.

They should be written using simple, concise language. The tone should be polite and business like, even if it is a letter of complaint.

A typical layout for a letter is shown in Figure 4.5.

The letter must be clearly constructed, with each fresh point contained in a separate paragraph for easy understanding. When you write a letter for any reason, remember the following basic structure:

- Your own address – Your address should be written in full, complete with the postcode.
- The recipient's address – This is the title of the person and the name and address of the organization to which you are writing. This should be the same as appears on the envelope.
- Date – Write the date in full, e.g. 14 April 2010.
- Greetings – Use 'Dear Sir/Madam' if you are unsure of the sex of the person. Use the person's name if you know it.
- Endings – Use 'yours faithfully' for all letters unless you have used the person's name in the greeting, in which case use 'yours sincerely'.
- Signature – Sign below the ending. Your name and status should be printed below the signature.

Planning your letter

Remember the simple rules in writing a 'good' letter:

- Think about what you want to say before starting to write.
- Think about the person who will receive your letter and what they will want to know.

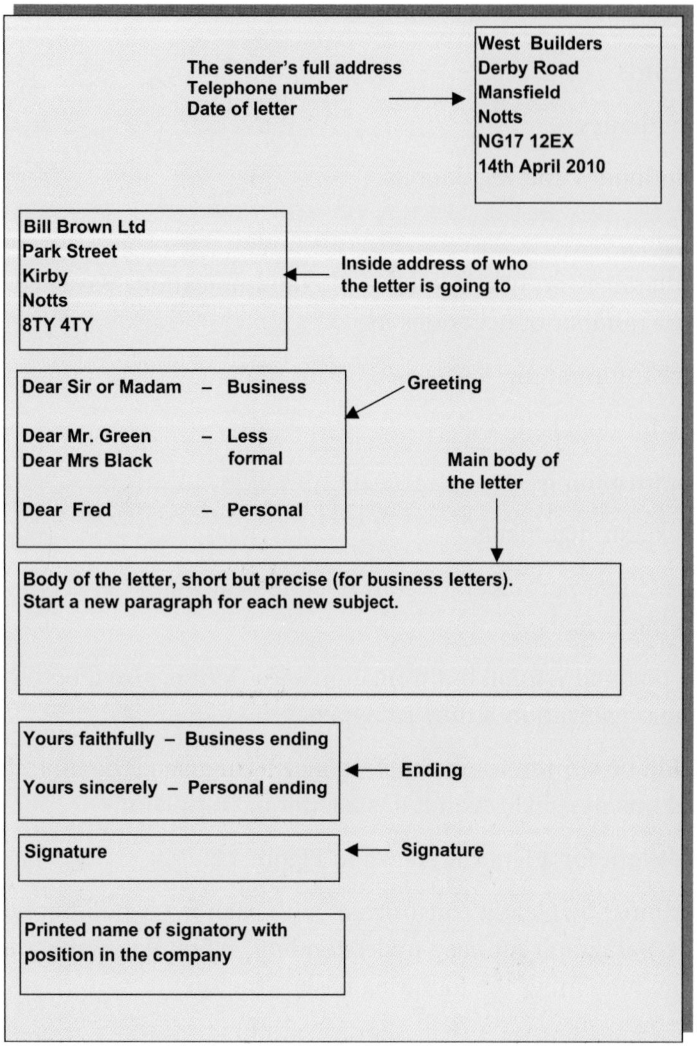

FIGURE 4.5
Typical letter

- Always jot down your ideas on paper and put them into some sort of order.

Remember that anyone who receives your letter will form their opinion of you by what they see.

When a job is advertised it may ask you to provide a CV. Always check what the advertisement is asking for. If you give the wrong information it will make you look inferior and careless.

The advertisement could ask you to do one or a number of the following:

- apply for an application form
- apply in writing
- apply in writing with full CV.

Report writing

A report is written to pass on information quickly and accurately to another party. It may be used to carry out investigations, witness an action or suggest something.

When writing a report, sufficient information should be included to allow the reader to understand fully what the report is intended to say.

A report should be divided into five parts:

- headings
- introduction to the report
- body of the report
- conclusion, with any recommendations
- any data, drawings, contracts, etc.

TELEPHONE

Telephones play a very important role within the communication system (Figure 4.6). Often, a telephone conversation may have taken place before a letter is written to confirm something. Its clear advantage over the written message is obviously the speed with which people are put in touch with one another.

A good telephone manner is essential. Remember, the person you are talking to cannot see you so you will not be able to use facial expressions and body language to help make yourself understood.

FIGURE 4.6
Types of telephone

The speed, tone and volume of your voice are very important. Speak very clearly at a speed and volume that people can understand. Always be cheerful and remember that someone may be trying to write down your message.

MEMOS

One of the most common methods of communication used to be the office memo (Figure 4.7). This has now been superseded by the e-mail.

Accommodation

On larger sites canteens and clean accommodation will be provided for taking meals and storing work clothes, along with washrooms, showers and toilets (Figure 4.8). As smoking is prohibited on most sites the areas are now cleaner and healthier than in the past.

Office

Administration can take place at any or all of the following places: office, workshop and/or site. Most of the administration takes place in the office. Site offices range from simple cabins to luxurious portacabins.

It is beneficial to both head office and the site supervisor that there is two-way communication to show how the work is progressing, what problems have developed, what information is required to ensure

MEMORANDUM	
TO:	FROM:
TIME OF MESSAGE:	DATE:
MESSAGE	

FIGURE 4.7
Memorandum

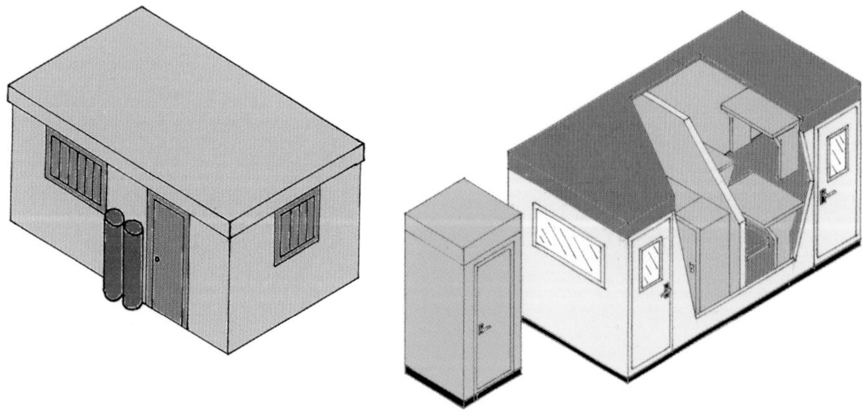

FIGURE 4.8
Site accommodation

continuity of flow of work, what materials and equipment are required to maintain progress, etc.

Most of the communication flows from site because of the need to meet deadlines, while the remainder is to keep management informed of every reasonable detail so that they are well placed to deal with any emergencies.

Compounds

The site should be made as secure as possible, but this will require a financial outlay, which many contractors feel is unjustified.

Compounds are generally associated with larger open sites and are erected to protect and safeguard valuable plant and equipment, fuel and materials.

The type of fencing used depends on that which is available to the contractor, and should be erected with careful consideration.

The site office is usually situated inside the compound for extra security. This is where many valuable items of stationery and equipment are retained for proper processing of information within offices, and conducting administrative or management duties.

Offering advice

Many site managers become a father-figure for some of the workforce. They are there to offer advice on all types of problems, not only regarding the work. They should be able to listen sympathetically and offer advice where possible.

Sometimes the site manager may have to speak to an employee about their standard of work, especially if it is dropping below the requirements set for the contract.

The site manager will also have to deal with suppliers if the standard of materials is falling below that required by the contract.

Dealing with alternative proposals

The site manager may, from time to time, have to deal with certain people regarding alternative proposals and to clarify these proposals.

Alternative proposals may be clarified orally, in writing, or using drawings or sketches (Figure 4.9).

It is an unfortunate fact of 'building life' that contracts often run behind schedule. There may be many reasons for a delay, but one of the most common is changes in design by the client.

Alterations to the design of the contract will be submitted in drawing form and these alterations will have to be explained to the various trades concerned.

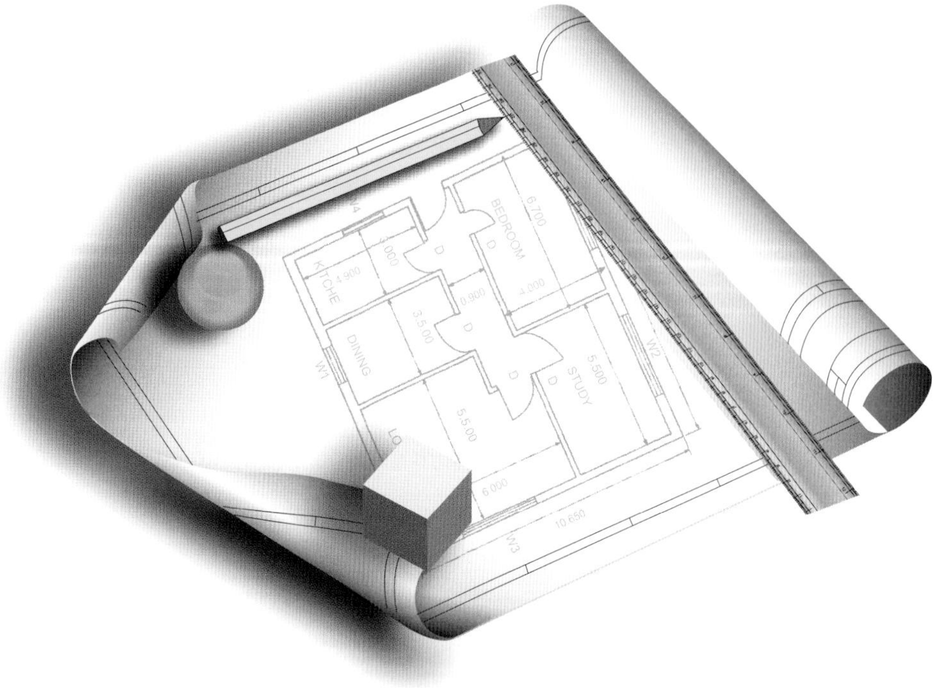

FIGURE 4.9
Site drawings

The site manager will have to alter the programme to accommodate the change and explain this to the workforce at the next site meeting.

Resolving conflicts

If all the previous notes on team spirit have been implemented, conflict should be kept to a minimum. But remember what was stated earlier:

> *... changes in conditions around the workforce have remarkable results.*

Most difficulties or disputes that arise on a construction site will likely fall under the following headings:

- wages
- bonus rates
- unhealthy working conditions
- discipline
- victimization by site managers or supervisors
- intimidation from other members of the workforce
- dangerous working areas.
- lack of welfare facilities

- lack of equipment or plant
- lack of promotion opportunities
- poor methods of working
- type of work allocated.

Handling complaints and site difficulties is one of the site manager's duties, and these should always be tackled while the problems are in their early stages.

Some individuals never appear to be satisfied and are obsessed with the desire to look for problems, and complain constantly about everyone and everything. These disgruntled individuals create a certain amount of depression among those with whom they work and affect morale on site. Site managers have to accept that there will always be such individuals in any society or work situation, with some sites having more than their share.

A supervisor must have an understanding of human behaviour, and needs to show patience with such behaviour, which could be the result of numerous causes. A certain amount of aggravation should be expected by supervisors at some stage. They should be able to accept such behaviour for what it is, although they should be on guard to prevent problems going beyond the acceptable norm.

Handling grievances

A site meeting between the site manager and the operatives can be useful for discussing all types of grievances.

In listening to an individual's grievance, the problem should be studied as follows:

- Listen to the aggrieved person's problems.
- Listen to the other person's side of the story.
- Collect any other facts from witnesses, if necessary.
- Study the facts and information.
- Seek assistance from other supervisors, if necessary.
- Make a decision, fairly and in an unbiased manner.
- State the decision and act on it.

Any problems of grievances outside the province of the site manager, e.g. wages and promotions, should be reported immediately to the personnel officer.

Where grievances develop regarding site duties, it is expected that a procedure would be outlined in the firm's handbook.

Where a problem of dispute develops, the supervisor and operative should remember that there is a standard procedure laid down in the National Working Rule Agreements.

Small grievances can escalate into large ones if not correctly dealt with. Suggestion or complaint boxes could be used to ensure anonymity and prevent victimization.

HANDLING INDIVIDUAL GRIEVANCES

The recommended method for an individual who has a grievance, no matter how trivial, should be verbally, by first approaching his or her site manager, who should then endeavour to solve the problem by whatever means to arrive at an amicable solution.

Failure of the immediate supervisor to solve the problem would entitle the aggrieved person to approach the site agent or manager, accompanied by the union steward.

The course of action open to the union steward if there is no satisfaction at site level would be to notify the full-time union officer, or the operative's regional joint secretary.

The full-time officer would next take up the case on behalf of the aggrieved operative with the firm's management to try to resolve the problem.

Terms of employment

The terms of employment (Figure 4.10) include details of:

- start date
- job title
- hours of work
- rates of pay

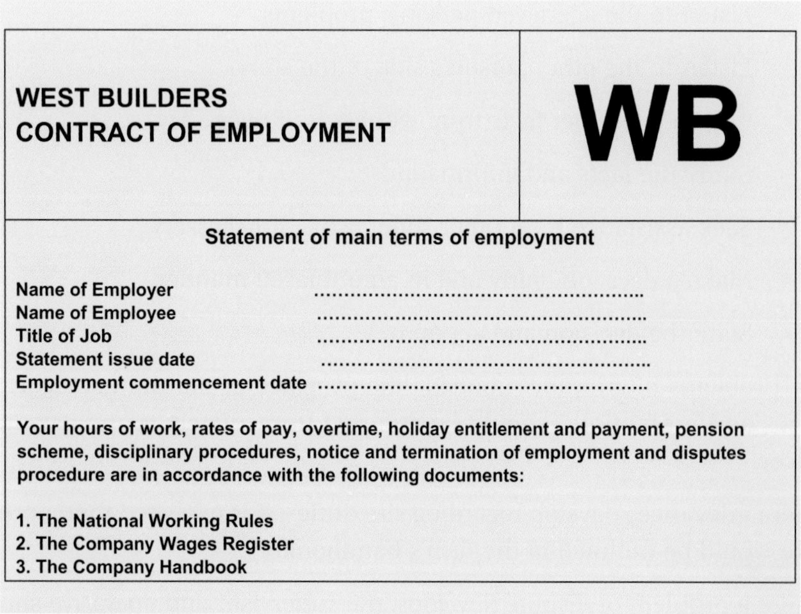

WEST BUILDERS
CONTRACT OF EMPLOYMENT

WB

Statement of main terms of employment

Name of Employer ....
Name of Employee ....
Title of Job ....
Statement issue date ....
Employment commencement date ....

Your hours of work, rates of pay, overtime, holiday entitlement and payment, pension scheme, disciplinary procedures, notice and termination of employment and disputes procedure are in accordance with the following documents:

1. The National Working Rules
2. The Company Wages Register
3. The Company Handbook

FIGURE 4.10
Typical contract of employment

- overtime
- payday
- holiday entitlement and pay
- sick pay
- pension scheme
- disciplinary procedures
- termination of employment
- dispute procedures.

Typical contract of employment

DISCIPLINARY RULES

This is a written statement outlining the firm's disciplinary rules and dismissal procedures. It should be issued to all employees to ensure that they are aware of the rules and procedures involved.

The rules normally provide for verbal warnings of unsatisfactory conduct, for example: poor attendance, timekeeping or production, and the failure to comply with working instructions or safety rules.

They are followed by the final written warning.

If this final written warning (Figure 4.11) is not heeded, dismissal may follow.

Note

Verbal and written warnings are not required in cases of gross misconduct, where instant dismissal can result.

These cases are defined as: theft from the company, falsification of records for personal gain, and acts placing people or property in danger.

WEST BUILDERS
CONTRACT OF EMPLOYMENT

WB

FINAL WARNING

To .... **Date** ....
Site ....

It is being brought to your attention that since the verbal warning given to you on
....
by ....
concerning ....

*** No significant improvement has been made / this conduct has been repeated.**

This is a final warning * failure to show improvement / repetition of this conduct will result in your employment being terminated.

Personnel Manager ....

*** Delete as appropriate**

FIGURE 4.11
Final warning

NATIONAL WORKING RULES

Most building operatives are employed using the wage rates, terms and conditions of employment as laid down in the National Working Rules for the building industry by the National Joint Council for the Building Industry (NJCBI).

The main exceptions are plumbing and electrical operatives, whose terms of employment are negotiated by the Joint Industry Board (JIB). Others are employed under terms negotiated by the Building and Allied Trades Joint Industrial Council (BATJIC), and others are self-employed.

The main functions of the council and their working rules are to:

- fix basic wage rates
- determine conditions of employment
- settle disputes referred to them by both employers and operatives.

Basic rates and main conditions of employment are determined on a national level by the NJCBI, which consists of representatives on the employer's side from the Building Employers' Confederation (BEC) and other associated employers' organizations.

There are also local and regional committees that negotiate regional variations and additions to the National Working Rules.

These rules are published in booklet form and should be available from your employer or a bookseller. The main contents of this booklet is the 27 National Working Rules, which are grouped under the following headings:

- wages
- hours, conditions and holidays
- allowances
- retirement and death benefit
- apprentices/trainees
- safety
- general.

DISCIPLINE

Discipline means orderly conduct, and it is essential in any group effort of employment. Without some form of discipline a day's pay would never be earned, and a day's planned workload never completed.

Today's construction industry has lost the old threat of dismissal in order to maintain discipline. The modern supervisor needs the support and interest of the workforce, and suggestions from them, and these will only be achieved by leadership and co-operation;

NOT BY FEAR.

Situations requiring discipline

These include:

- bad time keeping
- poor workmanship
- theft
- waste of materials
- insubordination
- carelessness with regard to safety
- ill-treatment of tools, plant and machines.

Limitations to discipline

These include:

- instant dismissal or the serving of notice
- a reprimand, or threat of dismissal on reoccurrence
- loss of bonus or incentive, if bad workmanship has been carried out
- loss of wages for time lost through bad timekeeping.

DECISIONS IN DISCIPLINE

The supervisor must realize that discipline may not be necessary if every operative is fully informed as to what can and cannot be done.

Conscience coupled with experience should tell the operative how far to go, but the example of the supervisor should be the final guide.

Before making any decisions on discipline the supervisor should base the actions on facts, obtain all the relevant information first hand and never jump to conclusions because of a worker's past.

Do not assume that dismissal will act as a threat to others to create better support; it may have the reverse effect. Discipline should be constructive and persuasive rather than destructive and aggressive.

Measures adopted should encourage better work rather than the possibility of any resentment or criticism. Any action taken should fit the crime, so that the culprit will realize their error.

Multiple-choice questions

Self-assessment

This section of the book is designed to allow you to check your level of knowledge. The section consists of revision questions for this chapter. The questions are all multiple choice and have four possible answers. The answers are to be found at the end of the book.

The main type of multiple-choice question will be the four-option multiple-choice question. This will consist of a question or statement, known as the stem, followed by a choice of four different answers, called the responses. Only one of these responses is the correct answer; the others are incorrect and are known as distracters.

You should attempt to answer the questions by choosing either (a), (b), (c) or (d).

Example

The person employed by the local authority to ensure that the Building Regulations are observed is called the:

(a) clerk of works
(b) building control officer
(c) council inspector
(d) safety officer

The correct answer is the building control officer, and therefore (b) would be the correct response.

Working relationships

Question 1 In an organizational structure, the relationship that starts at the top and stretches down to the trainee is known as:

(a) direct relationship
(b) formal relationship
(c) informal relationship
(d) staff relationship

Question 2 One of the most important reports site managers have to produce is the:

(a) site report
(b) site diary
(c) contract schedule
(d) contract programme

Question 3 A person employed to carry out functions not able to be completed by the main contractor is known as the:

(a) contractor

(b) secondary contractor

(c) specialist contractor

(d) subcontractor

Question 4 What is the document used to guide a chairperson through a meeting known as?

(a) programme

(b) agenda

(c) minutes of the meeting

(d) schedule

Question 5 A person who provides the materials for a contract is known as the:

(a) subcontractor

(b) client

(c) supplier

(d) specialist

Question 6 The secured area where building materials are stored is known as the:

(a) site compound

(b) site office

(c) site store

(d) site workshop

Question 7 Where can details concerning working arrangements, holiday and sickness entitlements and wage rates be found?

(a) personnel department files

(b) The Working Rule Agreements

(c) the contract of employment

(d) the contract programme

Question 8 What is the most common method of communication on a building site?

(a) telephone

(b) e-mail

(c) written

(d) oral

CHAPTER 5

Working Methods

This chapter will cover the following NVQ and Diploma units:

- NVQ VR 211
- CC 3003K

This chapter is about:

- Assessing project data to determine construction, installation and work methods
- Adopting safe and healthy working practices
- Selecting the methods of work
- Confirming the methods of work to the relevant personnel associated with the occupation

The following NVQ performance criteria will be covered:

- Assessment of project data
- Information sources for project data
- Identify work methods
- Communicate the method of work

The following Diploma outcomes will be covered:

This chapter has no comparable Diploma units but it gives the student an introduction to work methods.

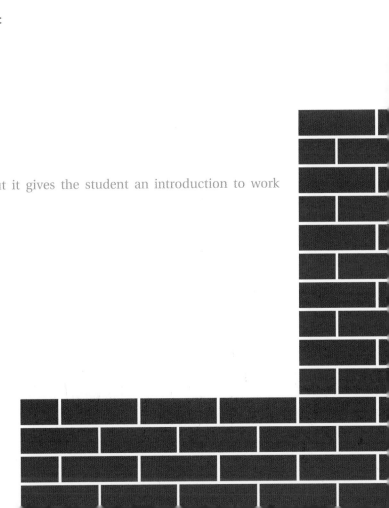

Assessment of project data

The construction of a building can be split into two stages:

- the preconstruction period – this involves the planning of the work

- the on-site construction period – this consists of the actual construction of the building and any administration that needs to take place.

The preconstruction period

This involves the collection of information to assist in a competitive price for the contract. Thought must be given to the way the contract is going to be carried out so that accurate quantities can be calculated to arrive at an acceptable estimate for the contract.

The contractor should use past contractual experience to assist the estimator to arrive at such a figure. The contractor must also look into the methods and timing of each operation and prepare an outline programme of work.

Some of the objectives of pre-tender planning are:

- to pool the contractor's past experience
- to delegate certain tasks to assist the estimating procedure
- to agree on method and output rates to avoid later problems
- to produce a realistic tender which could result in more future enquiries to tender.

When a contractor receives an enquiry or invitation to tender it should be classified as:

- necessary to keep staff employed
- only interested at the correct price
- not very attractive.

Decisions should have been taken on:

- the general order and procedure of work
- methods to be employed
- material requirements and usage
- work to be done by own labour
- work that has to be subcontracted out
- major plant requirements.

There may have been many alternative solutions; therefore the aim must be to select the most appropriate, bearing in mind *time* and *cost*. These are the influencing factors when planning decisions are made.

TIME

The majority of construction contracts must be completed within a certain period, otherwise financial penalties may be incurred. This contract period influences planning decisions.

It is pointless completing a contract weeks ahead of time if the production costs are going to be greater and there is no financial gain to the contractor; the aim must be to complete the work as quickly as possible but at the optimum cost.

COST

The costs of operations must be considered in relation to the time that is available for the operations to be performed.

Often the time available will indicate the method and cost.

Planning will provide information on:

- the dates when labour, plant and materials are required
- the methods to be adopted
- the dates when subcontractors will be required
- the dates when plans, details and information must be available.

The degree of detailed planning will obviously bear some relation to the type and magnitude of the contract, but even planning in its broadest sense will highlight problems and provide a way to overcome them.

PROJECT DATA

The amount and type of project data will alter according to the type of contract. More planning is required for larger, more involved contracts, and less for smaller contracts.

The data may include the following:

- quantities required
- specifications/bill of quantities

- detailed drawings
- health and safety requirements
- timescales
- scope of works.

This is not a complete list and could alter according to the contract.

Quantities required

The quantities of materials required for the contract can either be taken directly from the drawings for a small contract, or provided in a bill of quantities for a large contract.

The bill of quantities includes the materials measured net, which means that the contractor must make allowances for waste, etc.

When a supplier is asked to provide a quotation they must be supplied with as much information as possible to determine the exact specification of the materials. This information can be given by:

- extracts from the bill of quantities
- extracts from the specification
- a reference to British Standards
- detailed or sketch drawings
- reference to manufacturers' literature or catalogues.

Other information is also required which involves quantities, such as plant and labour.

Plant requirements will vary according to the type of contract. It is necessary to define the tasks they are to perform. There are three main areas where plant and equipment are required: excavation, cranage and ancillary equipment.

The type of contract will also show the type and amount of labour required. Some contracts will be very labour intensive, while others will not.

Specifications

As mentioned in Chapter 3, the specification is a precise description of all the essential information and job requirements that will affect the price of the work but cannot be shown on the drawings.

Figure 5.1 shows a typical specification document.

Bills of quantities

As mentioned in Chapter 3, the bills of quantities are produced by the quantity surveyor working for the architect.

This document gives a complete description and measure of the quantities of labour, material and other items requires to carry out the works, based on the working drawings, specifications and schedules.

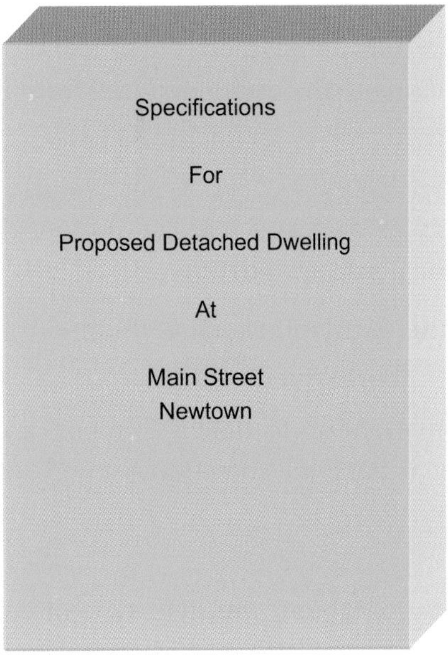

FIGURE 5.1
Specification document

Figure 5.2 shows a typical bill of quantities document.

Detailed drawings

As mentioned previously, the estimator should be able to take off the quantities of materials required for the contract directly from the working drawings.

Drawings that show the work to be carried out are drawn to scale, in one of the general scales used in the building industry.

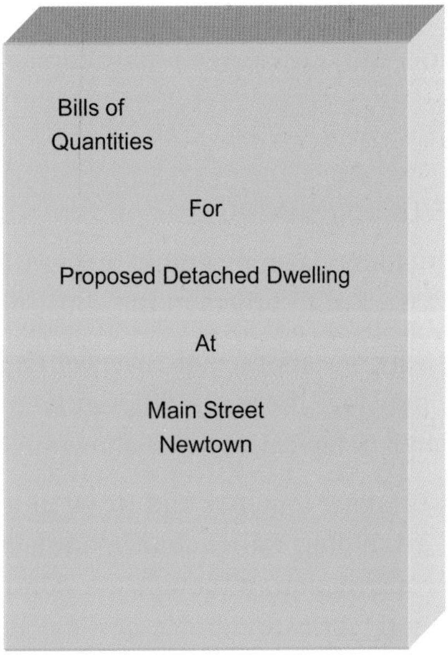

FIGURE 5.2
Bill of quantities document

When drawings are received on site they should be carefully studied so that the work to be done is fully understood. These can be numerous on a large site, so only those relating to the quote required from the supplier are sent.

Health and safety requirements

The Health and Safety at Work Act places a duty of care on employers or their agents to provide a safe system of work.

This in addition to other legislation such as the Control of Pollution and the Control of Substances Hazardous to Health.

It is vital that all the relevant up-to-date documents are at hand, along with other legislation such as the Fire Protection Act and various Highways Acts.

Timescales

To ensure efficient use of labour, the right type of labour should be contracted to do the work.

The various trades have different methods of working and different gang sizes. A bricklaying gang, for example, could be two and one, or four and two.

Some of the other trades may work individually.

A method statement is usually provided which will include the best method of carrying out the work.

Scope of works

The scope of the works will change according to the type of contract.

As mentioned in previous chapters, construction is divided into civil engineering and building. Each covers a very large area and can include the following:

- civils – roads, sewers, bridges, etc.
- building – living, working, recreational, storage, spiritual, transport, etc.

The on-site construction period

The construction of buildings is an assembly process. Whatever your trade, at some time you will work with others in the construction of a building.

Most buildings differ from one another and even where there are similarities in construction (e.g. housing) there are differences in siting, which often lead to unique problems. A typical house is shown in Figure 5.3.

Assembly means the bringing together and fixing at a certain point of the component parts of a building rather than manufacturing them. This is largely true for site construction, although the contractor is involved in some manufacturing and fabrication processes (e.g. reinforced concrete).

The efficient bringing together of the materials and components, plant and labour at a certain geological location for the purposes of erecting a building

FIGURE 5.3
House type

requires considerable planning and organization before and during the course of the contract.

Because of the unique nature of buildings and the sites on which they stand, a large section of this work must be related to providing an effective working environment by establishing a suitable layout organization for the site.

Information sources for project data

If the information received for a project is insufficient then other sources can be used. These could include the following:

- the client or the client's representative
- suppliers
- regulatory authorities
- manufacturers' literature.

Checking information

Before you commence work on site, ensure that all information received is accurate and current. This is usually in the form of charts, graphs, drawings, data sheets, diagrams and the written word.

All drawings should be dated when received and any that become out of date should be removed. A drawings register (Figure 5.4) could be created, to include:

- the number of drawings received
- from whom the drawings were received
- the date they were received
- the number of copies of each drawings received
- the date each drawing was amended or suspended
- the scale of each drawing.

JB BUILDERS LTD
DRAWINGS REGISTER

Contract: .... **Drawings from** ....

Contract No .... **Date commenced** ....

Item No	Description	Scale	Number supplied	Date of issue	Amendments								Comments
					A		B		C		D		
					No	Date	No	Date	No	Date	No	Date	

FIGURE 5.4
Drawings register

The client or the client's representative

The client is the most important person in a building contract.

The client has the responsibility for the initial formation of the project. He or she is also responsible for financing the project and in the design stage of project, and works in conjunction with the architect.

The client is accountable only to him or herself.

The architect

The architect is accountable only to the client and is responsible for the design of the project as required by the client, and ensuring that the contract is correctly executed.

The clerk of works

The clerk of works is accountable to the architect and the client, and is responsible for the inspection of the work, for issuing instructions to be confirmed by the architect, and for the recording of all work and labour on site.

The clerk of works is also responsible for ensuring that all work is carried out in accordance with contract documents.

Suppliers

These provide the various resources for the contract, including materials, plant and equipment.

The materials supplier can be nominated by the architect to ensure that the correct materials are priced by the possible contractor.

Suppliers should not be selected at random. Supplier evaluation should be a constant activity and the suppliers selected will have to meet minimum quality standards and delivery dates.

Legislation

Legislation, the main form of English law, is an Act of Parliament. Most legislation only provides a basis from which to operate.

The Health and Safety at Work Act is a typical example, where the agency carrying out the details and arrangements of the Act is the Health and Safety Commission.

Other building legislation is outlined below.

BUILDING REGULATIONS

Under the control of the Secretary of State for the Environment, the main purpose of the Building Regulations is to ensure that buildings are constructed so that they are safe and do not present a danger to those who occupy them. In simple terms, they describe how a building should be constructed.

TOWN AND COUNTRY PLANNING ACTS

There are many Acts in existence that affect planning and the control of development. The main aim of them all is to safeguard the public regarding buildings and developments. They are mainly dealt with by local government offices.

OTHER LEGISLATION

The following legislation also affects the construction industry:

- Building Control Act
- Clean Air Act
- Factories Act
- Guard Dogs Act
- Historic Building Act
- Housing Act
- Road Safety Act
- Water Acts, etc.

BRITISH AND EUROPEAN STANDARDS

These are issued by the British Standards Institution to lay down minimum standards for materials and components (e.g. walling blocks and doors) used in construction and other industries. They usually have the kite mark printed on them.

Standards greatly simplify the workings of construction and are quoted in:

- specifications
- Building Regulations.

BRITISH STANDARD CODES OF PRACTICE

Codes of good practice are issued by the British Standards Institution, to cover workmanship in specific areas, e.g. building drainage, and brick and block masonry.

Agrément Certificate

These certificates are granted by an independent testing organization, called the British Board of Agrément, stating that the manufacturer's products have satisfactorily passed agreed tests. Subsequent to the granting of the certificate strict quality control has to be continued.

Manufacturers' information

Technical information can be produced in several formats. When items of equipment are purchased there will always be manufacturer's information sheets with them.

The information may be:

- operating instructions – how to use the item
- safety guidelines – power supply, personal protective equipment (PPE) to be worn and recommended checks
- technical information – mechanical details and possible outputs.

Normal hand tools are not usually provided with manufacturer's instructions. Most textbooks explain hand tools in detail and the safety measures required while using them. Information on setting-out equipment is provided by the manufacturer to ensure correct usage. Manufacturer's instructions for a Cowley level will explain how to set up the instrument correctly and various types of readings which will give accurate and inaccurate levels.

Identifying working methods

Having obtained the contract the planning is worked out in greater detail to make the best use of resources and to meet the project, statutory and contractual requirements.

Preparation of the master programme for the work

STANDARD WORK PROCEDURES

The master programme will contain the following:

- starting dates and time allowed for each operation
- the sequence of the operations
- the phasing of operations
- the labour, plant and equipment requirements for various operations
- the dates when information such as drawings, details and schedules will be required
- the dates when nominated or own subcontractors will be required
- allowances for holiday periods.

Having obtained the contract, planning must be more detailed; site supervision, subcontractors, suppliers, etc., must be consulted at this stage.

Decisions must now be made on:

- site layout and staff requirements
- requirements of mechanical plant and tools (plant schedules)
- material requirements (advanced orders placed)
- labour requirements
- methods to be used, recorded for site information
- any bonus arrangements to be used on site
- standards to be used for programme calculations
- allowances for contingencies, weather, etc.

SEQUENCE OF WORK

A planned sequence of operations should be decided on according to the type of project. A construction sequence is the order in which the activities should be carried out. The sequence is usually obtained by asking the following three questions:

- What activity precedes?
- What activity can be done at the same time?
- What activity must follow?

The answer to the above questions is usually based on technical judgement by the programmer who will produce the main construction programme.

There are certain set answers to the above questions, such as the painter cannot paint the plastered walls until the plasterer has completed plastering the walls.

A traditional sequence for a domestic dwelling could be as follows:

- site set-up
- excavate – site strip – reduced level – foundations
- concrete foundations
- brickwork to DPC
- ground floors
- brickwork to first floor – door and window frames
- first floor
- brickwork to eaves – windows
- roof – roof coverings
- watertight.

These are placed into the programme as shown in Figure 5.5.

ORGANIZATION OF RESOURCES

One of the most important roles of site management is to ensure that the resources on site are sufficient, adequate, of the correct quality and type, and available at the correct time.

Resources consist of labour, plant and materials.

CONTRACT PROGRAMME											
	Job No. 123456										
	WEEKS										
OPERATION	1	2	3	4	5	6	7	8	9	10	REMARKS
Set up site											
Excavate foundations											
Concrete foundations											
Brickwork up to DPC											
Ground floors											
Brickwork to 1st floor											
First floor											
Brickwork to eaves											
Roof											
Watertight											

FIGURE 5.5
Main contract programme

Labour

The cost of labour represents about 40–60 per cent of the overall cost of building.

One of the site manager's responsibilities is to forecast future labour requirements so that labour is available as and when required to maintain the planned production rate.

The site manager should also be able to:

- ensure that the total labour force on site is of the right size at all times during the contract
- ensure that working gangs are of the optimum size in relation to the type of work and the associated plant
- communicate the necessary working instructions to first line supervisors and subcontractors' supervisors
- phase the work of different crafts
- ensure the correct plant/labour balance
- ensure the correct craftsperson/operative balance
- maintain timekeeping standards on site.

The site manager will have had the initial responsibility for making any necessary arrangements for site labour; therefore, he or she must be capable of forecasting labour requirements in advance.

Plant

Before any plant can be selected it is necessary to define the tasks they are to perform. This is never simple because the methods to be used are not always apparent from the information provided by the designer.

It is, therefore, necessary to quantify the work to be done. This will be divided into three areas: excavation, cranage and ancillary equipment.

During the planning stage of a contract the type of plant required to carry out particular operations should have been decided.

Arrangements should be made to ensure that the machines are brought onto site at the appropriate time (plant schedule).

The three key factors governing plant selection are:

- the type of work to be carried out
- the quantity of work to be done
- the time available for the work to be carried out.

Economic plant utilization depends on the following factors:

- a planned programme of work to keep the machines continuously employed to maximum capacity
- skilled machine operators

- the correct plant/labour balance
- efficient preventive maintenance to keep the machines working.

Plant and labour balance

Assuming that the necessary materials will be available, it is important that the correct balance is achieved between labour and plant. This balance is essential if low operating costs are to be achieved.

The site manager should aim to ensure that plant is operating to full capacity and that labour is kept fully occupied.

These are two basic essentials in carrying out construction work at low resource cost. The site manager cannot control the prices of materials and labour rates, etc., but he is in a position to control labour and material utilization and performance.

Labour and materials are the two main areas that differ greatly from company to company.

When work has been allocated by the site manager, each operative is responsible for achieving the day's target. If targets are not achieved other operations which follow are held up.

Results do not just happen. The main reason why some sites achieve better results than others is that they are better organized: the working procedure has been carefully worked out and not left to chance. Supervision is actively running the job.

Plant scheduling

Plant requirements should be ordered as soon as possible to allow the plant manager time to ensure availability from his or her own stock or by hiring from a reputable plant hire company.

There are times when the site manager is expected to use his or her own initiative and to select, order and use plant, the suitability of which depends entirely on their past experience.

Typical items of plant are shown in Figure 5.6, and a typical plant and equipment schedule in Figure 5.7.

Materials

The cost of materials used in the construction industry exceeds 50 per cent of the overall costs.

FIGURE 5.6
Typical site plant

JCB CONTRACTORS LTD PLANT AND EQUIPEMENT SCHEDULE Contract: Date: Job No: Prepared by:				
Plant type and equipment	Dates required		Plant hire firm	Remarks
	Date on site	Date off site		
Concrete mixer	October 22nd		Own plant	Call off when required
Excavator	October 22nd		Hire firm	Maintained by hire firm
Dumper	November 5th		Own plant	Call off when required

FIGURE 5.7
Typical plant and equipment schedule

The site manager has little or no control over the initial price of materials; every contractor has to pay approximately the same price. However, by employing strict methods of control in handling and using materials, the most efficient contractors will involve themselves with minimizing the costs in their utilization. In this respect the general supervisor plays a major role.

Effective planning and control procedures instigated by the site manager and management will result in considerable economies.

Types of materials and components

There are too many materials and components that could be used on site to mention all of them.

Materials and components could arrive on site by various modes of delivery, including loose, such as sand and gravel, boxed, such as wall tiles, and individual units, such as windows and doors.

Storage of materials

The storage of the material must be considered during early planning.

Depending on the availability of space and the size of the contract, properly fitted-out stores, with racks and bins of varying sizes into which materials can be placed for ready issue and easy handling, should be available under the charge of a responsible, competent storeperson.

The stores should lie within an enclosed stores compound. The compound should have adequate space and provision on which to place materials that cannot be accommodated within the stores and are not affected by the weather. These stores must be kept to a minimum for security reasons.

It is important that stores are strictly controlled and only authorized employees allowed to sign for receipt. The stores should have a receiving bay and entrance door, and an entirely separate entrance for the issuing of stores, complete with a counter and register for outgoing materials. This register should indicate for which section or part of work the materials are required, and under whose authority they are issued.

Tarpaulins, plastic sheeting or protective coverings should always be available for issue.

Cost of materials

It is very difficult to state the cost of each material. This information should be obtained directly from the manufacturer before ordering.

The cost of materials can be affected by the amount required – bulk buying can be cheaper. The distance from the works can cause the cost to rise.

Quantities of materials

On smaller contracts the amount of materials required is 'taken off' the drawings.

Taking off is the first stage in which the site manager may be involved in material control; company policy will determine who is responsible for taking off.

Whenever possible, taking off should be done directly from the main contract drawings, with cross-reference to bills of quantities and specifications.

Any discrepancies, especially those of major consequence, should be reported to the architect as his or her instructions will be required.

Preparation and checking of material delivery schedules

This follows taking off. A list of delivery dates, the amount of materials and the sequence in which they will be delivered is produced. This is called a materials delivery schedule (Figure 5.8).

A copy of the delivery schedule is held on site together with the copy orders. This allows the site staff to see when materials are due for delivery, allowing them to prepare for receipt and to adjust deliveries to suit the progress of the work.

WEST BUILDERS LTD
MATERIAL SCHEDULE

Contract: .... **No:** ....
Job No: .... **Date:** ....
Prepared by: ....

Bill reference	Description	Quantities	Delivery Dates	Comments

FIGURE 5.8
Materials delivery schedule

On receipt of materials it is essential that careful checks are made on quantity and quality, and the materials are not damaged. Any discrepancies should be advised in writing to the suppliers and the delivery ticket marked accordingly.

The material schedule sheet will have been completed before commencement of the site, and gives the site manager the information required for controlling the materials on site. If materials arrive on site very early they are more prone to damage or pilfering.

Material deliveries

The site manager should receive some notification of impending deliveries. This may take the form of an advice note when the order was made originally. The materials should be accompanied with a delivery note, presented by the driver as proof of delivery, quality and quantity.

Meticulous checks should be made of the materials delivered. Delivery notes should not be signed until the recipient is satisfied of the quality and quantity of the goods.

Each individual delivery ticket should be sent to the head office and a record of the delivery kept on site (Figure 5.9).

Material transfers

When materials need to be transferred between sites, correct documentation should be completed, as shown in Figure 5.10.

Security

It is important that materials are adequately secured on site to prevent loss.

Types of loss include:

- ordinary theft
- criminal activity

WEST BUILDERS LTD
MATERIALS RECEIVED

Contract: ....
Job No: ....
Prepared by: .... **No:**......................................

Date	Delivery No	Supplier	Materials	Rate	Total Value	Comments

FIGURE 5.9
Materials received sheet

WEST BUILDERS LTD
MATERIALS RECEIVED

No: ....
Date: ....

From site:	**To site:**
Description	**Quantity**

Issued by:	**Received by:**	**Driver's signature**

FIGURE 5.10
Materials transferred sheet

- pilfering
- vandalism
- short deliveries
- fraud.

Prevention of loss will vary according to where the site is situated and the type of site works to be undertaken.

Large housing estates tend to be more difficult to secure, while compact sites in the centre of towns require a minimum of effort and cost to keep out intending trespassers, as the general public usually act responsibly when they spot intruders.

SITE ACCOMMODATION

Site accommodation can be as simple as a single cabin or as elaborate as fully equipped multistorey offices (Figure 5.11).

Site fencing

The construction site should be fenced off to prevent the general public from entering the site. The type of fence depends on the type and position of the site. An example is shown in Figure 5.12.

REDUCTION OF WASTE

As mentioned before, construction sites vary considerably and the wastage of materials on these sites also varies considerably.

Expensive materials that are wasted, and in particular those in short supply, can make the difference between profit or loss.

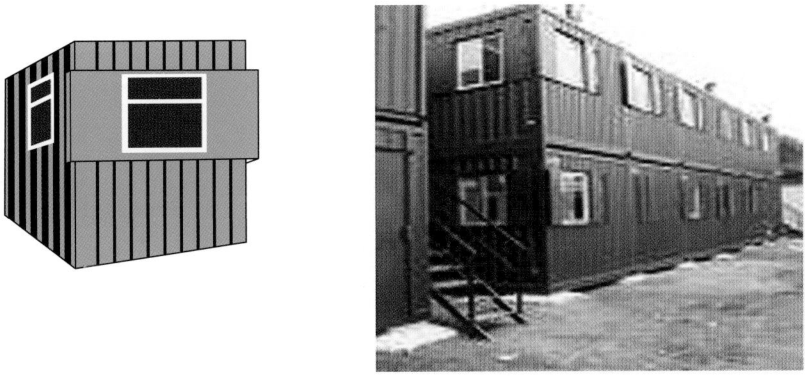

FIGURE 5.11
Typical site accommodation

FIGURE 5.12
Typical site fencing

Site management must, therefore, pay particular attention to the prevention of waste by instituting regular inspection and careful assessments of material requirements.

Waste occurs on site for a number of reasons, most of which can be prevented. Overestimating the quantities and poor storage and handling causes most wastage. Uneconomic cutting and faulty workmanship can also add a considerable amount of wastage.

With regard to timber and reinforcement, the preparation of cutting lists before actual ordering takes place, combined with correct storage, can avoid unnecessary wastage.

Working conditions

Every employer who has five or more employees must prepare a written safety policy. It must be revised as appropriate and brought to the attention of all employees.

Certain minimum health and safety considerations are laid down in relation to construction sites by the Factories Acts and Working Rule Agreements. These call for:

- canteens (places to eat food)
- drying and changing rooms

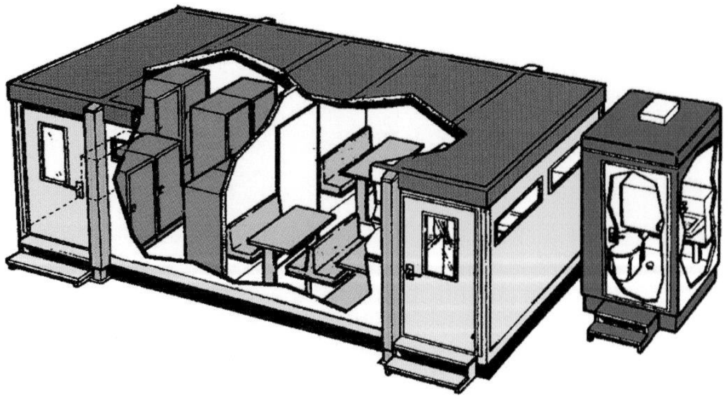

FIGURE 5.13
Site accommodation

- shelter from inclement weather
- washing and toilet facilities.

Portable welfare facilities can be provided, similar to the one shown in Figure 5.13.

When refurbishing existing premises it may be possible to use the facilities already in place. Table 5.1 shows a chart containing the facilities to be provided according to the number of site workers.

Whenever possible toilets should be flushed by water, but if not, chemical toilets may be used. Rooms containing sanitary conveniences should be adequately ventilated and lit. Men and women may use the same toilet, provided it is in a lockable room and is suitably positioned away from any urinals that have also been provided. A wash hand basin with water, soap and towels or dryers should be close to the toilets if the toilets are not near other washing facilities provided on the site.

First aid

Every employer and self-employed person on site must ensure that adequate first aid is available. It is sensible for all contractors to make arrangements with the main contractor to provide first aid (if possible).

First aid arrangements will vary with the degree of risk on the site, but should usually include as a minimum:

- adequately stocked first aid box(es)

Table 5.1 Required facilities

		No. of people employed by contractor on site: 0 5 10 20 25 40 50 100
First aid	Box clearly marked and in charge of named person	First aid boxes; Box + trained first aider
Stretcher ambulance		Stretcher provided. LHA informed
First aid room	To be used only for treatment and in charge of trained person	First aid room
Shelter and clothing	All people to have shelter and place for clothing	Warming workers and drying their clothes; Where possible warming workers and drying their clothes
Meals room	All people to have drinking water provided and facilities for boiling water and eating meals	Heating food if hot meals are not available
Washing facilities	All people on site for more than 4 hours to have washing facilities	Work lasting 6 weeks Hot and cold water soap and towels; Work lasting 12 months 4 wash places + 1 for every 35 more
Sanitary facilities	To be maintained and kept clean and well lit	1 for every 25 people; 1 for every 35 people

Washing facilities must be to be close to the meals room.
Protective clothing must be provided where people are required to work in inclement weather.
Subcontractors may use the facilities provided by another contractor.

- a trained first aider(s) (though for small sites it is sufficient to appoint a person to take charge of the first aid box and any situation where serious injury or major illness occurs (responsibilities should include phoning for an ambulance)
- information for workers on site about first aid arrangements, including the location of the nearest telephone.

Risk assessment

Under the Management of Health and Safety at Work Act regulations all work must have been subject to a risk assessment to detect and define hazards, and then suitable control methods put into place to remove or reduce them.

A risk assessment is nothing more than a careful examination of what, in your work, could cause harm to people, so that you can weigh up whether you have taken enough precautions or should do more to prevent harm.

The aim is to make sure that no one gets hurt or becomes ill. Accidents and ill-health can ruin lives, and affect a business too if output is lost, machinery is damaged, insurance costs increase or you have to go to court.

A risk assessment does not need to be complicated.

A hazard is defined as 'an activity with potential to cause harm or injury', which means that nearly all building operations are hazardous.

Risk assessment must be carried out and the likelihood of injury assessed on a rating scale (Table 5.2).

Table 5.2 Risk assessment

Effect of hazard		Likelihood of harm (risk)	
Category rating		Category rating	
Major	3	High	3
Serious	2	Medium	2
Slight	1	Low	1

Communicating the method of work

It is important the site manager can read and understand the contract programme and all the various legislative documents and drawings, and even more important that he or she can communicate these documents to the various employees on the site.

One of the first jobs when starting a new site will be to set out the site. This operation will depend on the type and position of the site.

Types of site

Owing to land shortages architects are being commissioned to design buildings on land that would otherwise be left barren.

There are numerous shapes and sizes of building plots, too many to mention, but which fall into certain categories.

CLOSED SITES

These are very restricted building areas with very limited, if any, storage facilities. Access could also be a problem.

Alterations to existing buildings, especially shops, fall into this category.

A typical example is shown in Figure 5.14.

OPEN SITES

This type of site has plenty of space for material storage and plant movement, and does not restrict in any way the method of excavation or construction.

This is traditionally a green field that has been released for development where no restrictions occur and access is readily available.

The main problem is with security. A typical open site is shown in Figure 5.15.

BROWNFIELD SITES

These are sites that have already been built on and are now being demolished and made available for new buildings.

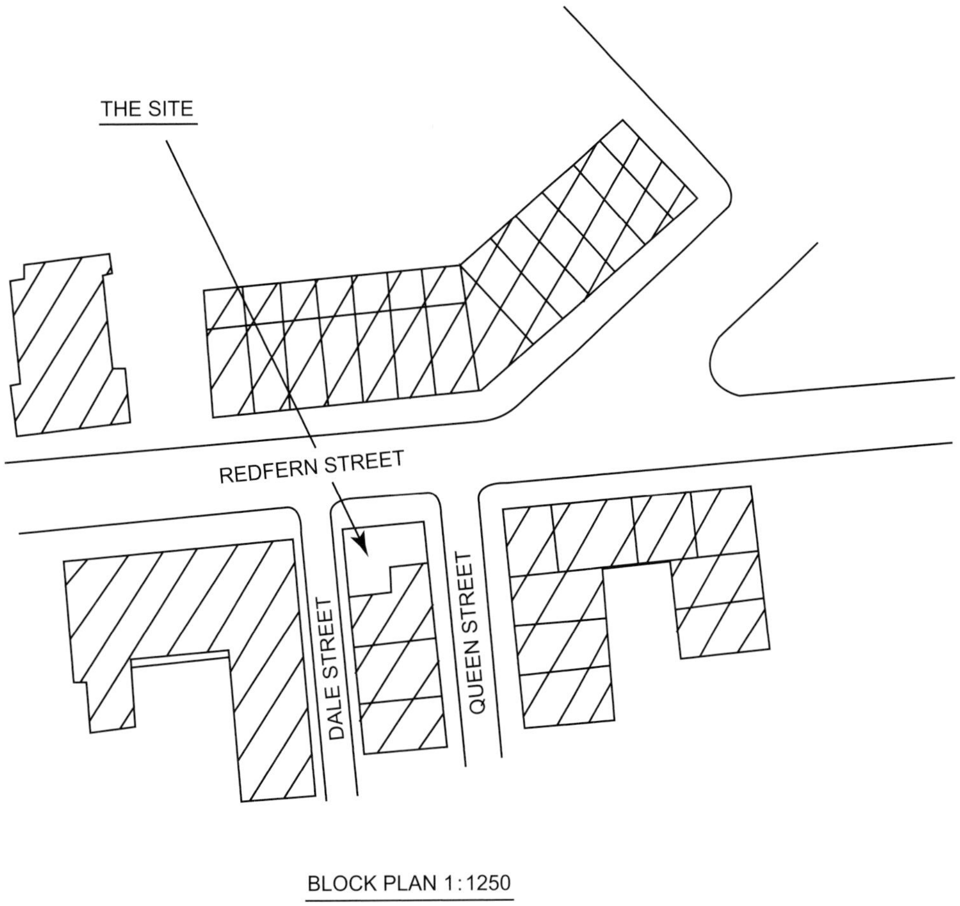

FIGURE 5.14
Closed site

These sites will almost certainly have existing services which will have to be identified and made safe. A risk assessment will have to be carried out to determine whether there is any contamination from the previous use of the building.

A typical brownfield site is shown in Figure 5.16.

OTHERS

Other types of construction site include refurbishment or alterations that take place inside existing buildings.

LAYOUT

Each of the above sites will present different problems for the site manager when planning out the site.

Technical criteria

The site manager will also have to identify various other technical criteria to ensure the best use of resources to be able to meet the contractual requirements.

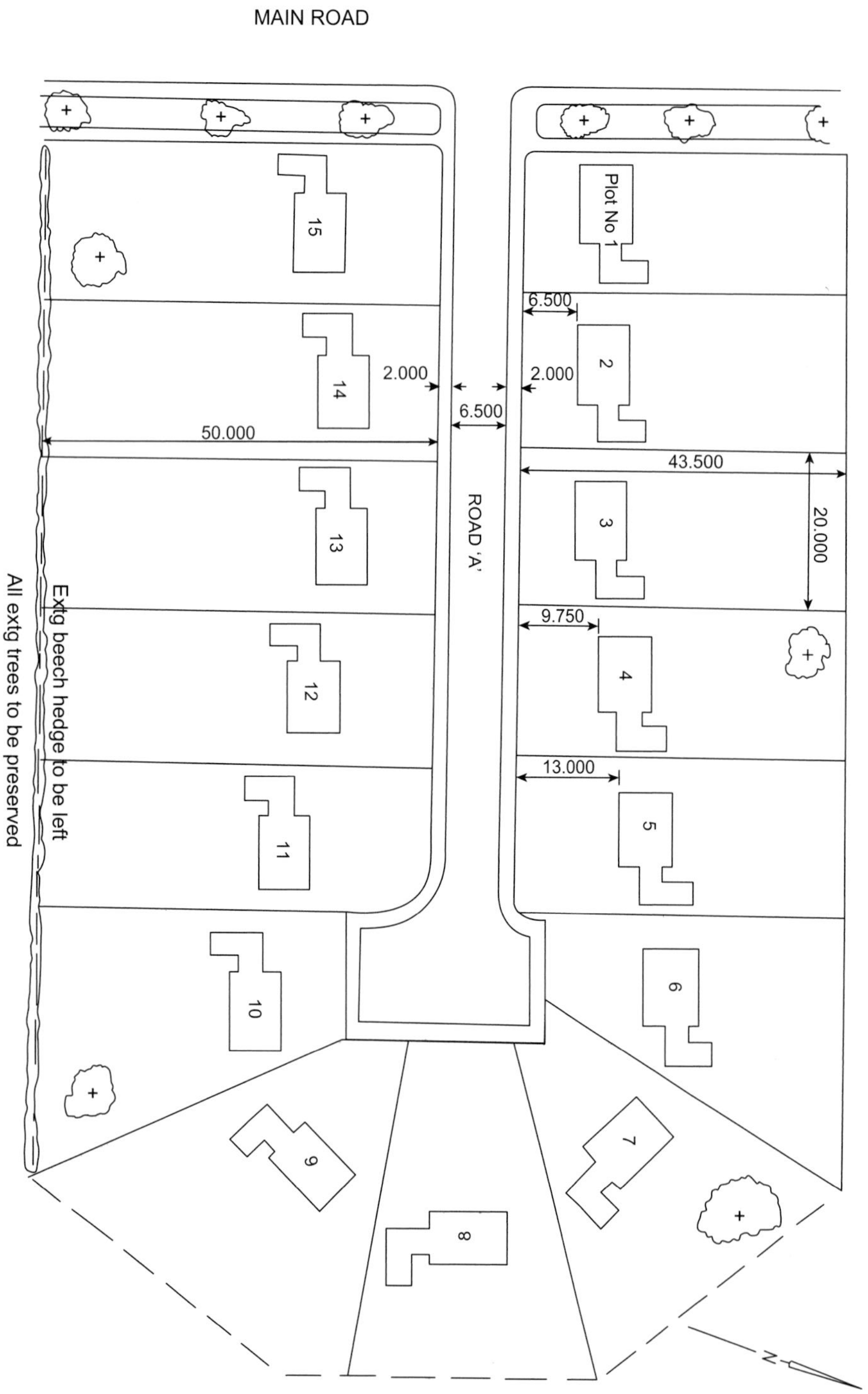

FIGURE 5.15
Open site

The various technical criteria could include the following:

- materials
- health, safety and welfare
- fire protection

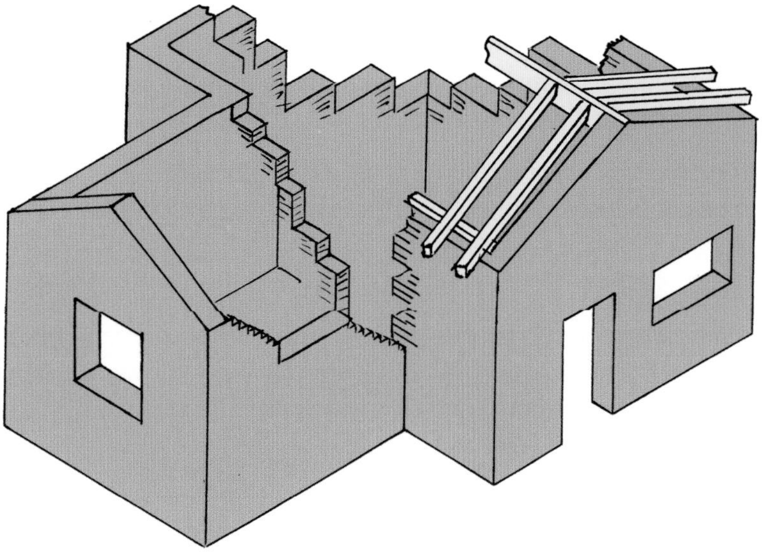

FIGURE 5.16
Brownfield site

- access
- equipment availability
- availability of suitable workforce
- pollution risk
- waste and disposal
- weather conditions
- site layouts.

Many site managers feel that as soon as they are given possession of a site they must start building straight away. However, a great deal of time and money can be lost with poor site preparation.

The site manager is responsible for setting up the site, and a good and efficient site layout can lead to more profits for the contractor.

Site visits

There are numerous site activities which require planning before the actual work should begin on the contract. The site manager should visit the new site as soon as possible to gather information that is not always shown on the drawings.

This information could include:

- access and egress
- external and internal roads
- position of site cabins
- materials and plant.

The efficient bringing together of the materials and components, plant and labour at a certain geographical location for the purposes of erecting a building requires considerable planning and organization before and during the period of the contract.

Therefore, to avoid delays due to poor site planning the following points should be considered.

ACCESS AND EGRESS

Access and egress must be provided for site traffic, to enable materials, equipment, etc., to be brought in and allow for the removal of spoil at the early stage of the contract as site clearance takes place.

- Breaking out the pavement – permission must be sought from the local authority, who may do it themselves for a suitable fee.
- Access through hoardings – doors must open inwards and not onto the pavement.

The site entrance should be near the administrative buildings for control purposes.

There is likely to be a constant flow of traffic around the site, and suitable roads, turning and unloading bays, entrance and exits must be provided.

TEMPORARY WORKS

Some temporary works are normally specified in the bill of quantities, but the contractor often adds to these as an additional safeguard or to help provide more effective working conditions.

These can include the following:

- hoardings and walkways to provide site security and separate the general public from working areas
- temporary roads and services (Figure 5.17)
- car parking, signs and notices.

EXCAVATIONS AND SUBSTRUCTURE WORK

The site layout required at this stage may be different from that required at a later stage in the contract. If the contract includes heavy plant and machinery for the excavations extra roads may be required until the ground work has been completed.

Any layout changes should be planned before commencement so that those items which are on site for the duration of the job do not need to be repositioned.

A large open site may be suited to open cut excavation, whereas a confined city site may require sheet piling. The excavation and spoil disposal methods and plant to be used must be determined and the output of each method balanced.

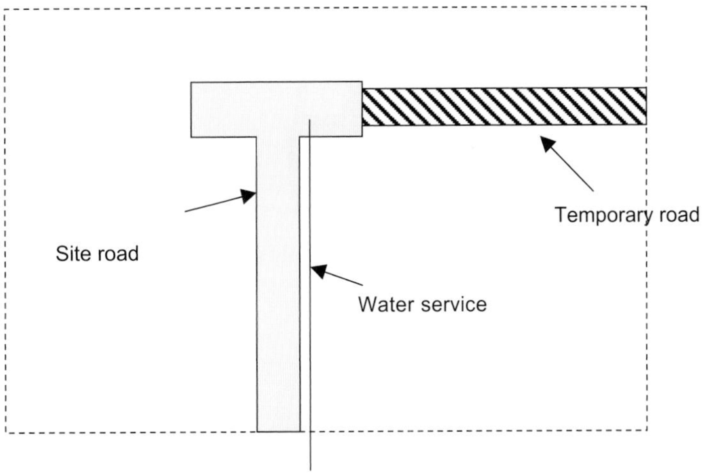

FIGURE 5.17
Temporary roads and services

Site layouts must be able to cope safely with the type and quantity of plant required.

PLANT

The correct choice, compatibility and positioning of plant are of great importance in the efficient working and profitability of the site.

A large variety of plant is available, which can be broadly classified into static and mobile plant. Examples are shown in Figure 5.18. A further subdivision can be made in terms of function: excavating, mixing, transporting and lifting.

Mobile plant requires adequate roads, whereas static plant requires a firm base and possible adequate connection to the building, as with cranage.

With static plant erection and dismantling times are additional considerations.

MATERIALS

Materials need to be brought on site, unloaded, stored or fixed directly into the building. The storage provided will depend on the nature of the material and its relative value.

The positioning of materials in relation to their final position must be considered.

Costs can be saved if transportation distances are reduced and if double handling is avoided.

Materials are delivered to site in various methods of packaging. Some are loose, such as sand and gravel. Other materials arrive in boxes, such as tiles and ironmongery. Others are palletted, such as bricks and blocks.

Materials may be stored in huts, in a compound either under cover or without cover, or in the building itself.

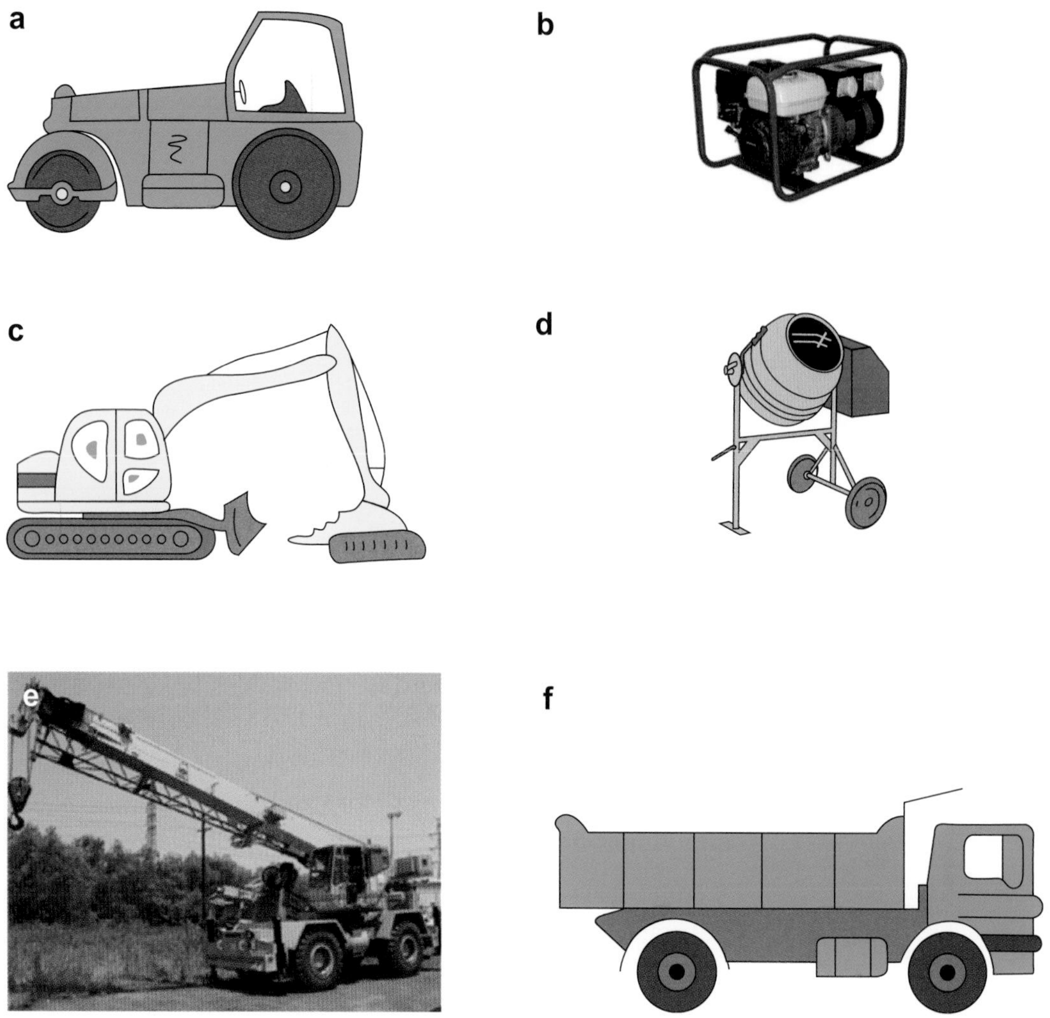

FIGURE 5.18
Site plant: (a) mobile plant – road roller; (b) static plant – generator; (c) excavating plant – backactor; (d) mixing plant – concrete mixer; (e) lifting plant – crane; (f) transporting plant – truck

TYPES OF STORAGE

Various materials require different storage:

- Loose materials such as sand and gravel require bins to prevent them spreading and being wasted.
- Cement requires protection from the weather.
- Components such as roof trusses require special attention.

The storage of liquids such as oil and petrol is subject to the following regulations:

- Any storage within 6 m of a building is governed by the petroleum regulations and the associated fire regulations. Where fuel is stored within the building, containers must have screw tops and not exceed 9 litres each in capacity, and the maximum number of containers in any store is two (Figure 5.19).

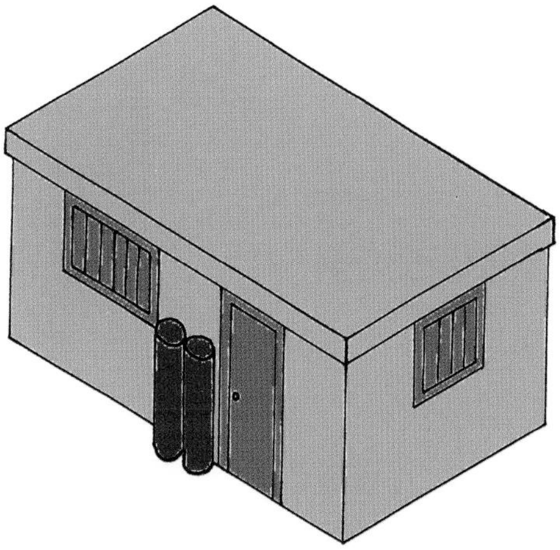

FIGURE 5.19
Storage for flammable items

SITE ACCOMMODATION

Site accommodation for staff and the workforce should be positioned a suitable distance from the construction works to avoid noise and dust. Accommodation should be provided according to the type and length of contract.

Typical accommodation should be provided for:

- contractor's site staff
- workforce, including subcontractors
- toilets and drying room
- canteen and rest areas
- first aid room
- storage of materials.

Various types of site accommodation are shown in Figure 5.20.

COMMUNICATIONS

The physical size of a contract has a considerable bearing on the need for adequate communications.

On a small site there is very little need for anything special to facilitate direct contact between the site manager and the operatives, mainly because the lines of communication are short and effective.

On a large site – both horizontally and vertically – the site manager will carry out a large proportion of their duties by observation and example, and will probably spend a large proportion of their time out of the office.

There are several methods and systems available to facilitate communications and the choice will depend on circumstances, conditions and the amount of money available. Examples are bells, loudspeakers, telephones, mobile phones, personal call systems, flashing lights, closed circuit television.

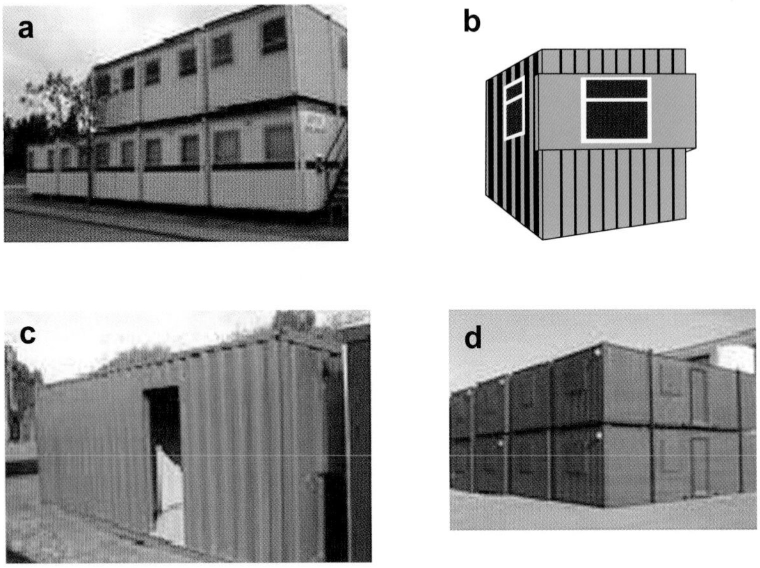

FIGURE 5.20
Site accommodation: (a) multi-storey offices; (b) steel site office; (c) steel storage cabin; (d) storage containers

The most common method of communications, particularly in the area of discipline and welfare, is the simple notice board, which is extremely effective providing that its use is controlled by a reasonable person.

Safety

Adequate safety measures have a marked effect on productivity in the building industry, because accidents are the cause of a large proportion of unproductive time.

It is the ultimate responsibility of site management to ensure the enforcement of all safety measures.

Large contracting companies have found that it is well worth employing safety specialists to educate and advise site management on the implementation of safety policy.

SIGN POSTING

This does not normally present a problem in urban or built-up areas, but when sites are remote then external sign posting will save time, particularly if the site is difficult to find.

On large sites internal sign posting will tend to assist in avoiding unnecessary movement on the site and reduce time wasting by people and transport looking for a particular person and place for unloading.

FIRE PROTECTION

All fires must be taken seriously and action taken immediately to prevent harm to people and damage to property.

If a small fire cannot be controlled quickly, the building must be evacuated.

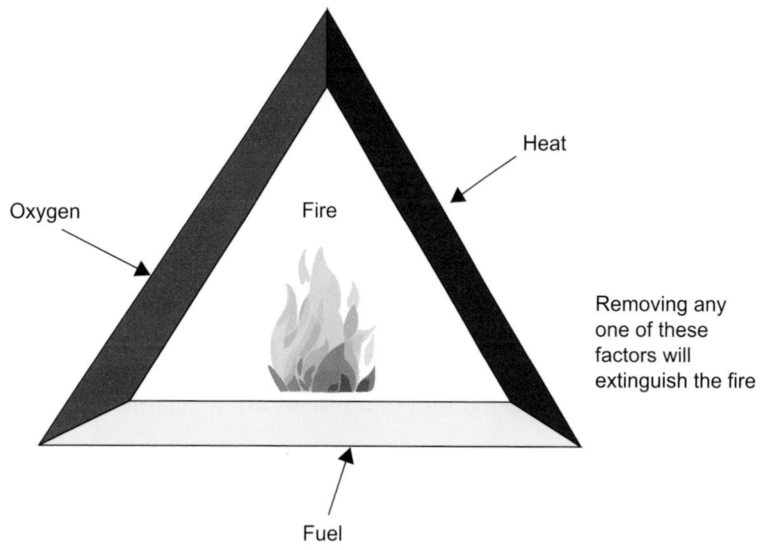

FIGURE 5.21
Factors in a fire

Fires require three factors to start. The removal of any one of these factors will extinguish the fire (Figure 5.21).

Fires, if they occur, need to be extinguished. Only tackle a fire if it is safe to do so or it is small enough to extinguish with a fire extinguisher.

The fighting of a fire should only be carried out by people who are fully trained. In some situations inexperienced site personnel are involved with the early stages of fighting a fire.

The fire-fighting should not continue if:

- the fire becomes too dangerous
- there is the possibility that the escape route might be cut off
- the fire continues to spread
- there are gas cylinders or explosives in the immediate area that cannot be removed or soaked continually with water.

Exits and oxygen

Never work in a position where a fire may block the exit.

Since all fires need an ample supply of oxygen, if a fire does break out try to close all windows and doors.

Fire-fighting equipment

To prevent fires the workplace should be kept clean and tidy. Rubbish should not be allowed to accumulate in rooms and storage areas. Fire prevention is therefore mainly good housekeeping.

When using flammable materials a suitable fire extinguisher must always be kept handy and ready for use. Before starting work on any job, make sure that the extinguisher operating instructions are fully understood.

Make sure that the extinguisher is the right type for the fire that may occur. Use of the wrong type can be disastrous. Fires can be classified according to the type of material involved:

- class A – wood, cloth, paper, etc.
- class B – flammable liquids, petrol, oil, etc.
- class C – flammable gases, liquefied petroleum gas, propane, etc.
- class D – metal, metal powder, etc.
- class E – electrical equipment.

The following types of fire-fighting equipment are available:

- Fire blankets (Figure 5.22) are useful for wrapping around a person whose clothing is on fire. They may also be used to smother a small, isolated fire.
- Sand (Figure 5.22) is also useful for dealing with small isolated fires, such as burning paint droppings.
- Water hoses (Figure 5.22) are fitted in some buildings.
- Fire extinguishers – pressurized extinguishers can be filled with various substances to put out a fire. Fire extinguishers are now all the same colour (red), but they have a colour band which identifies the substance inside:
  - Water: the colour band is *red* and the extinguisher can be used on class A fires.
  - Foam: the colour band is *cream* and the extinguisher can be used on class A fires. A foam extinguisher can also be used on class B fires if the liquid is not flowing and on class C fires if the gas is in liquid form.
  - Carbon dioxide: the colour band is *black* and the extinguisher can be used on class A, B, C and E fires.
  - Dry powder: the colour band is *blue* and the extinguisher can be used on *all classes* of fire.

A range of extinguishers is shown in Figure 5.23.

FIGURE 5.22
Sand bucket, fire blanket and hose

FIGURE 5.23
A range of pressurized extinguishers

SITE TIDINESS

An untidy site is primarily caused by waste materials.

This is most common when a lot of broken bricks and blocks are lying around. A large number of undamaged bricks may be mixed in with the broken ones and also become waste.

DISPOSAL OF MATERIALS FROM HEIGHTS

Careful disposal of materials from heights is essential. They should always be lowered safely and never thrown or dropped from scaffolds and window openings, etc. (Figure 5.24a). Even a small bolt or fitting dropped from a height can penetrate a person's skull and almost certainly lead to brain damage or death.

Waste can be collected (Figure 5.24b) and taken to a suitable tipping area (Figure 5.24c).

HAZARDOUS AND NON-HAZARDOUS WASTE AND DEBRIS

Waste resulting from work materials can include: sand, gravel, cement, concrete, bituminous, sealants, bricks, stone, steel, mesh, nails, timbers, fuels, oils, etc.

Waste resulting from work activities can include: soils, vegetation, cuttings, inorganic materials, etc.

Depending on the sector in which the operative works, the materials will differ.

Some operatives will work in a specific trade occupation, relating to bricks or timber, while others will be involved in general construction activities, relating to excavation materials.

Most substances and materials are safe provided they are handled or worked on sensibly and with proper precautions. Almost anything can be dangerous, of course, if handled or used irresponsibly.

Some materials require extra care; a few need extreme caution. It is important to know what the hazards are, when they occur and how they can be prevented. Prevention will usually involve the use of protective clothing and equipment.

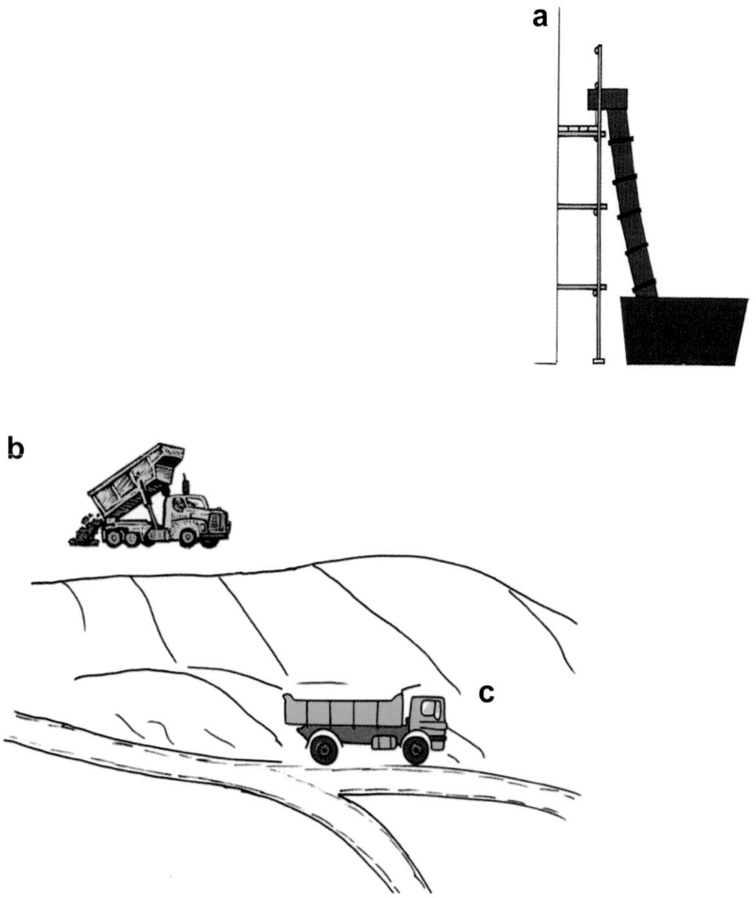

FIGURE 5.24
(a) Chute, (b) skip, and (c) disposal of waste

DUST AND FIBRE

These are released when the materials are worked on or handled. They can harm eyes, lungs and, in certain cases, the skin.

Brick, masonry, concrete, timber, plastics and other hard materials release dust when worked on with cutting or drilling tools.

If there is a danger of materials entering the eyes or lungs, protective equipment and clothing will be required.

Excavating in some soils can be hazardous, depending on the type and situation. Brownfield sites require some form of demolition which will involve many hazardous materials. Other sites are developed on land that has been filled, such as former council tips.

ASBESTOS

The harmful effect of asbestos dust on the lungs is now well known. Blue asbestos is considered so harmful that it is no longer used in building work or building products in the UK.

However, it was widely used before the dangers were recognized, particularly for lagging and insulation of pipework. Everyone, especially those engaged in demolition or maintenance work, should be alert to the hazards

FIGURE 5.25
Asbestos danger sign

of this material and should not deal with it on their own initiative but seek professional advice.

The correct personal protective equipment should be worn at all times when dealing with hazardous waste materials

The main dangers occur when asbestos is cut or drilled, or when it is otherwise disturbed, such as during repair or demolition work. It is then that harmful dusts are released.

The handling and use of asbestos is controlled by law under the Asbestos Regulations 1969. Anyone involved in this work should be fully trained and protected against dust hazards. Signs should be used (Figure 5.25).

Some of the most common materials containing asbestos are boiler and pipework coatings and laggings, sprayed coatings providing fire or acoustic insulation, insulation board, cement-based boards, sheets and formed products, ceiling and some floor tiles, gaskets and brake linings.

Removal of waste

Since 1 April 1992, the law on waste has included a duty of care that means you must take all reasonable steps to look after any waste you have and prevent its illegal disposal by others.

If you break the law, you could be fined an unlimited amount.

'Controlled waste' is any household, commercial or industrial waste, such as waste from a house, shop, office, factory, building site or any other business premises.

HOW DO I KNOW WHAT TO DO?

The law requires you to complete certain paperwork and to take all reasonable steps to meet the duty of care. What is reasonable depends on who you are and on the circumstances. *The Duty of Care, A Code of Practice*, is published by HMSO (ISBN 011 75 25 57 X).

STEPS TO BE TAKEN IF THE DUTY OF CARE AFFECTS YOU

- When you have waste you have a duty to stop it escaping. Store it safely and securely.
- If you hand the waste to someone else:

First, secure it.

- Most waste should be in suitable containers.
- Loose material loaded into a vehicle or skip should be covered.

Secondly, check that the person taking your waste away is legally authorized to do so.

People allowed to take your waste, and the ways to check on them are explained as follows.

Council waste collectors

The council will collect your waste from most shops and small offices. In this case you do not have to do any more checking. You will have to fill in some paperwork though, as will be explained later.

Registered waste carriers

Carriers of waste (unless exempt) have to be registered. Look at the carrier's certificate of registration. Check with the council that issued it that the carrier is properly registered. Ring up the council, or go to their offices and ask to see their register.

Holders of waste disposal or waste management licences

Some licences are only valid for certain kinds of waste or certain activities. Ask to see the licence. Check that it covers your type of waste.

In Scotland only, council waste disposers

Check with the council that its resolution (the equivalent of a licence) covers your type of waste.

Exempt waste carriers

Not all carriers of waste have to be registered. The main people who are exempt are charities and voluntary organizations. If someone tells you that they are exempt, ask them why.

There are exemptions for very specific activities and types of waste. If someone claims not to need a licence, check that the exemptions apply to their case.

Thirdly, hand over a written description of the waste, and fill in and sign the transfer note. The description and note can be the same document. An explanation of the paperwork is given next.

FILLING IN THE PAPERWORK

When waste changes hand a transfer note must be completed and signed by both parties and a written description of the waste handed over. These two may be on a single piece of paper. The government has published a model form with the code of practice, but any design may be used if it contains the correct information.

Repeated transfers of the same kind of waste between the same parties, e.g. weekly collections from shops, can be covered by one transfer note for up to a year.

The transfer note

The transfer note, to be completed and signed by both parties, must include.

- what the waste is
- the amount of waste
- what sort of container it is in
- the time and date the waste was transferred
- where the transfer took place
- the names and addresses of both parties
- details of which category of authorized person each one is
- if either or both parties, as a waste carrier, has a registration certificate, the certificate number and the name of the council that issued it must be included
- if either or both parties has a waste licence, the licence number and the name of the council that issued it must be included
- reasons for any exemptions from the requirement to register or to have a licence.

The written description

The written description must provide as much information as someone else may need to handle the waste safely.

Keeping the information

Both parties must keep copies of the transfer note and the description for two years. They may have to prove in court where the waste came from and what they did with it.

Waste can be collected directly into a skip which can be collected when full (Figure 5.26), or it can be removed from the site by lorry, ensuring in both cases that the material is covered to prevent loose materials from falling while in transit.

FIGURE 5.26
Disposal of waste

Weather conditions

The weather is possibly the only criterion that cannot be controlled on construction sites. Poor weather conditions can have a detrimental effect on the completion of a contract on time.

The site manager can do more about the effects of weather than anyone else.

Building in the UK can be disturbed in winter periods by uncertain weather conditions, and during this period many construction workers are laid off. Bad weather is often blamed for rising unemployment and falling production in the winter.

WINTRY CONDITIONS

Rain for prolonged periods may entail the transfer of people from outdoor work to inside work.

Frost may occur any time between late October and early March, and could bring to a halt concreting, brickwork or plastering.

High winds may hinder the use of cranes, or damage newly laid brickwork or partly constructed roofs.

Fog may cause a delay of labour and materials and is probably worst around the industrial areas from November to January.

Snow is usually the danger from December to January and with frosts could cause the biggest hold-ups on construction sites.

PREPARATION

It is vital to have a supply of protective clothing and portable heaters for drying out. A supply of tarpaulins, sacking and polythene should be kept for protection of the work and materials.

In countries where weather is continuously poor, such as Sweden and Finland, a huge tent made from scaffold and tarpaulins is erected before the work begins. This prevents any stoppages during the external work. The cover stays on until the building is watertight.

PRECAUTIONS

During the construction period certain precautions can be taken such as extending the curing time or accelerating the hardening of mortar and concrete.

PLANNING FOR CONTINUITY OF WORK

It is difficult to generalize what should be done to plan in advance for continuity of work under wintry conditions, especially as any form of preparation will vary according to the different trades, some of which are more affected than others.

For wet trades – especially concrete and brickwork – keep a check on the temperature levels. Work should stop at 6°C on a falling thermometer and start when the temperature reaches 4°C on a rising thermometer.

The following temperature levels provide a useful guide:

- 0°C – water freezes, and hardening of concrete and mortar stops
- 4°C – minimum temperature for placing concrete or mortar
- 10°C – rate of hardening is reduced below this temperature
- 13–25°C – the best range for concrete or mortar
- 30°C – too high a temperature for concrete or mortar, which may result in flash setting.

The normal working day can be lengthened during winter with the use of site lighting.

Remember

A hot, dry summer can also present difficulties such as premature drying out of concrete, resulting in cracking, and dust caused by site transport.

Environmental considerations

CONTAMINATED SITES

It is a statutory requirement for premises holding dangerous or toxic substance to be licensed. In addition, the company has to complete a risk assessment identifying the risks of the chemicals that they use and the precautions required to minimize the risk.

The Building Regulations Approved Document C also details land contaminants, which may be naturally occurring or the result of previous land or building use.

Special precautions may be needed to treat these hazards and monitor potential long-term problems.

Some landfill sites produce large quantities of methane gas, which should be adequately vented to atmosphere in a suitable way to prevent explosion.

Radon gas created in ground rock such as granite can permeate into buildings, and special precautions may be required to prevent this if development is anticipated on such ground. The geology of the area should be known when dealing with radon, and gas discharges should be monitored, if possible, over long periods.

Sites where the ground to be covered by the building is likely to contain contaminants may be identified at an early stage from planning records or from local knowledge of previous uses.

Examples of such sites are shown in Table 5.3.

DERELICT LAND

Disused, abandoned and spoilt land is described as being derelict.

This land results from demolition or dilapidation, as slums and old industrial buildings are left abandoned. It can give an area an air of desolation and neglect.

Table 5.3 Possible contaminants and actions

Signs of possible contaminants	Possible contaminant	Relevant action
Poor growth or absence of vegetation	Metals, metal compounds	None
	Organic compounds, gases	Removal
Unusual colours and contours may indicate wastes and residues	Metals, metal compounds	None
	Oil and tarry wastes	Removal, filling or sealing
	Asbestos (loose)	Filling or sealing
	Other mineral fibres	None
	Organic compounds including phenols	Removal or filling
	Combustible materials including coal and coke dust	Removal or filling
	Refuse and waste	Total removal
Fumes and odours	Flammable, explosive and asphyxiating gases including methane and carbon dioxide	Removal
	Corrosive liquids	Removal, filling or sealing
	Faecal animal and vegetable matter	Removal or filling
Drums and containers	Various	Removal, along with all contaminated ground

The appearance of derelict land often deters new development and can cause economic and social depression (Figure 5.27).

SLUM AREAS

These are areas of overcrowding with very squalid and substandard buildings within the built environment. Slums are created by poor planning and design and buildings being left abandoned.

FIGURE 5.27
Derelict land

POLLUTION

An important goal of any project should be to maintain or improve the quality of life. It is essential that the natural environment is seen as a vital part of this quality of life.

Designers and constructors must therefore look in detail at the effects that their projects and the materials used in those projects are having on the environment.

Types of pollution

Pollution may be defined as the release by humans of substances or energy to the environment in quantities that will damage the environment or resources.

An example of this is the leaking of engine oil into the road or a leaking pipe. In itself this is very minor, but the oil will be washed down the storm water drain and out to sea, joining other 'minor spillages'.

Legislation to control pollution was first passed with the Alkali Act of 1863.

Today, environmental issues are a major concern. However, up to the 1980s, 'green issues' were almost exclusively dominated by pressure groups and a few public and private organizations.

Increased concern about the quality of the environment has influenced property and construction. The impact of the built environment extends from global factors, e.g. the depletion of the ozone layer, to the quality of the environment inside the property. The impact arises from all stages of a building's life from all parties involved, not just the owner or occupier. These stages include site selection, design, construction, occupation and even demolition.

There must be a more cautious approach to the future of the built environment, as the construction and occupation of buildings will continue to use resources, but with much greater attention to the environment.

Site pollution

It is sometimes easy to forget the possible inherent environmental pollution of the site.

Following large-scale industrial restructuring over the past 20 years, many industrial and commercial developments now take place on or adjacent to land that was once put to contaminative use (landfill).

This is especially so in the south-east of England, where the amount of available land is so small that sites are invariably redeveloped rather than developed for the first time.

To avoid any health risks caused by a polluted site a thorough environmental site investigation (ESI) is essential. The major aim of the ESI is to identify any hazardous substances present, their distribution over the site and their concentrations below or at the surface.

Guidance on the assessment of potentially contaminated land is provided by the Department of the Environment, the British Standards Institute and the Institution of Environmental Health Officers.

Landfill

The problem is that Britain produces 25 million tonnes of rubbish every year. This presents a tremendous disposal problem.

Some local authorities incinerate their rubbish. This yields energy that can be used to make small amounts of electricity. The residue is separated into ash, which is buried, and metal, which is recycled.

Unfortunately, all too often the rubbish is simply buried in landfill sites. These are large holes that are usually the result of the extraction of resources such as clay or gravel. The material gradually rots and, as this happens, produces gases. The two most important gases are carbon dioxide (35 per cent), which suffocates, and methane (65 per cent), which can explode.

There is a scarcity of building land in many areas and this has led to the development of many former landfill sites. Unfortunately, the gases can migrate up through the soil and collect in the buildings. This leads to a great risk of explosion. Even buildings hundreds of miles from the landfill site can be affected because the gases can move horizontally.

The solution

Sealing the floors and lower sides of the houses with a special membrane will prevent gases from entering the building. In addition, if a space or layer of granular material is placed underneath the floor slab, this will allow the gasses to be vented into the atmosphere.

In the case of buildings next to landfill sites, the answer is to dig a 1 metre wide vent trench along the side of the landfill and fill it with stones of about 50 mm diameter. The trench, which must be at least as deep as the landfill, is sealed with a membrane barrier on the side farthest away from the landfill to prevent horizontal movement. The gas is then released into the atmosphere through vent pipes. The composition of the gas is regularly monitored.

The ground conditions that result from landfill normally require piles to be driven into the soil to provide a firm foundation for construction.

By careful treatment of the situation, the construction industry is able to build both homes and industrial units that are safe from landfill gases.

Multiple-choice questions

Self-assessment

This section of the book is designed to allow you to check your level of knowledge. The section consists of revision questions for this chapter. The questions are all multiple choice and have four possible answers. The answers are to be found at the end of the book.

The main type of multiple-choice question will be the four-option multiple-choice question. This will consist of a question or statement, known as the stem, followed by a choice of four different answers, called the responses. Only one of these responses is the correct answer; the others are incorrect and are known as distracters.

You should attempt to answer the questions by choosing either (a), (b), (c) or (d).

Example

The person employed by the local authority to ensure that the Building Regulations are observed is called the:

(a) clerk of works
(b) building control officer
(c) council inspector
(d) safety officer

The correct answer is the building control officer, and therefore (b) would be the correct response.

Working methods

Question 1 Which contract document details the quality of materials and standard of workmanship?

(a) bill of quantities
(b) specification
(c) manufacturer's details
(d) contract drawings

Question 2 The person responsible for financing the work is known as the:

(a) client
(b) contractor
(c) supplier
(d) specialist

Question 3 What is the following logo known as?

(a) British Standard

(b) European Standard

(c) safety mark

(d) kite mark

Question 4 A building site that has existing buildings demolished and made available for new buildings is known as:

(a) an open site

(b) a closed site

(c) a brownfield site

(d) a greenfield site

Question 5 The three factors required to produce a fire are:

(a) oxygen, heat and fuel

(b) oxygen, electricity and fuel

(c) oxygen, heat and carbon dioxide

(d) gas, heat and fuel

Question 6 What is the most common method of removing waste from a building site?

(a) skip

(b) barrow

(c) fork-lift truck

(d) dumper

Question 7 How would you recognize a hazardous substance?

(a) by its smell

(b) by a symbol on the container

(c) by the type of container

(d) by the colour of the container

Question 8 Disused, abandoned and spoilt land is known as:

(a) slum areas

(b) polluted areas

(c) contaminated land

(d) derelict land

CHAPTER 6

Setting Out

This chapter will cover the following NVQ and Diploma units:

- NVQ VR48
- CC 3027K, 3029K

This chapter is about:

- Interpreting building information
- Adopting safe and healthy working practices
- Selecting materials, components and equipment
- Setting out complex brickwork and blockwork on level and sloping ground

The following NVQ performance criteria will be covered:

- Interpretation of information
- Safe work practices
- Selection of resources
- Minimize the risk of damage
- Meet the contract specification
- Allocated time

The following Diploma outcome will be covered:

- Know how to set out structural and decorative brickwork

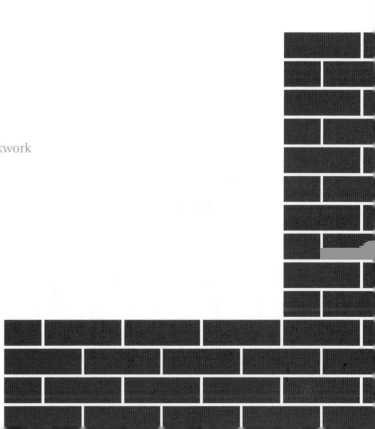

Types of instructions

This chapter will deal with the setting out of small buildings.

The information gained from this chapter will assist the student with the practical requirements of the following chapters.

Information can be gathered from numerous sources such as drawings, specifications and schedules.

Drawings

Trainees should be able to extract the information required for setting out from drawings. Drawings have been dealt with in Chapter 3. Please refer back if necessary.

When drawings are received on site they should be carefully studied so that the work to be done is fully understood.

Groups of individual measurements should be added up and checked against overall dimensions.

It is also very important to find out how the measurements have been taken, e.g.

- over all the walls
- centre to centre
- between the walls.

Drawings that show the work to be carried out are drawn to scale, in one of the general scales used in the building industry. The larger the scale, the more detail can be shown on a drawing.

APPLICATION OF SCALES

It is impracticable to draw most parts of buildings, construction sites and components, etc., to full life size. It is therefore necessary to present them to a smaller size which bears a known ratio to the real thing. This is known as drawing to scale.

For example, if it is decided to reproduce an object at half life size, this can be described as drawing to a scale of 1:2 or half size.

IDENTIFYING AND TAKING OFF DIMENSIONS

Taking off

This is a term used in the construction industry meaning identifying from drawings the type and amount of materials required to carry out the task.

This chapter will discuss taking off dimensions, datum positions and levels from the drawings.

Very often the site supervisor will be expected to be able to understand drawings and extract setting-out information. On larger site a site engineer will carry out all the setting out and levelling.

SITE LOCATION

When it is planned to erect a new building, one of the first considerations: is 'where are we going to build on the plot of land?'

To determine this a site location plan is drawn. See Figures 3.12 and 3.13 in Chapter 3. This drawing, along with other plans and documentation, is submitted to the local authority for approval. Using this drawing you should be able to extract sufficient information to be able to set out the building. There may also be information regarding the drainage runs.

EXAMPLES OF SETTING-OUT DRAWINGS

The design team will be required to produce working drawings for the builder to use on the site.

Drawings, schedules and specifications will have to be prepared, explaining how the design team requires the building to be constructed. To be able to read these drawings it is essential that the trainee is able to understand them.

Drawings should be produced according to British Standard Recommendations for Drawing Office Practice BS 1192. These recommendations apply to the sizes of drawings, the thickness of lines, dimension of lettering, scales, various projections, graphical symbols, etc. The person carrying out a task should be able to read drawings and extract the required information.

Information concerning a project is normally given on drawings and written on printed sheets.

Drawings should only contain information that is appropriate to the reader; other information should be produced on schedules, specifications or information sheets.

WORKING DRAWINGS

These have been explained in Chapter 3. This chapter will deal with the setting out of the building shown in the site plan (see Figure 3.13).

Using the information explained previously, the building has a front dimension of 12 m and a depth of 10 m.

This is calculated as:

Plot width = 25 m

$8\text{ m} + 5\text{ m} = 13\text{ m}$

$25\text{ m} - 13\text{ m} = 12\text{ m}$ width

Plot depth = 25 m

$8\text{ m} + 7\text{ m} = 15\text{ m}$

$25\text{ m} - 15\text{ m} = 10\text{ m}$ depth

Specifications

Except in the case of very small building works the drawings cannot contain all the information required by the contractor, particularly the required standard of materials and workmanship.

Therefore, the architect will prepare a document known as the specification to supplement the working drawings.

The specification is a precise description of all the essential information and job requirements that will affect the price of the work but cannot be shown on the drawings.

Typical items included in the specification are:

- description of materials, quality, size and tolerance
- description of workmanship, quality, fixing and jointing
- other requirements, site clearance, making good on completion, nominated suppliers and subcontractors
- inspection of the work.

BILLS OF QUANTITIES

The bills of quantities are produced by the quantity surveyor working for the architect.

These documents give a complete description and measure of the quantities of labour, material and other items requires to carry out the works, based on the working drawings, specifications and schedules.

Schedules

These are used to record repetitive design information about a range of similar components.

The main areas where schedules are used include:

- doors
- windows
- ironmongery
- sanitary ware
- radiators
- finishes
- floor and wall tiling.

The information that schedules contain is essential when preparing estimates and tenders.

Schedules are also extremely useful when measuring quantities, locating work and checking deliveries of materials and components.

MANUFACTURER'S INFORMATION

Technical information can be produced in several formats. When items of equipment are purchased there will always be manufacturer's information sheets with them.

The information may be:

- operating instructions – how to use the item
- safety guidelines – power supply, personal protective equipment (PPE) to be worn and recommended checks
- technical information – mechanical details and possible outputs.

Normal hand tools for setting out are not usually provided with manufacturer's instructions.

Information on setting out equipment is provided by the manufacturer to ensure that the item of equipment is used correctly.

For example, manufacturer's instructions for a Cowley level will explain how to set up the instrument correctly and various types of readings which will give accurate and inaccurate levels.

Selection of resources

Setting-out equipment

Before setting out any work the equipment should be carefully checked for accuracy.

TAPES

Linen tapes tend to stretch after they have been used for some time, so it is always better to use a steel tape (Figure 6.1).

There are several other types of measuring equipment such as hand tapes, wooden rules and fibreglass tapes (Figure 6.2).

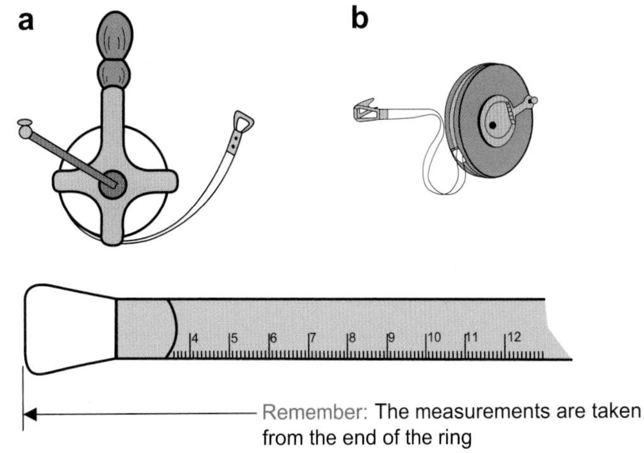

FIGURE 6.1
Types of tape: (a) steel tape; (b) metallic linen tape

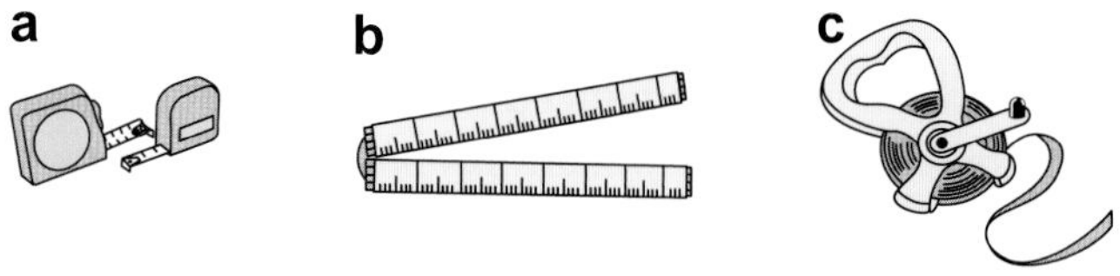

FIGURE 6.2
Other types of measuring equipment: (a) hand tapes; (b) wooden rule; (c) fibreglass tape

SPIRIT LEVELS

Spirit levels have a dual purpose as they are used for checking the horizontal and vertical accuracy (Figure 6.3).

STRAIGHT EDGE

A straight edge is normally made from timber but aluminium ones are also available. They are usually 2–3 m long and tapered at either end (Figure 6.4). The straight edge has to be checked from time to time for straightness and accuracy.

Remember that when transferring levels over a distance the straight edge and level should be reversed each time. This will counteract any discrepancies in either level or straight edge.

RANGING LINES

There are numerous types of lines available for setting out the building, ranging from cotton and hemp to nylon. It is most important to avoid ravelling the lines.

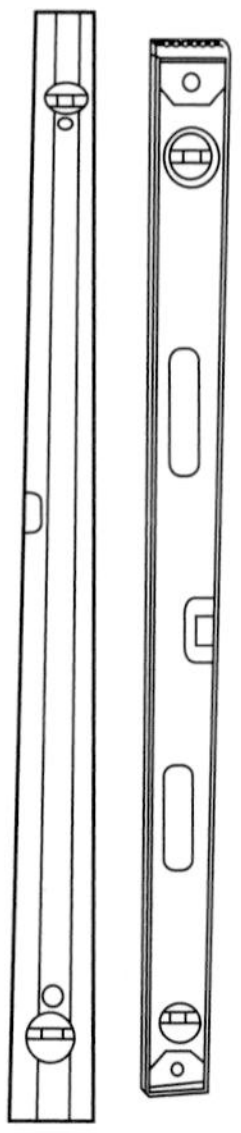

FIGURE 6.3
Spirit levels

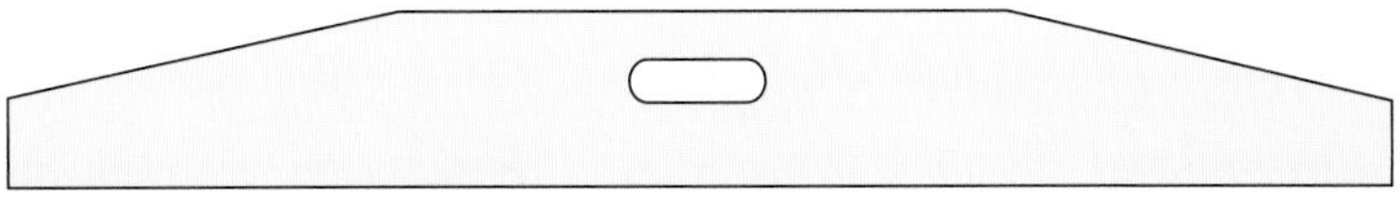

FIGURE 6.4
Straight edge

HAMMERS

Various types of hammers are required, from large ones for driving in the setting-out pegs to small ones for knocking nails into the pegs and profile boards.

RIGHT ANGLES

Most buildings have square corners; that is, corners set at 90 degrees or a right angle. The builder's square (Figure 6.5) is the most commonly used method of setting out a right-angled corner.

There are two methods of setting out a right angle which the trainee should know at this stage:

- the builder's square
- the 3:4:5 method.

Builder's square

The builder's square can be constructed of 75 mm × 25 mm timbers, half-jointed at the 90-degree angle with a diagonal brace, tenoned or dovetailed into side lengths.

The builder's square is laid to the previously fixed front line and the end wall line is placed to the square to produce a right angle (Figure 6.6).

3:4:5 method

A right angle can also be set out using the 3:4:5 method (Figure 6.7).

A peg should be fixed exactly 3 m from the corner peg on the fixed frontage line. A tape is then hooked on to the nail on the corner peg and another tape hooked on to the peg on the front line.

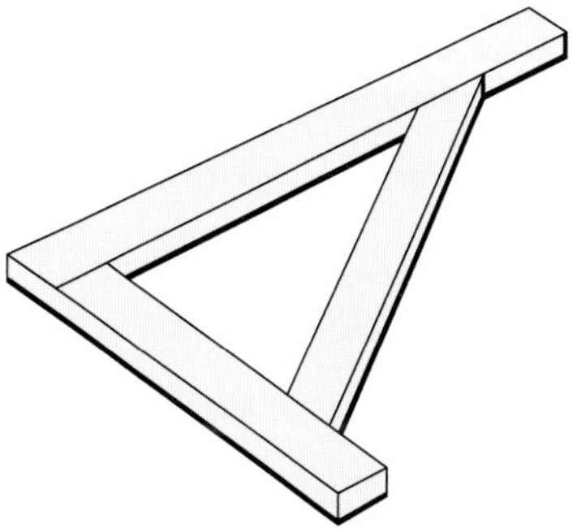

FIGURE 6.5
Builder's square

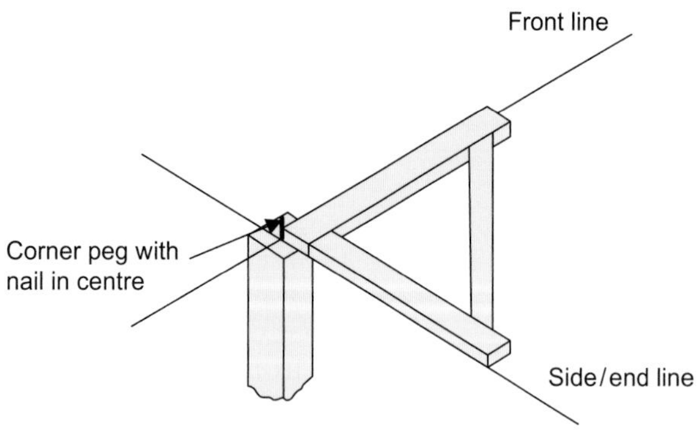

FIGURE 6.6
Setting out a right angle with a builder's square

Stretch both tapes towards the side wall line. At the crossing point of 4 m on one tape and 5 m on the other, a third peg should be fixed.

This will establish the end line at 90 degrees to the front line.

Unless using a builder's square with at least 2 m long sides, the 3:4:5 method is the most accurate method of setting out right angles on the building site.

Maintenance of setting-out equipment

There are several hidden dangers when setting out on the building site or in the workshop.

Tools used for setting out should always be maintained in a safe manner, kept clean and stored away safely until required again.

Wooden-handled hammers should have their heads fastened on correctly and never have split shafts.

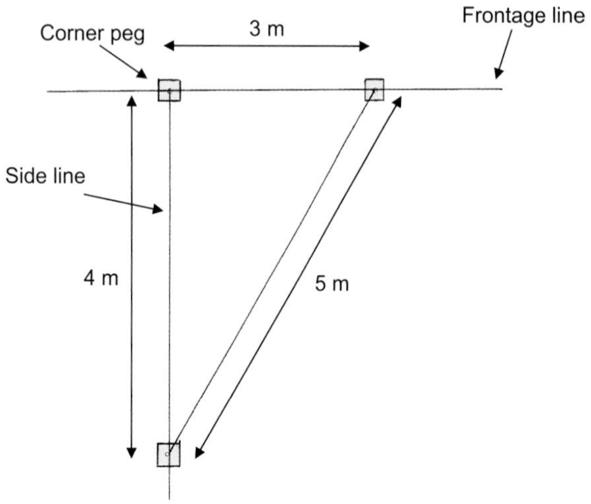

FIGURE 6.7
Setting out a right angle with the 3:4:5 method

Take care when using knives and saws as they are very sharp. Saws should always be sharp and stored correctly when not being used. Never cut directly towards the body: stand at the side of the cut and keep the hands behind the blade.

Always wear the correct PPE.

Workers should be instructed to report any faults immediately and stop using the item of equipment or cable as soon as any damage is seen.

Faulty equipment should be taken out of service as soon as the damage is noticed.

> ***DO NOT*** *CARRY OUT MAKESHIFT REPAIRS.*

TAPES

These should always be cleaned and oiled immediately after use and stored correctly.

BUILDER'S SQUARES, ETC

Any home made items of equipment such as boning rods and builder's squares should be checked before use.

The dimensions of builder's squares should be checked regularly to ensure that they are true.

AUTOMATIC LEVEL

It is essential that any type of automatic level is checked regularly to ensure that all setting out on site is accurate. Inaccurate setting out on site can cause problems and delays which could be very expensive.

The camera section should be sent away to the makers periodically for calibration. This cannot be done on site!

CHECKING SPIRIT LEVELS

Great care should be taken of spirit levels as they are expensive and if ill-used can lead to inaccurate levelling and plumbing. It is therefore important to check the spirit level occasionally to ensure its accuracy (Figure 6.8).

Checking for level

1. Set two screws into a bench, equal to the length of the level apart.
2. Turn one of the screws until the bubble reads level.
3. Reverse the level and replace it on the screws, and if the bubble is between the lines the level is accurate.
4. If the bubble is not in the centre of the lines, adjustment must be made to the bubble tube.

Remember that some spirit levels cannot be adjusted.

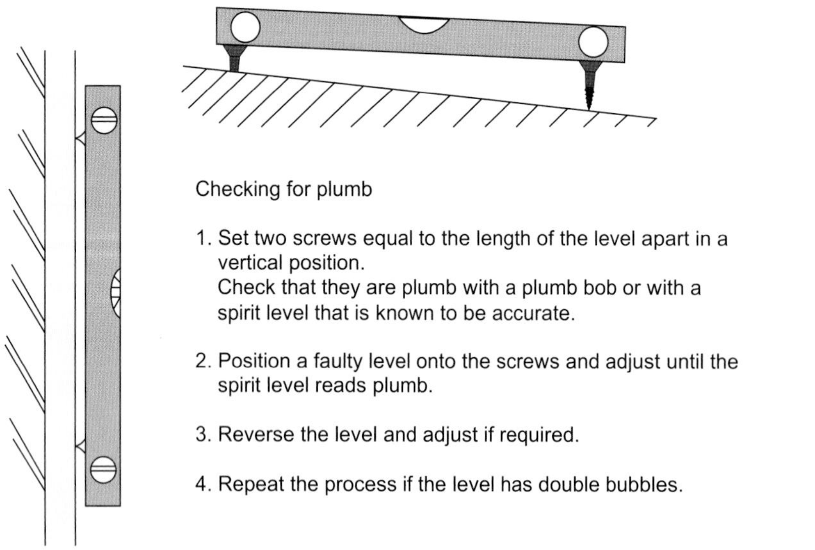
Checking for plumb
1. Set two screws equal to the length of the level apart in a vertical position.
Check that they are plumb with a plumb bob or with a spirit level that is known to be accurate.
2. Position a faulty level onto the screws and adjust until the spirit level reads plumb.
3. Reverse the level and adjust if required.
4. Repeat the process if the level has double bubbles.

FIGURE 6.8
Checking the accuracy of the spirit level

Site clearance

Before setting out can begin it is necessary to clear the site of all rubbish and top soil.

There could be hidden dangers in the rubbish to be cleared and care must be taken.

Services

If the site has been developed before there could be underground services such as water, gas and electricity.

Care should be taken when driving pegs into the ground that these pipes are not penetrated.

> **Remember**
>
> If in doubt ask your supervisor.

Setting-out calculations

Calculations required for setting out involve diagonals and the 3:4:5 method of producing a right angle.

3:4:5 RIGHT ANGLE

Pythagoras states that the square on the hypotenuse (longest side) is equal to the sum of the squares on the other two sides of a right-angled triangle.

Written as a formula, this is:

$$A^2 = B^2 + C^2$$

This is the basis of the 3:4:5 triangle.

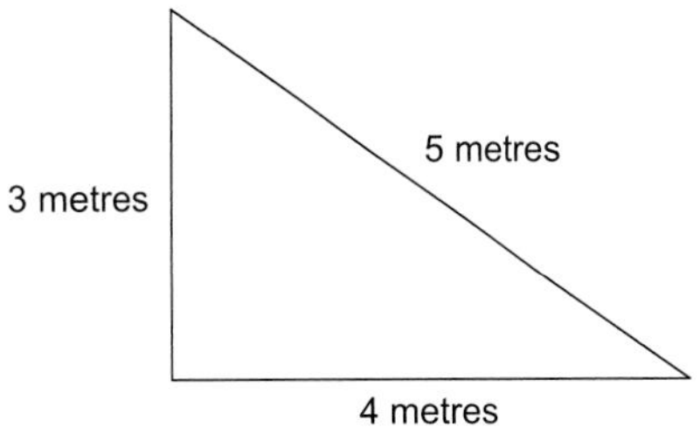

e.g. $5^2 = 4^2 + 3^2$

$25 = 16 + 9$

$25 = 25$

By transposing this formula it is possible to find the length of any side of a right-angled triangle provided we know the length of the other two sides.

Example 1

To find the length of side A:

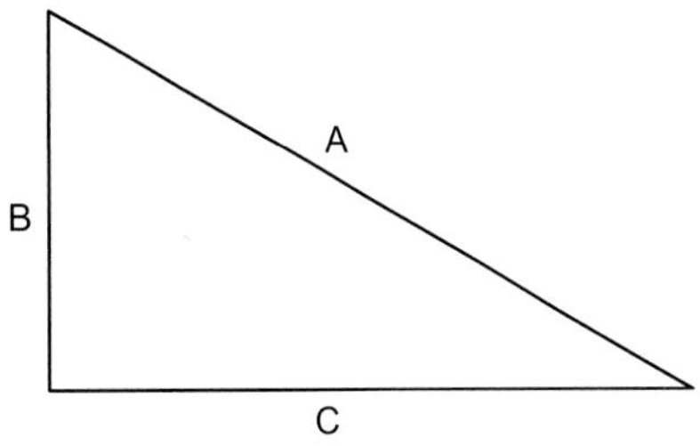

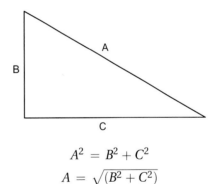

$$A^2 = B^2 + C^2$$
$$A = \sqrt{(B^2 + C^2)}$$

To find the length of side B:

$$A^2 = B^2 + C^2$$
$$A^2 - C^2 = B^2$$
$$B^2 = A^2 - C^2$$
$$B = \sqrt{(A^2 - C^2)}$$

To find the length of side C:

$$A^2 = B^2 + C^2$$
$$A^2 - B^2 = C^2$$
$$C^2 = A^2 - B^2$$
$$C = \sqrt{(A^2 + B^2)}$$

Example 2

To find the length of the sides of a triangle given the base and the perpendicular height:

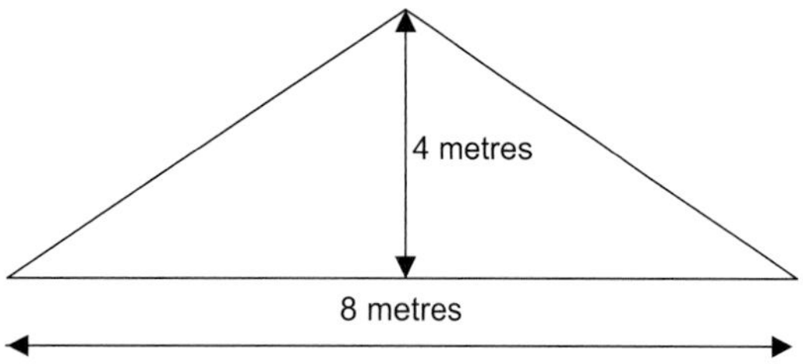

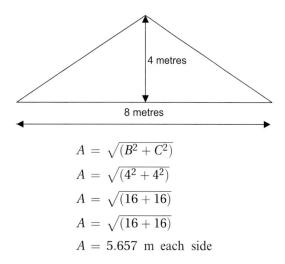

$$A = \sqrt{(B^2 + C^2)}$$
$$A = \sqrt{(4^2 + 4^2)}$$
$$A = \sqrt{(16 + 16)}$$
$$A = \sqrt{(16 + 16)}$$
$$A = 5.657 \text{ m each side}$$

Example 3

To find the length of the diagonal of a rectangle given the length of the two sides:

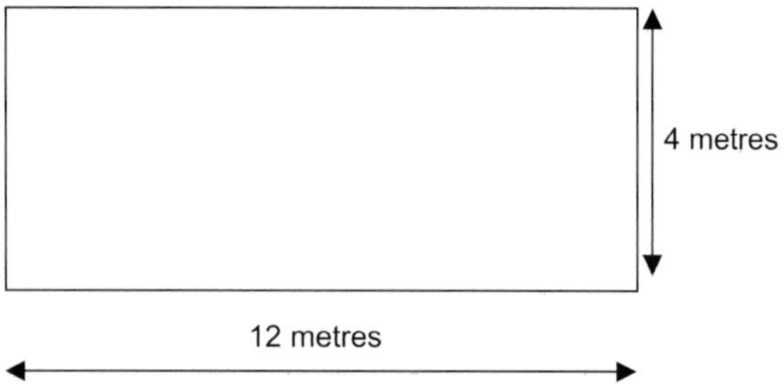

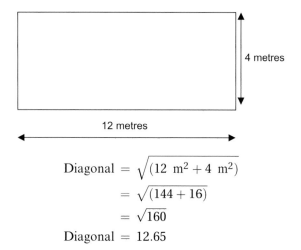

$$\text{Diagonal} = \sqrt{(12 \text{ m}^2 + 4 \text{ m}^2)}$$
$$= \sqrt{(144 + 16)}$$
$$= \sqrt{160}$$
$$\text{Diagonal} = 12.65$$

Minimizing the risk of damage

As stated before, there are several hidden dangers when setting out on the building site or in the workshop.

Setting out is usually the first item of work to take place on a building site.

It is usual for the top soil to be removed first. At this point the site is usually clear of all obstacles and dangers. If the setting out takes place before removal of the top soil then there could be hidden dangers on the site.

It is therefore essential to be aware and to ensure that correct PPE is worn, especially safety boots.

Accident reporting

Every accident should be reported – there should be an accident report book on every site and in every workshop, usually with the general supervisor or whoever is in charge of the site or workshop.

Make sure that you report any accident in which you are involved as soon as possible.

Site security

It is the responsibility of everyone on the work site to ensure that security of that site is maintained.

Security can take many forms and they are all equally important.

VISUAL SECURITY

- alarms – positioned in an accessible place within view of the general public
- bars, mesh, locks – fitted to glass-panelled doors and windows
- padlocks and padlock and chains – fitted to compound gates, pieces of plant and machinery
- lighting – flood lights and movement-activated lights
- security firms.

INDIVIDUAL SECURITY

It is the responsibility of all employees to contribute to the overall security of the firm. For example:

- Tidiness – do not invite crime by leaving tools and equipment where they may be easily seen. If there is a secure store, lock them away.
- If necessary immobilize large plant.
- Unauthorized access – it is the responsibility of all employees on site to challenge anyone who they feel has no authorization to be within a particular area. (Politeness is the best approach.)

If, despite the security measures taken, your site is breached, there are certain procedures you should follow. These should be given to you by your line manager, and may include:

- reporting the incident to the site supervisor
- reporting the incident to the police

- checking the inventory to find out what has been taken
- recording damage done to the premises and/or equipment.

Safety on site and in the workshop

Maintain a clean and tidy work station.

It should be the aim of everyone to prevent accidents.

The main contribution you as a trainee can make towards the prevention of accidents is to work in the safest possible manner at all times, thus ensuring that your actions do not put you, your workmates and the general public at risk.

A safe working area is a tidy working area.

All unnecessary obstructions that many create a hazard should be removed, e.g. offcuts of material, unwanted materials, disused items of plant, bricks, etc., and nails should be flattened or extracted from discarded pieces of timber which have been used when setting out and levelling.

Clean your work area periodically as offcuts of any materials are potential tripping hazards.

Remember

Vandalism and theft does not always happen at night.

Always take care of tools and equipment with which you are working and report any suspicious people or events to your line manager.

Remember

You are required by *law* to be aware and full your duties under the Health and Safety at Work Act.

STACKING AND STORING REUSABLE MATERIALS AND COMPONENTS

When clearing the work area there may be many materials that can be salvaged and used again.

Materials used for setting out consist mainly of timber. Any materials left should be stored away safely and securely. These materials should be properly cleaned before storing.

Any item that is damaged and cannot be reused is classed as waste and should be correctly disposed of.

HAZARDS

It is important that all materials are checked for any potential hazards, such as nails in timber or loose tin lids.

Always report any potential hazards to your supervisor.

Remedial actions

Any actual or potential hazards should be correctly dealt with.

Remedial action may involve:

- correcting – removing a nail from a piece of wood
- reporting – telling your line manager about the hazard
- warning others – placing a sign to warn others about the hazard.

DISPOSAL OF WASTE

There should be very little waste to be disposed of after completing setting out.

Most substances and materials are safe provided they are handled or worked on sensibly and with proper precautions.

Almost anything can be dangerous if handled or used irresponsibly. Some materials require extra care, and a few need extreme caution.

It is important to know what the hazards are, when they occur and how they can be prevented.

Usually prevention will involve the use of protective clothing and equipment.

REMOVAL OF WASTE

Since 1 April 1992, the law on waste has included a duty of care that means you must take all reasonable steps to look after any waste you have and prevent its illegal disposal by others.

If you break the law, you could be fined an unlimited amount.

'Controlled waste' is any household, commercial or industrial waste, such as waste from a house, shop, office, factory, building site or any other business premises.

PROTECTION OF THE SURROUNDING AREA

It is necessary to protect not only the site you are working on but also the neighbouring land or property. The surrounding properties could be affected by noise, smoke or dust.

Try to arrange parking so as not to upset the surrounding properties and cause traffic problems throughout the duration of the work.

A little understanding can prevent problems during the works.

Setting out

The first task in setting out a building is to establish a base line to which all the setting out can be related.

The base line is very often the building line; this is an imaginary line that is established by the local authority. You may build behind the building line and even up to the building line.

Note

NEVER build in front of the building line.

It is usual for the building line to be given as a distance from one of the following:

- the centre line of the road
- the kerb line
- existing buildings.

The frontage line of the building must then be on or behind the building line, never in front of it.

Degree of accuracy

A high degree of accuracy is required when setting out and this can be achieved if a steel tape is used and supported to avoid any sag.

Procedure for setting out

For this example we will use the site plan (Figure 3.13) in Chapter 3.

In this example, the building line has been set by the local authority 7 m from the front of the plot.

Two square lines are set from the kerb to the building line, at a distance apart larger than the required building, and two pegs are knocked in.

A line is stretched between the two pegs to establish the building line.

STAGE 1

The building line is fixed to two pegs on either side of the plot (Figure 6.9).

STAGE 2

In this example the frontage line of the building is on the building line (Figure 6.10).

After taking off the dimensions from the drawing the frontage line is calculated at 12 m.

Remember

The building cannot be in front of the building line, but it can be behind it.

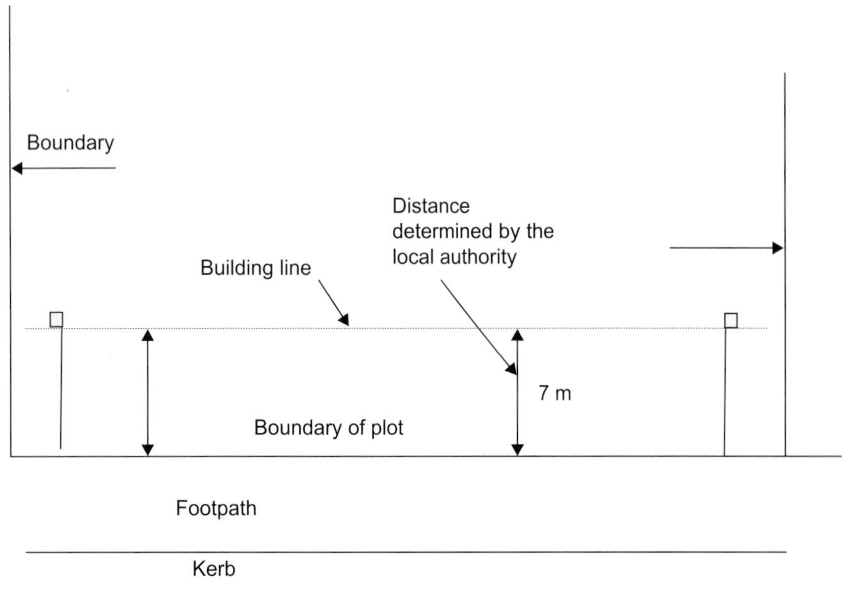

FIGURE 6.9
Stage 1

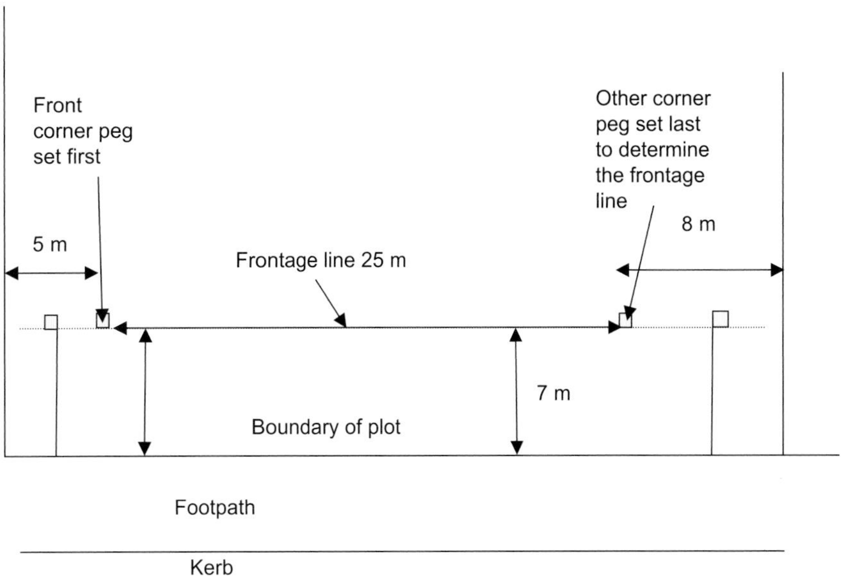

FIGURE 6.10
Stage 2

The distance from the boundaries should be read from the drawing. In this case the left-hand dimension is 5 m and 8 m from the right-hand boundary. Pegs should be knocked in to represent both points.

This will determine the front corners of the building.

In our example the two front corner pegs should be on the building line

STAGE 3

End wall lines can now be set out at right angles to the frontage line (Figure 6.11).

The depth of the building is calculated from the drawing and is 10 m.

The lengths of the end walls are pegged out.

Produce a right angle on both front corners of the building using a builder's square, the 3:4:5 method or an optical square.

Extend the lines back farther than the side wall dimensions of 10 m and knock two pegs in to hold the side wall lines.

STAGE 4

Knock two pegs in on each of the side walls at the required distance, i.e. 10 m.

Connect a line to the two back pegs to give the back wall line (Figure 6.12).

STAGE 5

The building is square if the diagonals are equal.

When completed check all dimensions and then check the diagonals.

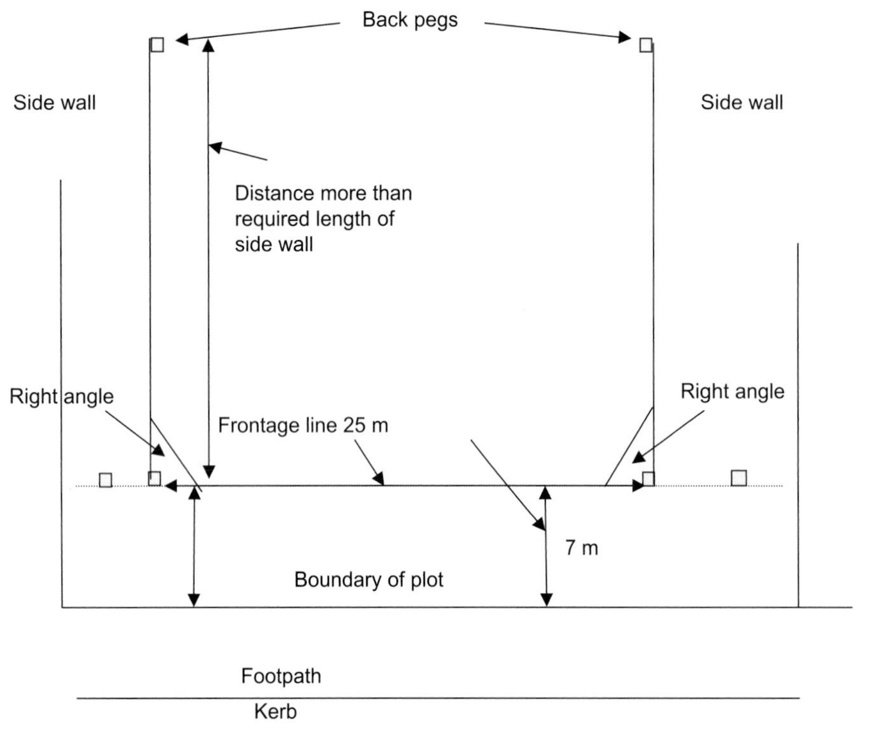

FIGURE 6.11
Stage 3

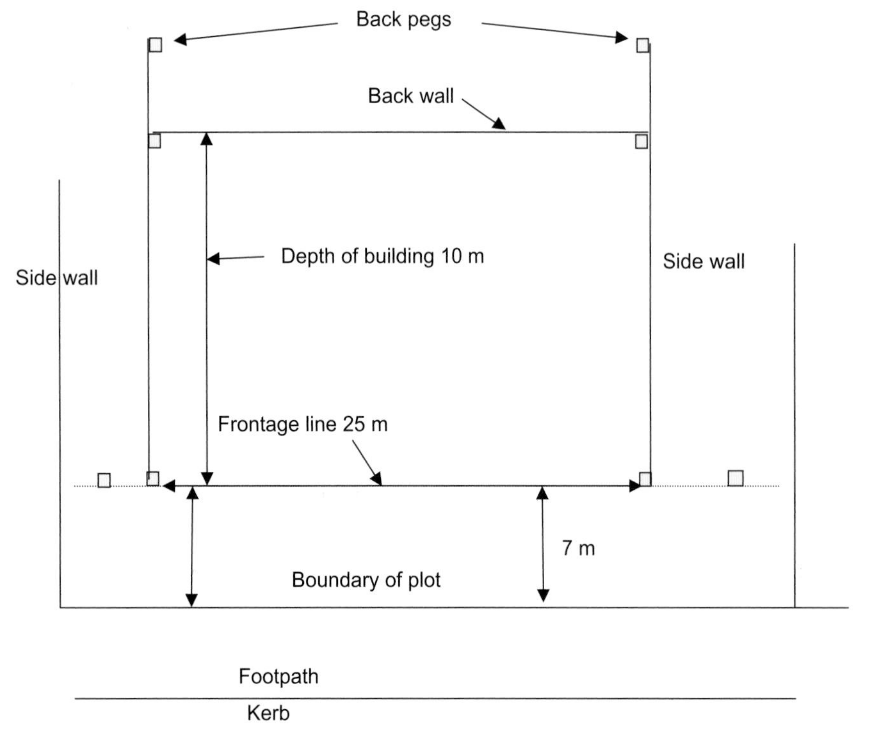

FIGURE 6.12
Stage 4

Measure each diagonal and if they are the same then the building is square (Figure 6.13). If not, then an adjustment has to be made.

Remember

Do not alter the frontage line.

Depending on which diagonal is out adjustments have to be made to make the diagonals equal.

Always check the dimensions of the walls after any adjustments.

Profiles

When the building has been set out and proved by checking the diagonals, profiles can be erected.

At the moment, the setting-out pegs are in the foundation trench, so they have to be repositioned approximately 1 m away from all the wall faces.

Profiles consist of 75 mm × 75 mm pegs with 150 mm × 25 mm boarding nailed to them (Figure 6.14). They can be either single (Figure 6.15) or corner type (Figure 6.16).

SINGLE PROFILE TYPE

Profiles are erected to enable the corner setting-out pegs to be removed to allow excavation to take place without disturbing the pegs.

One of these corner profiles could be set to a given level which is known as the 'datum', and may relate to the finished floor level or damp-proof course (DPC). This datum peg should be protected with concrete.

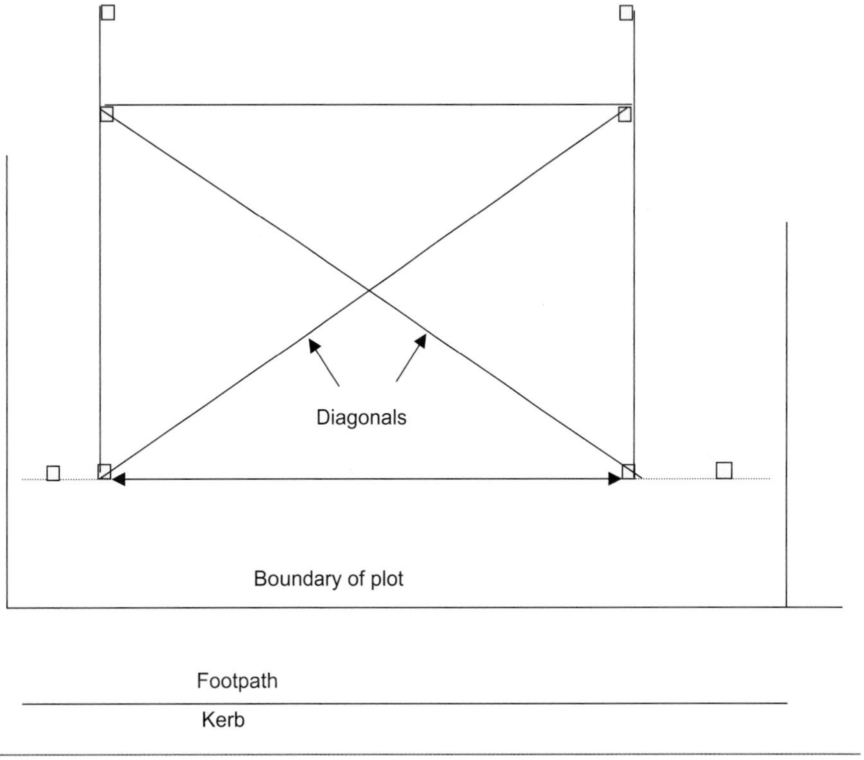

FIGURE 6.13
Stage 5

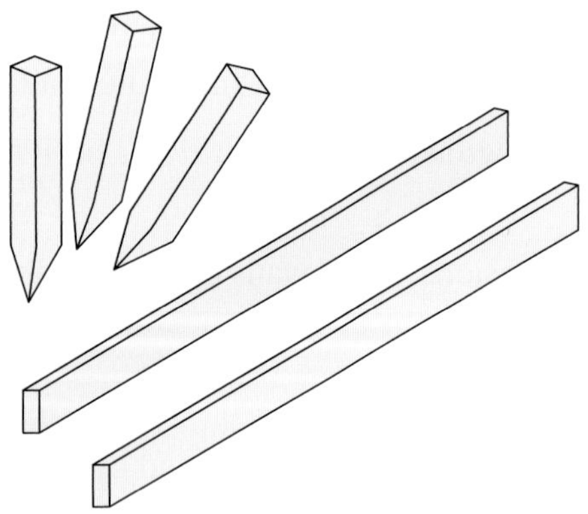

FIGURE 6.14
Material for making profiles

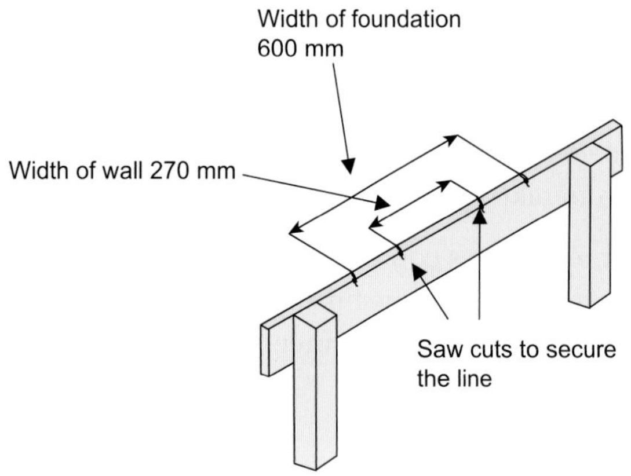

FIGURE 6.15
Single profile

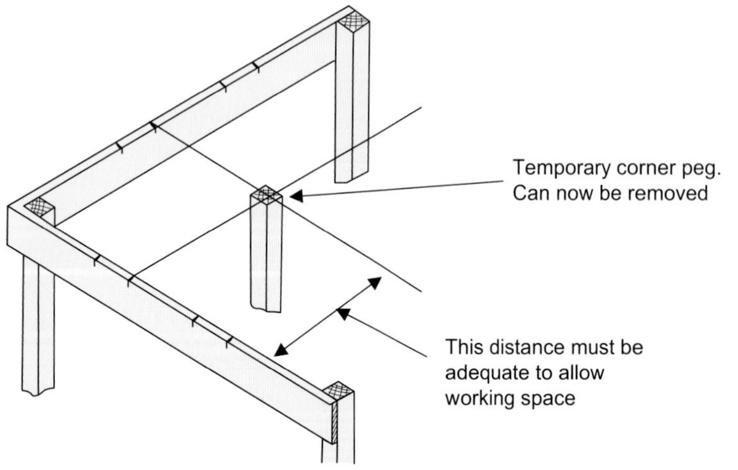

FIGURE 6.16
Corner profile

The datum peg could also be positioned away from the setting pegs but close enough to be accessible, and should also be protected with concrete and a small barrier of pegs and rails (Figure 6.17).

The profiles should be positioned approximately 1 m away from the face of the building to allow working space for the excavation.

The completed profiles can have foundation and wall widths marked on them in one of various methods. Saw cuts are best, as nails could be accidentally removed (Figure 6.18).

Once the profiles have been accurately constructed the dimensions should all be checked again, as should the diagonals.

The original setting-out pegs can now be removed.

Building lines can now be fastened to the profiles and the trench be marked out ready for excavation (Figure 6.19).

Lime or sand can be used to mark out the trenches ready for excavation (Figure 6.20).

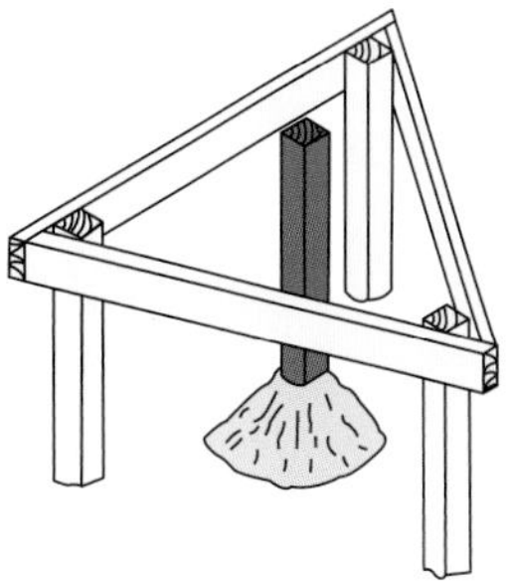

FIGURE 6.17
Datum peg protected

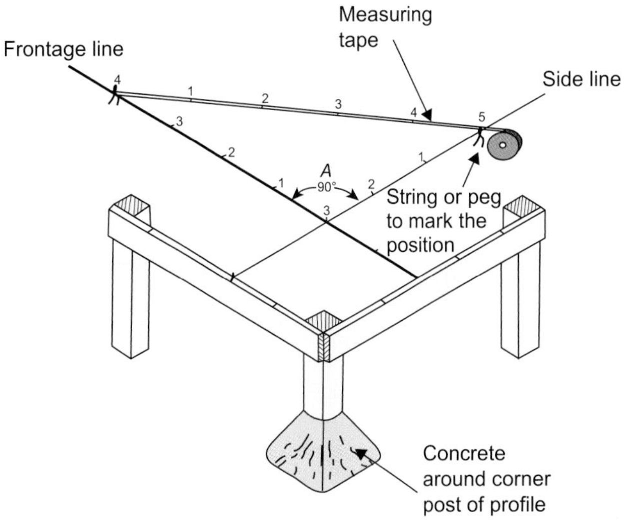

FIGURE 6.18
Checking a right angle by the 3:4:5 method

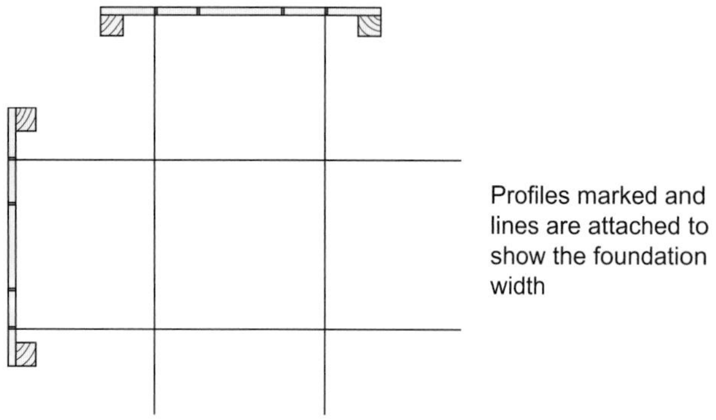

Profiles marked and lines are attached to show the foundation width

FIGURE 6.19
Fixing lines to the profiles to mark out the trench

Transferring levels

There are many methods of transferring levels on the building site, such as:

- straight edge and spirit level
- boning rods
- water level
- Cowley level
- site square
- telescopic levels
- laser level.

STRAIGHT EDGE AND SPIRIT LEVEL

Remember to reverse both level and straight edge for each reading (Figure 6.21).

BONING RODS

Boning rods (Figure 6.22) can also be used to transfer levels between two known points.

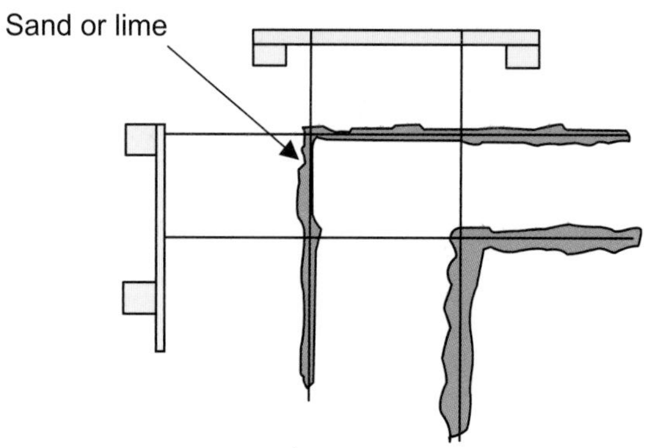

FIGURE 6.20
Marking out the ground ready for excavation

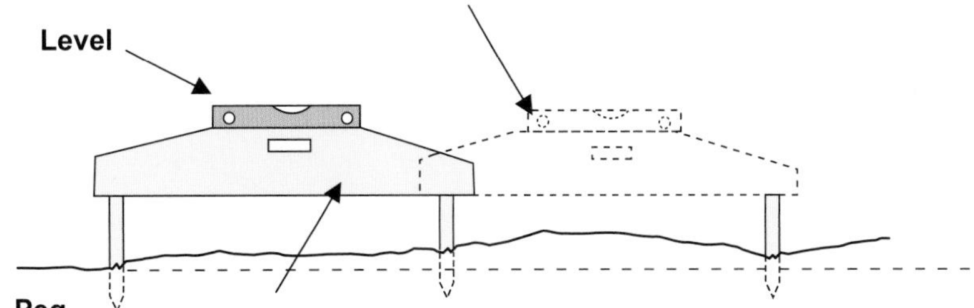
Remember to reverse both level and straight edge for each reading.
Level
Peg
Straight edge

FIGURE 6.21
Use of straight edge

The operation can be speeded up with the use of boning rods, as spirit levels are not used.

After the main level pegs have been set out at either end of a foundation, the extra levels between these can be set out with the use of boning rods.

Eye sight is used, instead of the spirit level, to sight in the top of the rods.

Boning rods are also used to transfer level points on large areas of hardcore, paving, etc., which are not necessarily level points.

The boning rod that is moved around the area or trench to produce intermediate level points is called the 'traveller' (Figure 6.23).

WATER LEVEL

The water level is simply a length of rubber tubing with a glass tube attached at each end (Figure 6.24).

The tube is carefully filled with water at one end only, to ensure that no air bubbles are trapped in the tube.

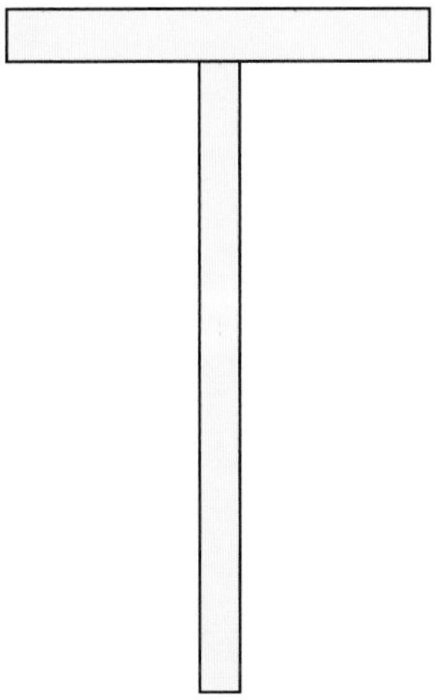

FIGURE 6.22
Boning rod

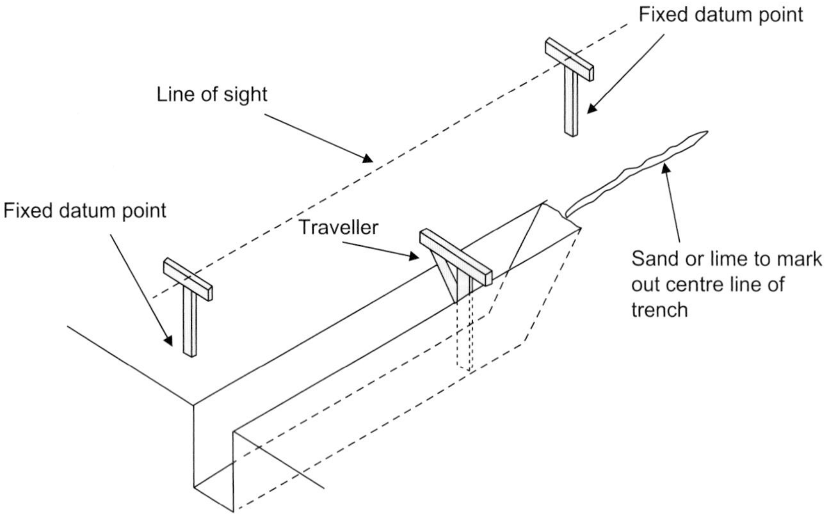

FIGURE 6.23
Use of boning rods and traveller

When the two glass tubes are held at the same level, the height of the water will be the same level in each tube.

If one tube is significantly lower than the other, then water will pour out of the lower tube, until once again the water finds its own level. To prevent this happening, both tubes are fitted with stoppers.

The water level is particularly useful for transferring levels from room to room or around corners.

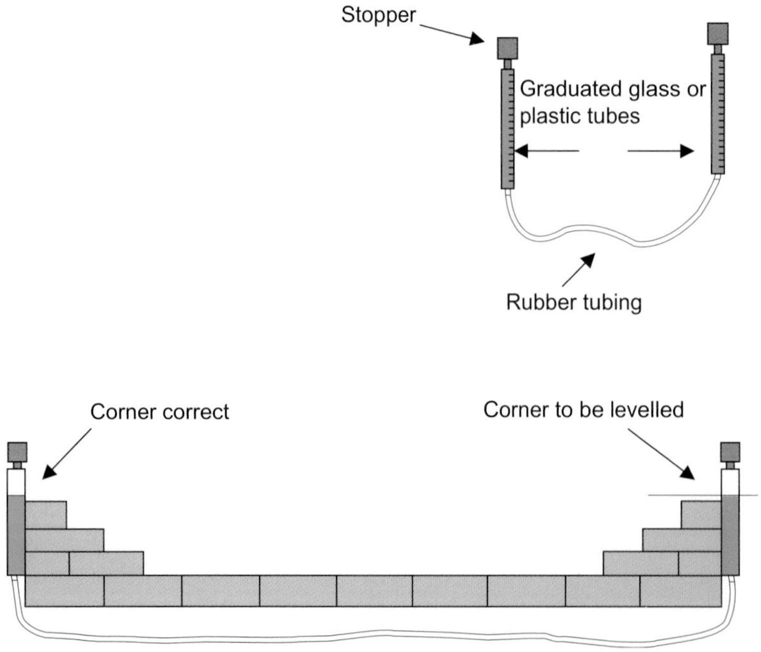

FIGURE 6.24
Use of water level

To transfer a level, hold one of the glass tubes so that the middle is about level with the required mark. Take the other glass tube to the position where you require the level to be transferred to.

Unplug both ends and raise or lower the glass tube nearer the level to be transferred until the water comes to rest level with the mark. The new level mark can then be marked off equal to the water level in the other glass tube.

COWLEY LEVEL

When levels have to be transferred over a long distance the method of straight edge and level can be very time consuming and rather inaccurate. A more accurate method is the Cowley level (Figure 6.25).

It can be used to transfer the site datum peg to all corners of the building very quickly and very accurately.

It is an automatic device working a system of mirrors, some of which are controlled by pendulums so that the sighting is always along the same line.

The target is 450 mm long and 50 mm wide and slides up and down a staff, which is marked out on the back in metre and millimetre graduations.

Set up the tripod as level as possible.

There is a rod on the top of the tripod which slides into the base of the Cowley level. This releases the two mirrors.

Place the instrument over the rods on the tripod, noting the release of the glass mirrors (Figure 6.26).

Rotate the camera around until you bring the target into vision.

Sight through the eyepiece: the view is of two mirrors, one the correct way up and the other upside down. Various views are shown in Figure 6.27.

Signal the assistant to slide the target up or down as required. As soon as both sides of the target are level the staff should be fastened off.

Note

The camera should never be moved while attached to the tripod for fear of damaging the mirrors.

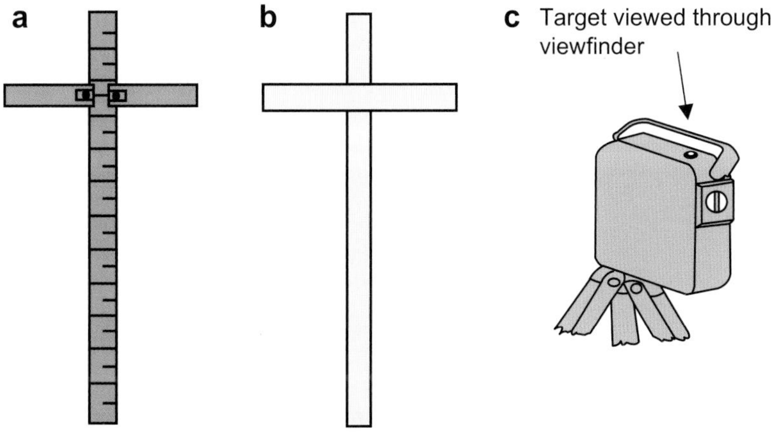

FIGURE 6.25
Cowley level equipment: (a) front of target; (b) rear of target; (c) Cowley level

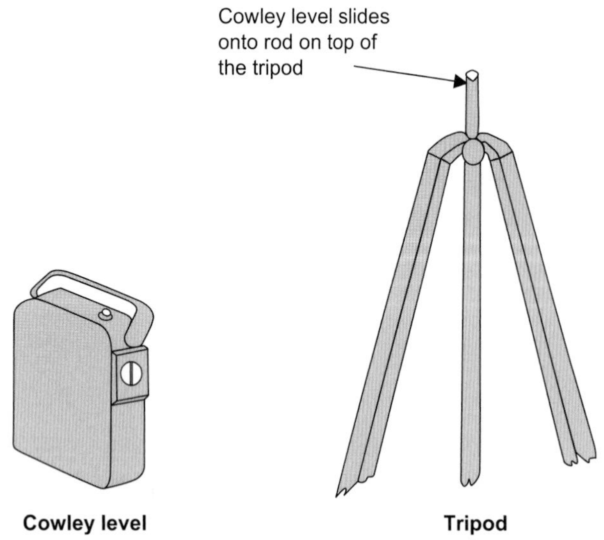

FIGURE 6.26
Setting up the Cowley level

This can then be transferred to other pegs. Each peg being knocked in to the required depth so that the view through the eyepiece reads level at every new peg.

THE OPTICAL SITE SQUARE

The site square is an accurate setting-out instrument (Figure 6.28).

The optical square consists of two telescopes mounted one above the other and with their lines of sight set at 90 degrees to each other. The instrument is fixed on a tripod (Figure 6.29), which is set up over a known corner peg.

Once the site square has been set up on a known corner peg, look through the bottom telescope and a view along one side of the building is shown until a view of the nail on the peg is in sight. Three views are shown in Figure 6.30. The correct view is when the nail in the top of the peg is in the centre of the hair lines.

When looking through the top telescope, without moving the equipment, it will trace out a line at right angles to the original line and then a further site peg can be positioned along the new line (Figure 6.31).

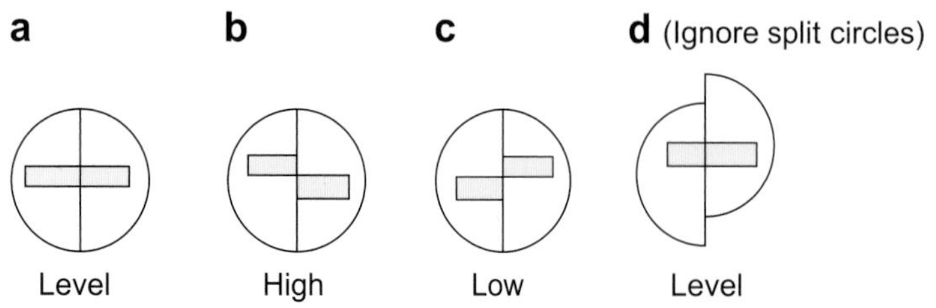

FIGURE 6.27
Various views of the staff

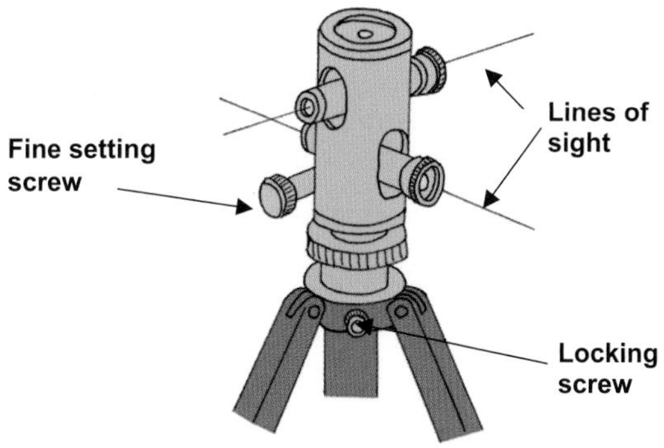

FIGURE 6.28
Site square

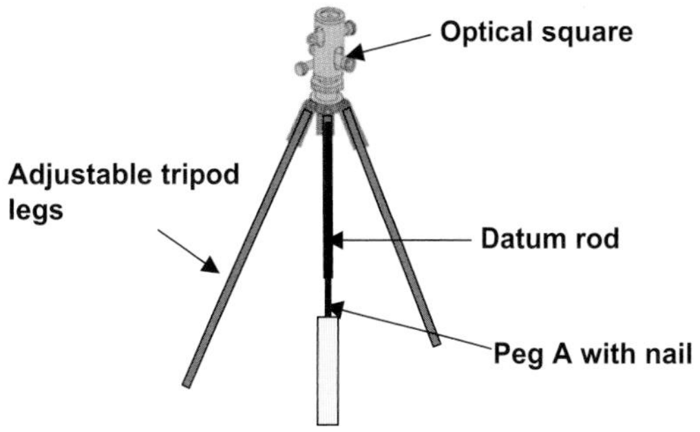

FIGURE 6.29
Site square set on tripod over corner peg

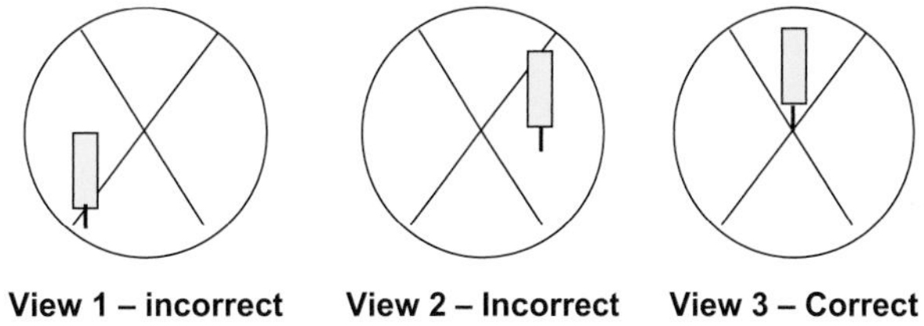

FIGURE 6.30
Views through the bottom telescope

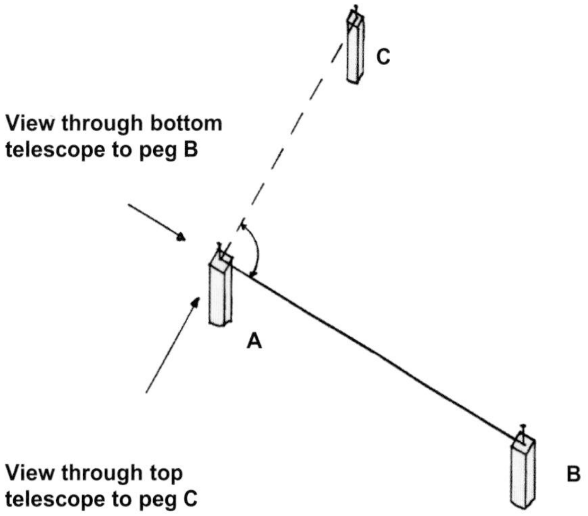

FIGURE 6.31
Producing a right angle

TELESCOPIC LEVELS

Telescopic levels are a series of instruments each using the same basic principles of a telescope set on a tripod that is fixed to read on a horizontal plane. The difference between the various optical instruments lies in the manner in which the telescope is levelled.

Component parts of levels

The telescope

This is usually a tube with a large object lens at one end and a small eyepiece at the other Between these two lenses are the focusing lenses, as shown below. The telescope magnifies the image to ease readings. Some telescopes produce an inverted image while others show the image the correct way up (Figure 6.32).

The diaphragm

This positions the cross-hairs (Figure 6.33).

Stadia lines

The vertical cross-hair normally has two small horizontal lines called stadia lines.

The distance between these two horizontal lines, when measured on a staff and multiplied by a constant (100 on modem instruments), gives the distance between the level and the staff.

Collimation line

This is the axial line passing through the centre of the telescope. For a level to be correct, the collimation line must be set exactly horizontal.

Figure 6.34 shows readings being taken from points A and B.

Bubble tubes

Some instruments have a spirit level tube which is used to level the instrument. These spirit levels have different configurations – some are plain spirit levels while others are 'split' or coincidence bubbles.

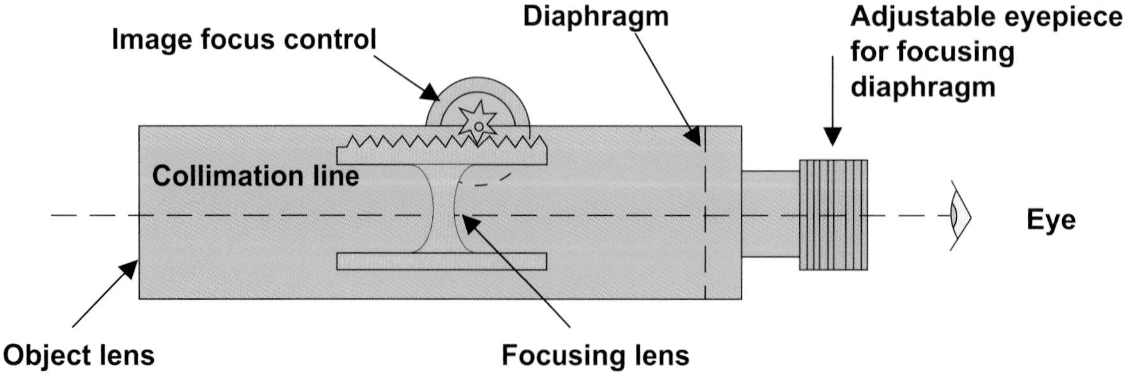

FIGURE 6.32
The telescope

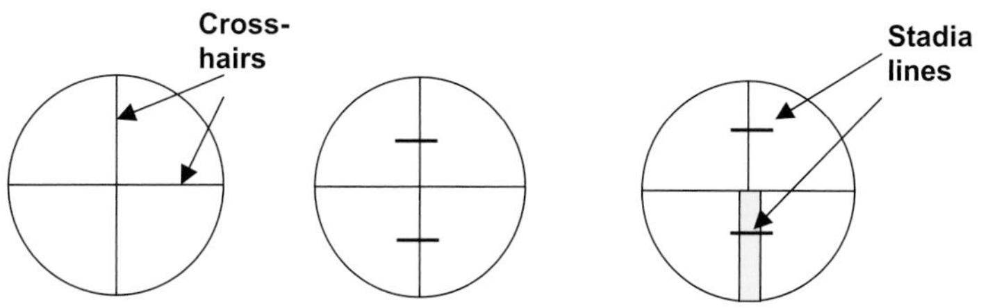

FIGURE 6.33
The diaphragm

Fish-eye/circular/pond bubble

Most surveying instruments nowadays have a fish-eye bubble (Figure 6.35). This assists in making the instrument approximately level.

Classification of levels

Different levels are designed for different purposes depending on the degree of accuracy required. The three groups are as follows:

- builder's level – low magnification but rugged; suitable for site use
- engineer's level – high magnification quality and accuracy and consequently more expensive
- precise levels – used for geodetic surveying with great accuracy over long distances; have optical micrometers for reading staffs and these are faced with Invar® steel to minimize thermal movement.

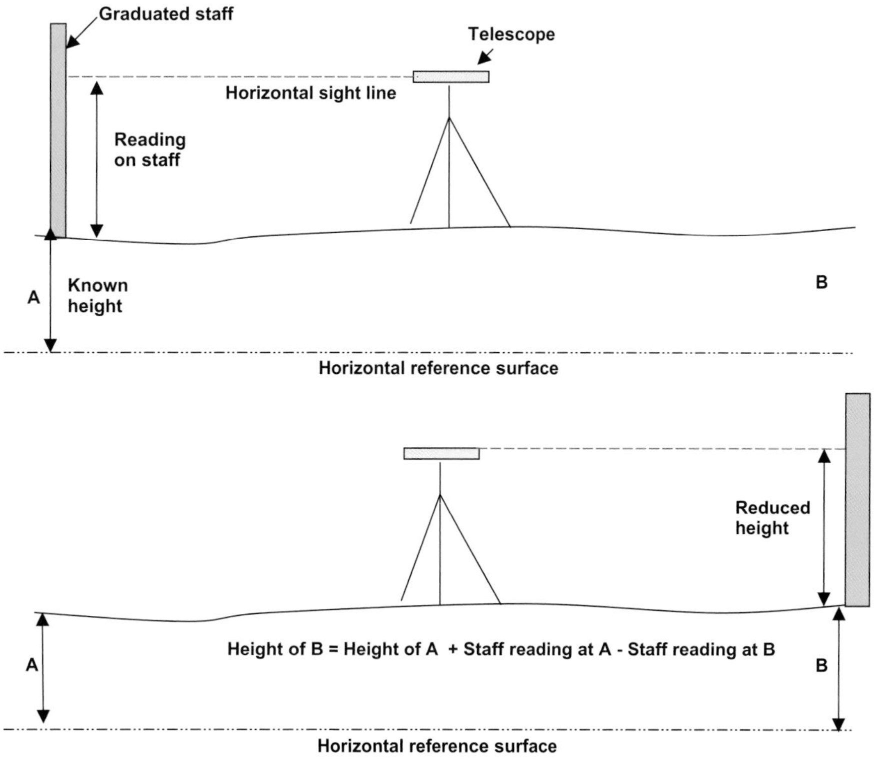

FIGURE 6.34
Collimation line

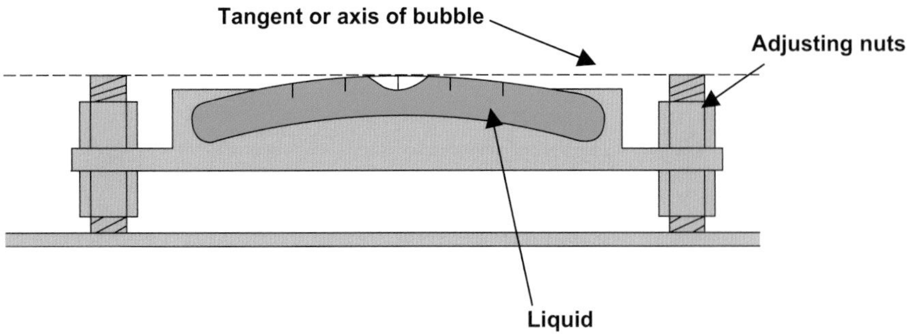

FIGURE 6.35
Bubble

Types of level

There are four basic types of instrument:

- dumpy level
- tilting or quickset level
- automatic level
- laser level.

All have a locking clamp or friction stop and slow-motion screw to direct the telescope centrally onto a staff. Most have a horizontal circle graduated in degrees so that horizontal angles can be set out.

These instruments are less accurate than a theodolite.

Dumpy level

The method of levelling a dumpy level, which distinguishes it from other levels, uses a spirit level in two directions. It has three footscrews between the bottom and top plates (Figure 6.36).

1. Position the telescope so that the bubble is parallel with two of the three footscrews.
2. Rotate the two footscrews in opposite directions until the bubble is showing level.
3. Turn the telescope 90 degrees and rotate the third footscrew until the bubble is again showing level.

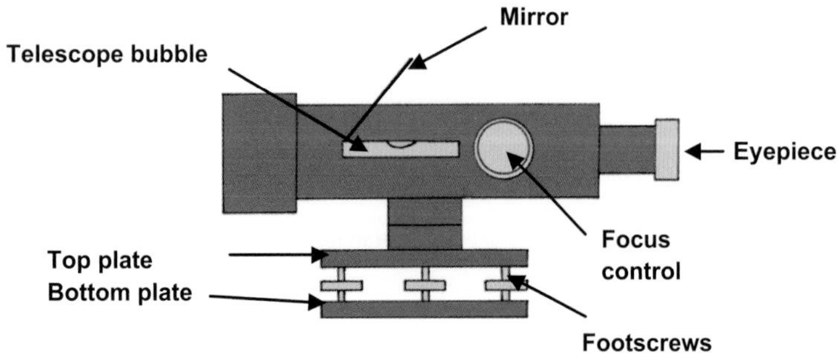

FIGURE 6.36
Dumpy level details

4. Turn the telescope in all directions to check that the bubble remains level at all times.

This is rather a laborious method and so the dumpy level is now virtually obsolete. We shall therefore consider in more detail the other instruments that are in common use in the industry.

Tilting or quickset level

This instrument was developed from the dumpy level, and it is set up for reading as follows.

1. Attach the instrument to the tripod.
2. Using the domed head joint or footscrews (depending on the design), centre the fish eye.
3. Sight onto the staff.
4. Level the bubble tube.
5. Take the readings.

Two levels are built into the instrument, a tubular one on the telescope and a circular one fixed to the top plate (Figure 6.37).

It is very easy to set up and this is the main advantage over the dumpy level.

Automatic level

This is a further development from the quickset level and need only be approximately levelled; then, a compensating device, usually based on a pendulum inside the telescope, corrects itself in case of any inaccuracy.

There are no precise levels to set; they need only be roughly levelled to centralize the circular bubble using the dome and cup joint or three footscrews.

It is set up for reading as follows.

1. Attach the instrument to the tripod.
2. Using the domed joint or the footscrews (depending on the design), centre the fish eye.

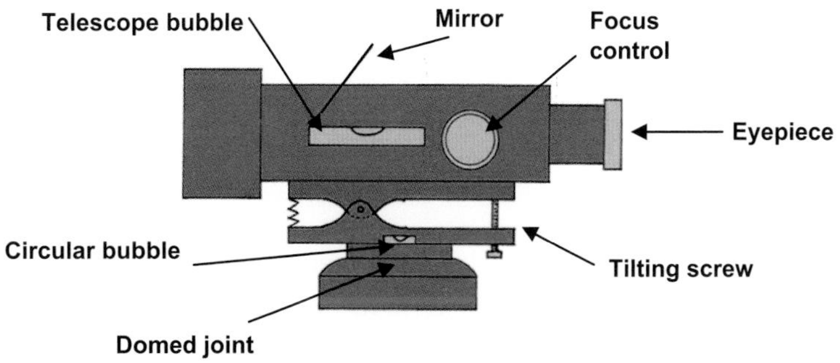

FIGURE 6.37
Tilting level details

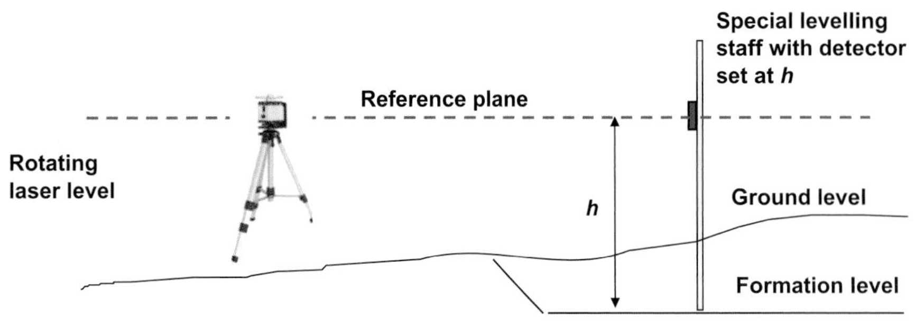

FIGURE 6.38
Laser level

3. Sight onto the staff.
4. Take the reading.

Care must be taken when using these levels in situations where there is a chance of any vibration, such as in heavy winds, owing to their precise compensating device.

Laser level

All the instruments described so far are optical levels and require two people for their operation. The laser level is not an optical level and can be operated by one person.

1. Set up the level using the fish-eye bubble as for the automatic level.
2. Press the set button and it automatically finds its own level (the plane of collimation) (Figure 6.38).
3. A special target receiver is then used to transfer this level to a target at the position where the level is to be taken or set out.

Laser levels are the latest method for transferring levels around a construction site. As long as they are looked after correctly they are very accurate and easy to set up.

They are a very versatile item of equipment and can be used for all kinds of levelling, both internal and external.

The range of most laser instruments is between 100 and 300 m and some units have a self-levelling facility.

Most instruments operate from a 12 V rechargeable battery, which produces an invisible beam of light.

OTHER LEVELLING EQUIPMENT

Tripod

Tripods are usually made of aluminium or wood. They should always be set up as level as possible and at a convenient height.

The British Standard metric staff

These are available constructed in wood, glass fire or aluminium.

Staffs can be telescopic or folding and are available in various lengths, e.g. 3, 4 or 5 m. The drawing in Figure 6.39 shows the common E-type face marking. The marks are 10 mm deep with spaces of 10 mm between them.

The numbers show the reading at their base. Different colours, normally red or black, are used for graduation marks in alternate metres.

Direct readings can be made to 0.01 m and estimated readings to 0.001 m.

DATUM

A datum is a fixed point, positioned on the building site, to which all other levels are related. It should be established at an early stage.

Wherever possible the datum should be positioned where it will not be disturbed but can be easily reached. A peg is usually driven into the ground near the site cabin and protected with concrete, as shown in Figure 6.17.

These are known as temporary bench marks.

Other fixed points could be used, such as a manhole cover or a kerb stone (Figure 6.40).

ORDNANCE BENCH MARKS

These are levels above sea level and are taken from a location at Newlyn, Cornwall (which is exactly at sea level) (Figure 6.41).

These types of levels are often shown on block plans, cut into stonework or brickwork, and usually on the corner of old buildings and stone gate posts, so that it is convenient for levels to be transferred from them.

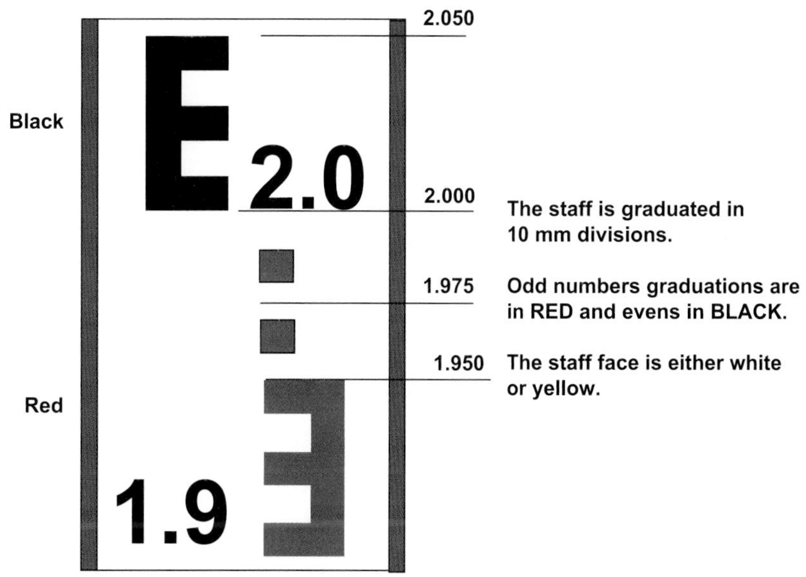

FIGURE 6.39
Metric staff

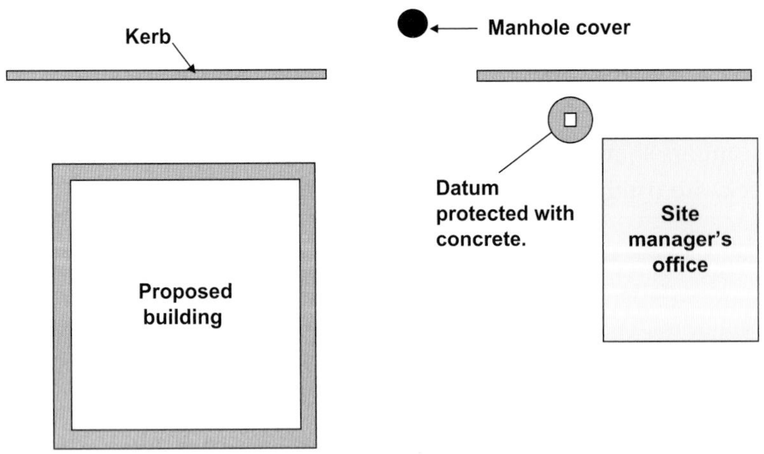

FIGURE 6.40
Site datums

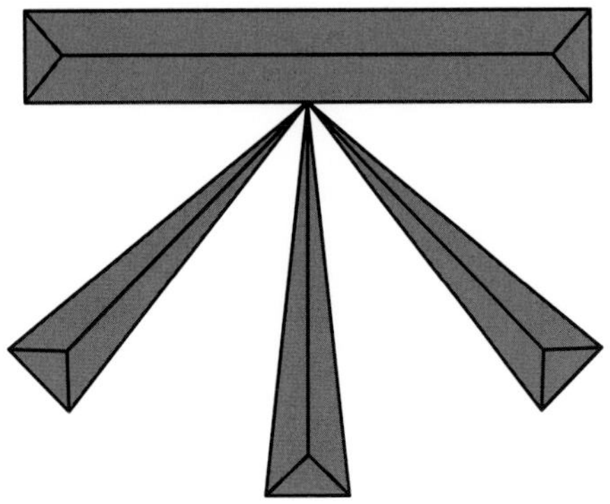

FIGURE 6.41
Ordnance bench mark

READING LEVELS

It is important to understand the various terms used when reading and recording levels.

- back sight – the first staff reading
- fore sight – the last staff reading
- intermediate sight – any reading between the back sight and fore sight
- change point – when the position of the level is moved
- reduced level – the height of any point in relation to the datum used.

There are two methods of recording the readings: rise and fall and height of collimation. 'Reduced levels' is the process of calculating the required reduced levels from staff readings by either of these methods.

For example, assume a set of readings consisting of six levels to be taken between an ordnance bench mark (OBM) and a manhole (Figure 6.42).

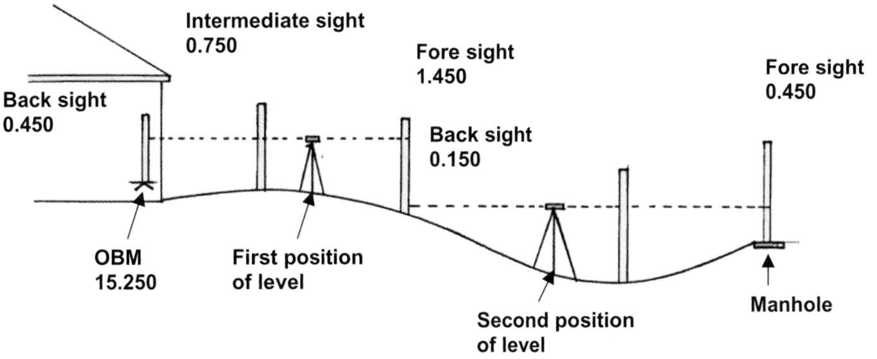

FIGURE 6.42
Staff readings. OBM: ordnance bench mark

Rise and fall method

Readings should be entered in the appropriate column as shown in Table 6.1.

It should be noticed whether the ground has risen or fallen between readings. If the second reading is larger than the first the ground level has fallen. If the second reading is less than the first reading the ground has risen.

The first reading is 0.450 and the second 0.750; therefore the calculation is:

$$0.450 - 0.750 = 0.300 \text{ (fall)}$$

The next calculation is as follows:

$$0.750 - 1.450 = 0.700 \text{ (fall)}$$

Each reading should be either added to or subtracted from the reduced level to give a new reduced level.

If the original bench mark is not known a figure usually around 10.000 should be used, to avoid negative readings.

The second set of readings is added to the chart. The first back sight is inserted on the same line as the last fore sight and the same reduced level accepted.

Table 6.1 Rise and fall method

Back sight	Intermediate sight	Fore sight	Rise	Fall	Reduced level	Remarks
0.450					15.250	OBM
	0.750			0.300	14.950	
0.150		1.450		0.700	14.250	
	1.550			1.400	12.850	
		0.450	1.100		13.950	Manhole
Check:						
0.600		1.900		2.400	15.250	
		0.600		1.100	13.950	
		1.300		1.300	1.300	

OBM: ordnance bench mark.

Once all the readings have been completed the calculations must be checked.

Add up all the rise and fall figures and subtract the smaller from the larger.

Total the back sights and fore sights and again subtract the smaller from the larger. Note the first and last reduced levels and subtract the smaller from the larger.

All three answers should be the same. In the example in Table 6.1 they are all 1.300.

Height of collimation method

In the example the first sight at the OBM gives a staff reading of 0.450 above the bench mark.

Adding the two together, 15.250 + 0.450, gives a new reading of 15.700. This is the height of collimation.

The intermediate staff reading would be:

$$15.700 - 0.750 = 14.950$$

The fore sight reading would be:

$$15.700 - 1.450 = 14.250$$

At this point the level is moved to the second position and a back sight reading is taken onto the last staff position.

The readings at a changeover point have the same reduced level and are both recorded on the same line.

The new height of collimation is:

$$14.250 - 0.150 = 14.400$$

The remaining readings are calculated and inserted into the chart, and a check is made.

Total the back sights and fore sights and subtract the smaller from the larger. Note the first and last reduced levels and subtract the smaller from the larger.

Both answers should be the same. In the example in Table 6.2 they are all 1.300.

Completing the work on time

It is important that all work on site, even the setting-out programme, is kept on time.

If one part of the programme is delayed then it has a knock-on effect on following programmes. Any problems with keeping to the programme should be reported to the site manager so that allowances can be made if possible.

Table 6.2 Height of collimation

Back sight	Intermediate sight	Fore sight	Collimation	Reduced level	Remarks
0.450			15.700	15.250	OBM
	0.750			14.950	
0.150		1.450	14.400	14.250	
	1.550			12.850	
		0.450		13.950	Manhole
Check:					
0.600		1.900		15.250	
		0.600		13.950	
		1.300		1.300	
OBM: ordnance bench mark.					

One of the most common reasons for delays is poor drawings. When drawings are difficult to understand or there is a lack of clear information it can cause a time delay.

Another cause is the weather. Poor weather during winter months can affect the duration of setting out.

Setting out brickwork and blockwork

Once the initial setting out has been completed and the foundations have been excavated and concreted the brickwork can commence.

Brickwork up to DPC level is laid first and the correct bonding must be sorted out before any brickwork is exposed.

Multiple-choice questions

Self-assessment

This section of the book is designed to allow you to check your level of knowledge. The section consists of revision questions for this chapter. The questions are all multiple choice and have four possible answers. The answers are to be found at the end of the book.

The main type of multiple-choice question will be the four-option multiple-choice question. This will consist of a question or statement, known as the stem, followed by a choice of four different answers, called the responses. Only one of these responses is the correct answer; the others are incorrect and are known as distracters.

You should attempt to answer the questions by choosing either (a), (b), (c) or (d).

Example

The person employed by the local authority to ensure that the Building Regulations are observed is called the:

(a) clerk of works
(b) building control officer
(c) council inspector
(d) safety officer

The correct answer is the building control officer, and therefore (b) would be the correct response.

Setting out

Question 1 Identify the item of setting out equipment shown:

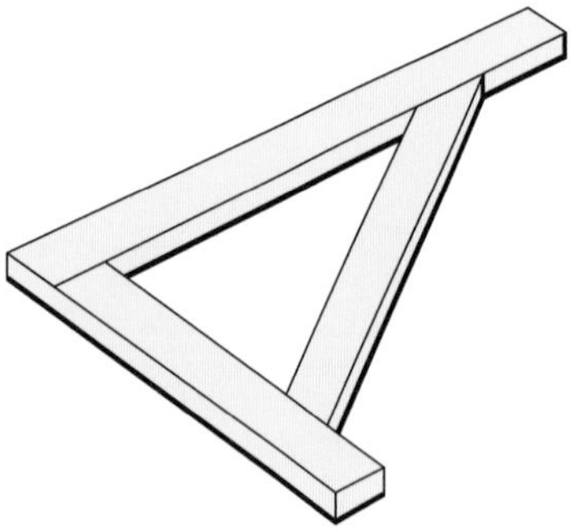

(a) builder's square
(b) builder's level
(c) site square
(d) site level

Question 2 Identify the item of levelling-out equipment shown:

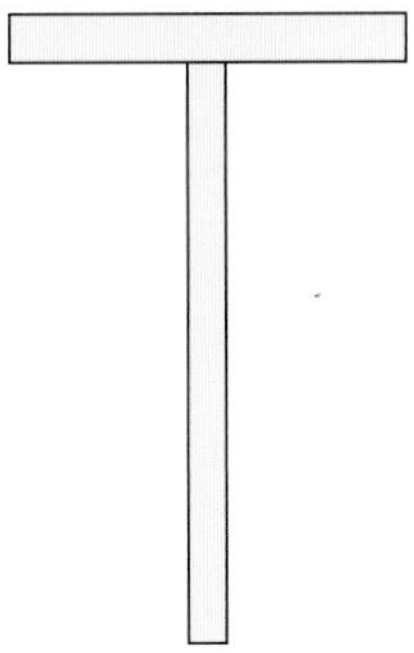

(a) profile

(b) ranging rod

(c) level

(d) boning rod

Question 3 Which of the following is a correct 3:4:5 ratio?

(a) 2:4:6

(b) 6:8:10

(c) 8:10:12

(d) 9:10:15

Question 4 Which of the following is the correct dimension for the diagonal x?

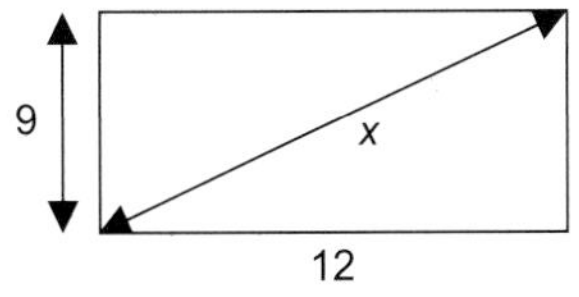

(a) 12

(b) 14

(c) 15

(d) 21

Question 5 Who is responsible for setting the building line?

(a) the local authority

(b) the local council

(c) the site manager

(d) the client

Question 6 If, when checking the diagonals after setting out, they are found to be wrong, which line should not be altered?

(a) the back line

(b) the side line

(c) the frontage line

(d) the diagonal line

Question 7 Why is it necessary to reverse both the level and straight edge when transferring levels?

(a) to save using two levels

(b) to correct any deficiency in the equipment

(c) to achieve the correct fall

(d) to be able to read the level correctly

Question 8 Name the item of equipment used for setting intermediate-level points when transferring levels using boning rods:

(a) datum point

(b) straight edge

(c) range rods

(d) traveller

CHAPTER 7

Chimneys, Flues and Fireplaces

This chapter will cover the following NVQ and Diploma units:

- NVQ VR49
- CC 3027

This chapter is about:

- Interpreting building information
- Adopting safe and healthy working practices
- Selecting materials, components and equipment
- Preparing and erecting brickwork and blockwork chimneys and decorative features

The following NVQ performance criteria will be covered:

- Interpretation of information
- Safe work practices
- Selection of resources
- Minimize the risk of damage
- Meet the contract specification
- Allocated time

The following Diploma outcomes will be covered:

- Know how to set out and build fireplaces and flues
- Know how to set out and build decorative chimney stacks

Building information

This chapter will deal with the construction of chimney stacks, flues and fireplaces.

Information can be gathered from numerous sources such as drawings, specifications and manufacturers' information sheets.

Information from drawings

The student should be able to extract information required for setting out from drawings. Drawings have been dealt with in Chapter 3. Please refer back if necessary.

Information from specifications

The specification is a precise description of all the essential information and job requirements that will affect the price of the work but cannot be shown on the drawings.

Manufacturers' information

Technical information can be produced in several formats. When items of equipment or components are purchased there will always be manufacturers' information sheets with them.

Safe work practices

As stated before, there are several hidden dangers when working on the building site, especially when working on chimney stacks at height. It is therefore essential to be aware and to ensure that correct personal protective equipment is worn, especially safety boots.

The scaffold should have been erected by a trained scaffolder and all ladders fixed in position.

It is vital that the highest standard of workmanship and the current interpretation of the current Building Regulations are maintained throughout the construction of chimney breasts and stacks. Failure to do so can have devastating results.

The regulations are mainly concerned with preventing the spread of fire and the discharge of dangerous gases given off by burning fuel, ensuring that they are safely transported to the atmosphere without endangering health.

A chimney stack passing through the roof must also be properly protected against downward penetration of dampness.

When the stack penetrates a pitched roof the junction between the roof and the chimney stack requires the services of a plumber. An apron is needed at the front of the stack and a gutter at the rear. Both sides will need stepped flashings.

It is the responsibility of the bricklayer to rake out the bed joints to receive these flashings. The joints should be raked out to a depth of about 25 mm. They should be raked out cleanly and not with a sloping section. When the flashing is in position and plugged into the joints, the joints should be pointed.

Selection of resources

The most important resources when constructing chimney stacks are the bricks and mortar.

Chimney stacks are constructed at the highest point in the building and are subject to attacks from the weather on all faces.

Good class engineering bricks are recommended, with a cement mortar equal to the strength of the brick. The chimney stack should be terminated with good solid capping, well weathered to allow water to fall away quickly.

Construction of chimneys, flues and fireplaces

So that the apprentice can clearly understand this important craft operation, this chapter has been divided into four subsections. These cover:

- The types and positioning of fireplaces, flues and chimneys, as desired by the architect.
- The Building Regulations controlling construction.
- The construction of a fireplace chimney through the ground floor, first floor and roof of an ordinary domestic dwelling. By following the sequence of operations, the reader should have no difficulty when confronted with a problem.
- The fixing of fireplaces.

Types of chimney breasts and flues

Single fireplaces

The fireplace opening is formed on one side of the wall only, by the formation of attached piers called jambs. Figures 7.1 and 7.2 illustrate the plans of various jamb arrangements, in solid and cavity walls, at the ground floor level or wherever the base or a chimney breast begins.

Figure 7.1(a) shows a simple arrangement where the main wall forms the back of the fireplace opening. Figure 7.1(b) illustrates a similar example where the main wall is one-and-a-half bricks thick and the back of the

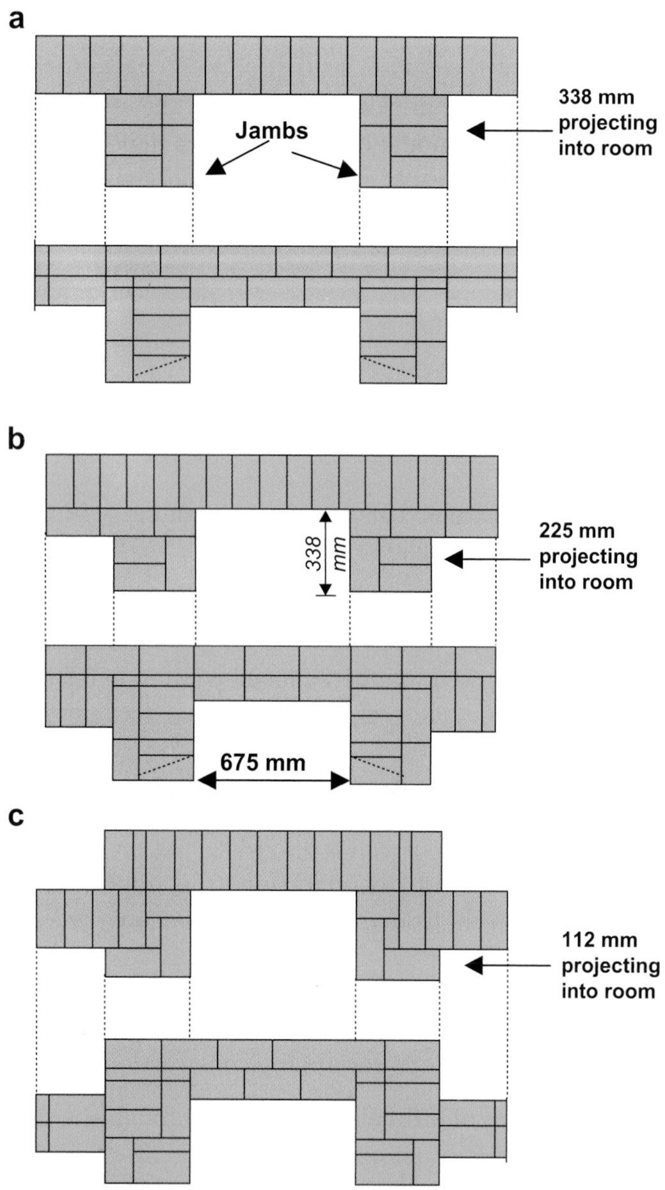

FIGURE 7.1
Single fireplaces, solid walls in English bond

fireplace is one brick thick. Figure 7.1(c) shows a typical example of a fireplace placed on external walls; the projection of the jambs into the room is lessened by the formation of an external breast or the breaking of the wall line on the opposite side of the wall. The advantage gained is the larger area available in the room in which the fireplace is situated.

Figure 7.2(a, b) shows various details of chimney breasts in cavity walls with building blocks on the inner leaf. Figure 7.3 shows a chimney breast constructed on an internal wall.

The width and height of the fireplace opening depend on the type of stove or grate to be inserted, while the width of the jambs depends on the width of the chimney breast required on the upper floors.

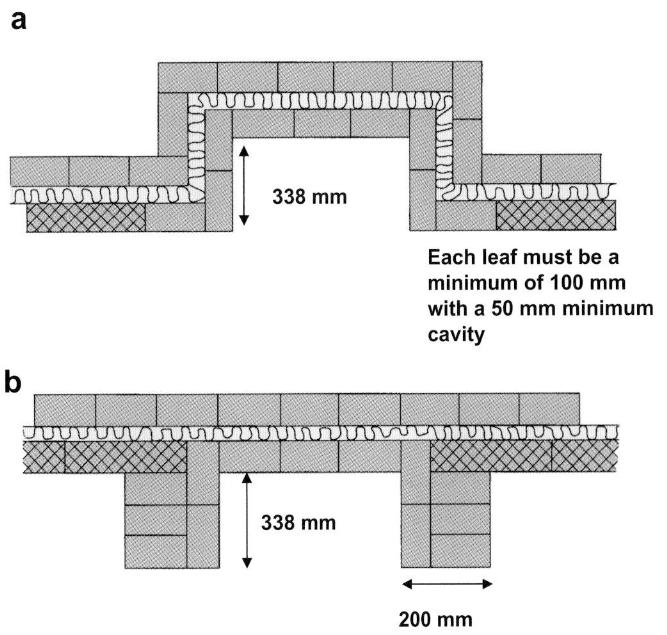

FIGURE 7.2
Single fireplaces in cavity walls: (a) external chimney breast; (b) internal chimney breast

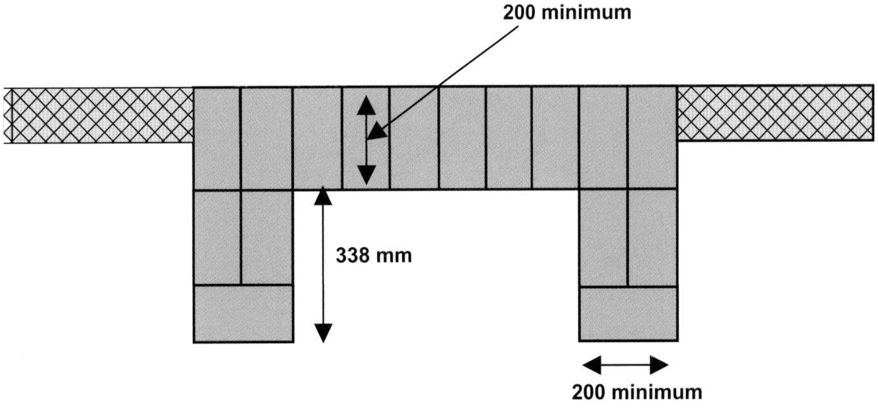

FIGURE 7.3
Single fireplaces on an internal wall

The minimum depth of the fireplace opening is 338 mm, to allow a 225 mm flue to be formed together with a covering of 112.5 mm of brickwork when the chimney breast is being constructed above the lintel or arch level of the fireplace opening.

Double or back-to-back fireplaces

Back-to-back fireplaces are the typical fireplace construction in terraced or semi-detached dwellings (Figure 7.4). The fireplaces are formed on the party wall.

Figure 7.5 shows a method of bonding adopted by some bricklayers to save the labour of cutting. This practice should be discouraged as bricks are often

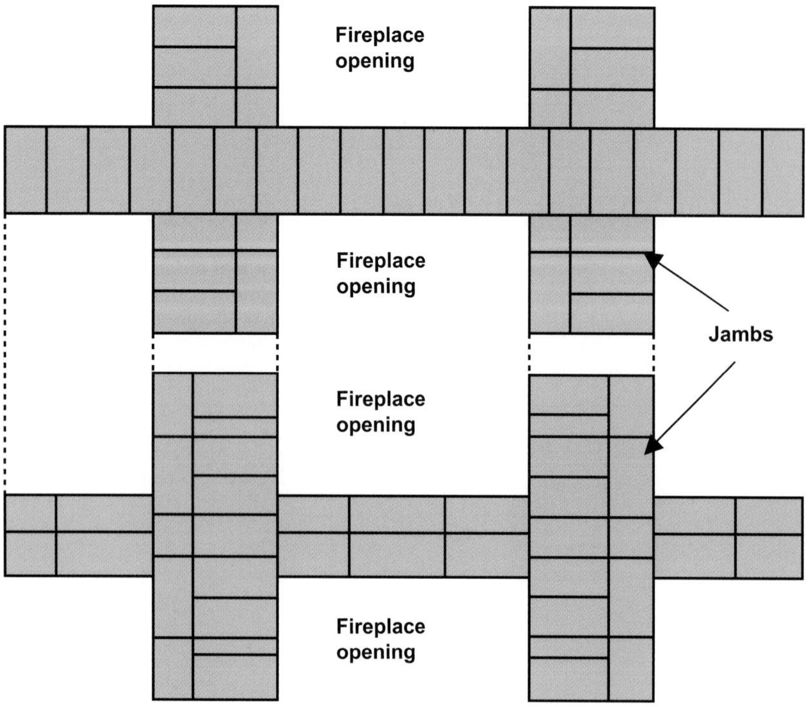

FIGURE 7.4
Back-to-back fireplaces

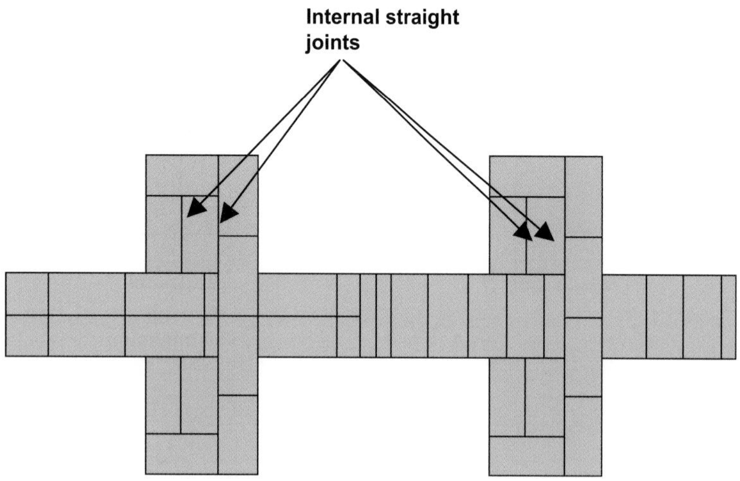

FIGURE 7.5
An example of bad bonding practice

omitted where the straight joints occur and pockets are formed, which are filled with rubbish – an example of bad workmanship.

Interlacing fireplaces

These are fireplaces constructed on an internal wall and placed side by side. This arrangement lengthens the chimney breast, but the projection of the jambs into the room is reduced, giving a greater room area (Figure 7.6).

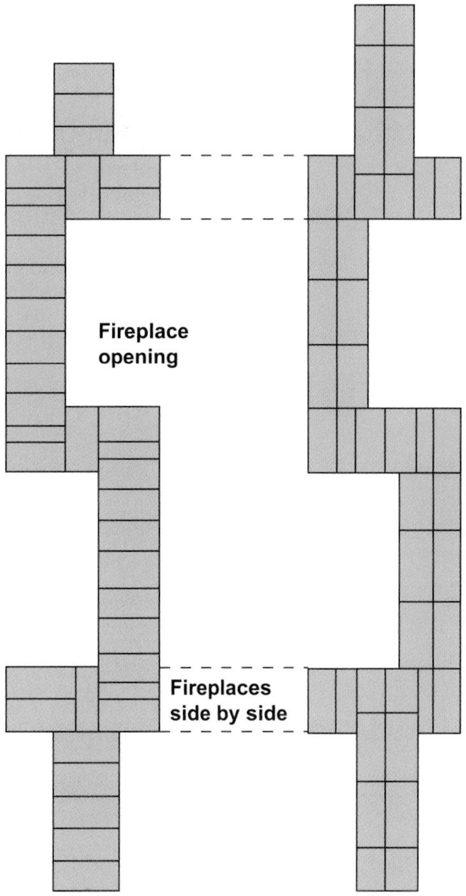

FIGURE 7.6
Interlacing fireplaces, ground floor

Angled fireplaces

Figure 7.7 shows the plans of the alternate courses of brickwork bonding at ground level. Angled fireplaces are difficult to construct because of the need for a number of twists of the flue to obtain correct positioning. Their construction will therefore not be shown at this stage.

Upper floor fireplaces

Figures 7.8–7.12 illustrate the bonding arrangements of upper floor fireplaces in the various types of construction.

The alternate arrangement of flues in the interlacing fireplaces should be noted; this will be better understood when the grouping of fireplaces and chimneys has been considered.

Grouping of fireplaces and chimneys

The details in the following section will mainly be found on existing buildings and have been included for those who are involved in the maintenance of such buildings. Repairs and maintenance of buildings will be covered in Chapter 14.

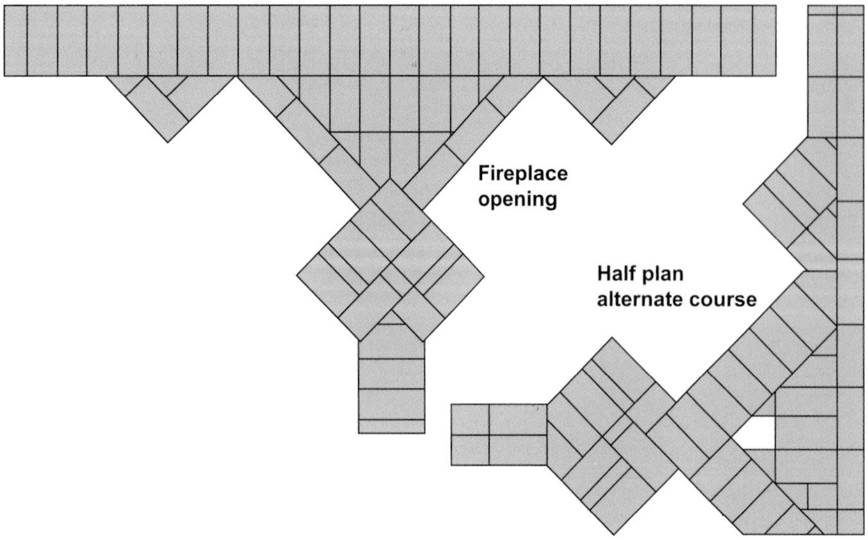

FIGURE 7.7
Plans of angled fireplaces, English bond

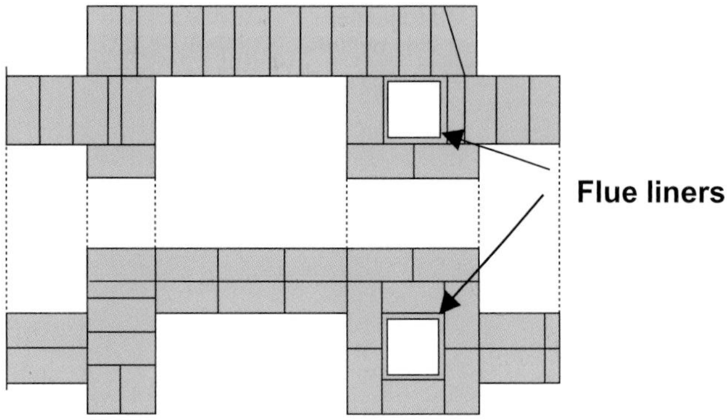

FIGURE 7.8
Single fireplaces, first floor

To achieve sound construction and to reduce the number of chimney stacks emerging through the roof, fireplaces and chimneys were usually grouped to a central position.

The fireplaces of each successive floor were positioned one above the other, and where possible the flues from fireplaces in adjacent rooms were gathered together before emerging through the roof.

The base of a chimney breast was designed to be wide enough to support the fireplace construction that occurred in the floors above.

Where buildings extended to four or five floors this was not always practicable, and in this case it was permissible to extend the length of the upper floor chimney breasts by corbelling. In simple domestic construction the latter point did not arise.

Figure 7.13 shows single-breast fireplace construction on an external wall. The passage of the flues has been marked by dotted lines, and it will be noted that the flue from the ground floor fireplace has been gathered over

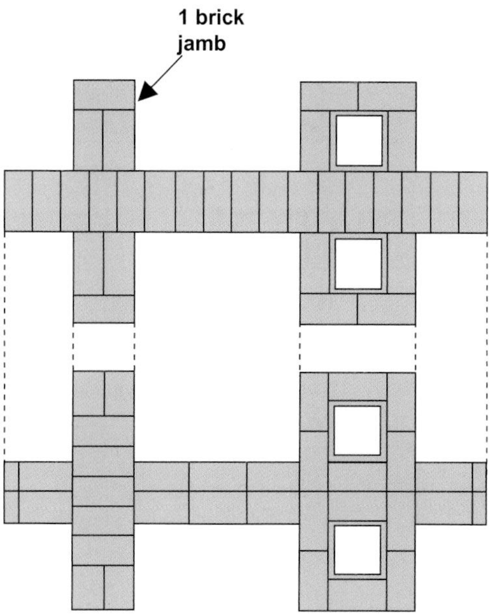

FIGURE 7.9
Back-to-back fireplaces, first floor

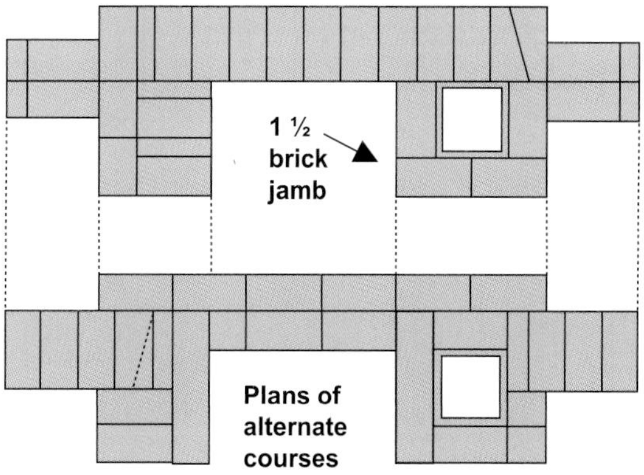

FIGURE 7.10
Single fireplace at first floor, showing wider right-hand jamb

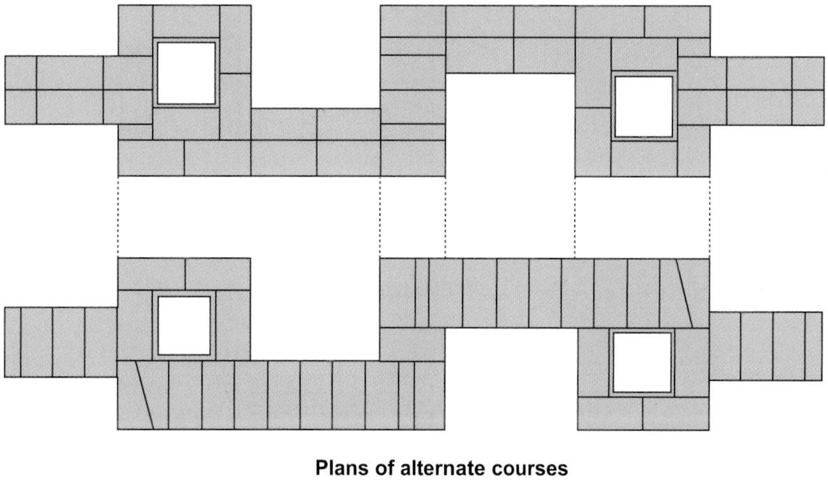

FIGURE 7.11
Interlacing fireplaces, upper floor

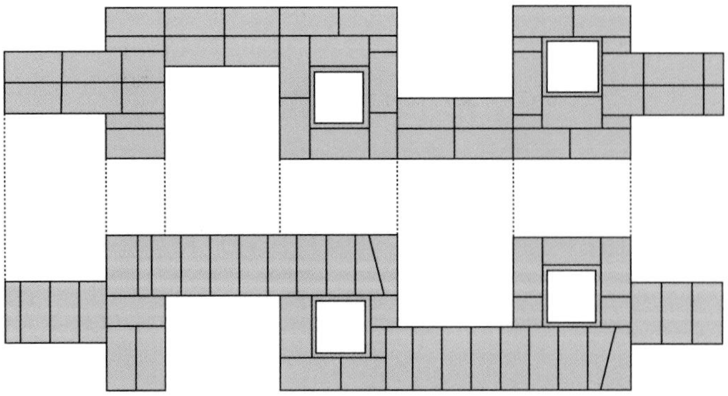

FIGURE 7.12
Interlacing fireplaces, upper floors, with central flue position

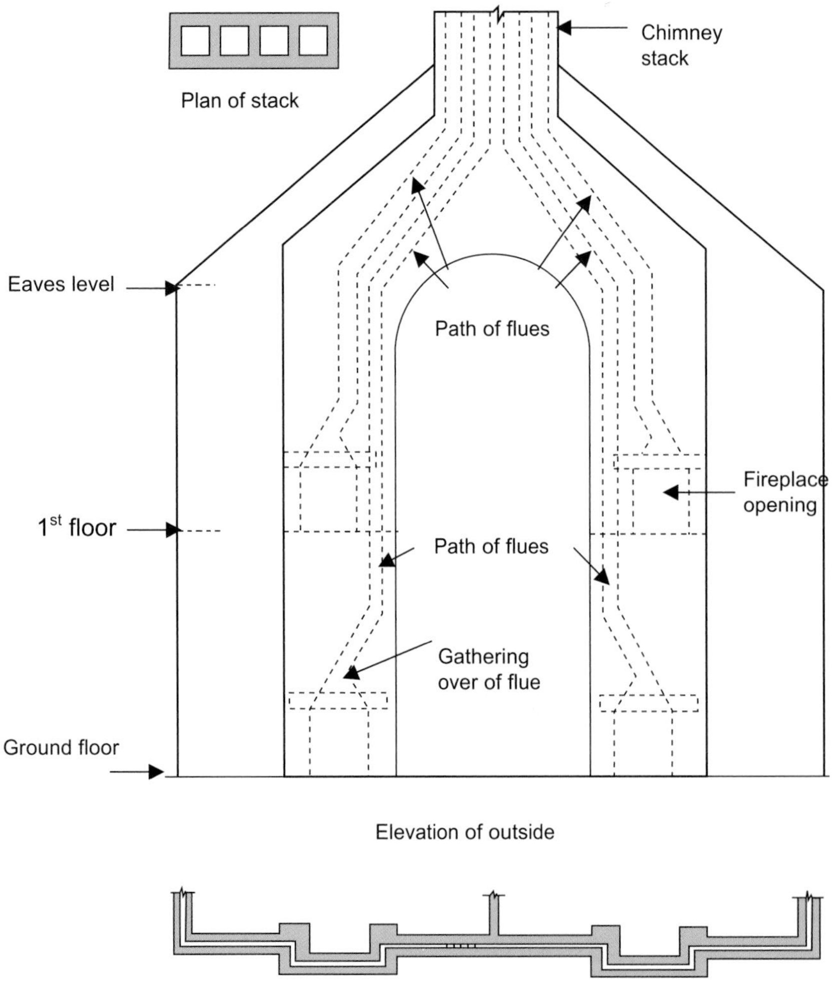

FIGURE 7.13
Single fireplaces on an external wall

to miss the first floor fireplace, while at roof level the flues from adjacent rooms have been grouped to form a single stack of four flues. The external chimney breasts have been connected by a face-brickwork semi-circular arch. The alternative is to continue the single chimney breast and to reduce it to stack size by means of circular ramps (Figure 7.14) or by tumbling (Figure 7.13), where it will develop into a two-flued stack.

Figure 7.15 shows the grouping of chimneys and fireplaces on a double-breast party wall. The grouping of fireplaces from adjacent rooms takes place within the roof space and does not require the decorative finish as in the external chimney breast (Figure 7.14), which accounts for the set-off a few courses above first floor ceiling level. The chimney stack emerging from the centre of the roof consists of eight flues, but the chimney breasts can be carried up individually, thus emerging one on each side of the ridge of the roof as a four-flue stack.

Figure 7.16 shows the grouping of interlacing fireplaces. In the description of types of fireplaces two methods of arranging the upper floor fireplaces

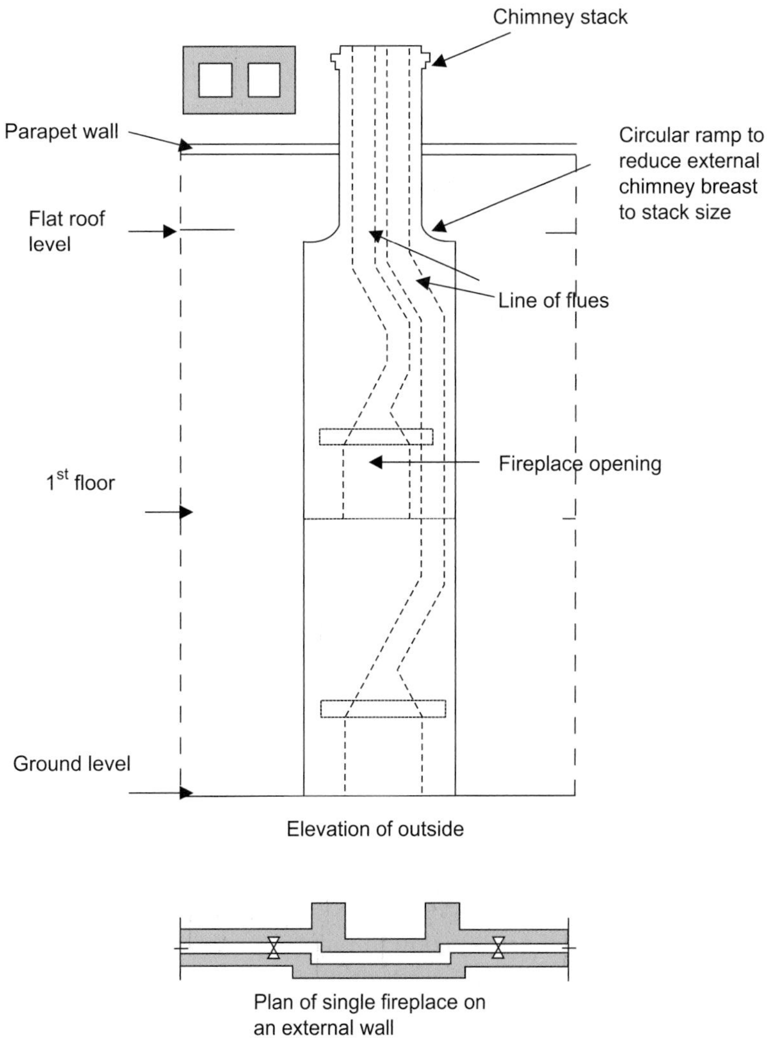

FIGURE 7.14
External chimney breast reduced to stack width using brick ramps

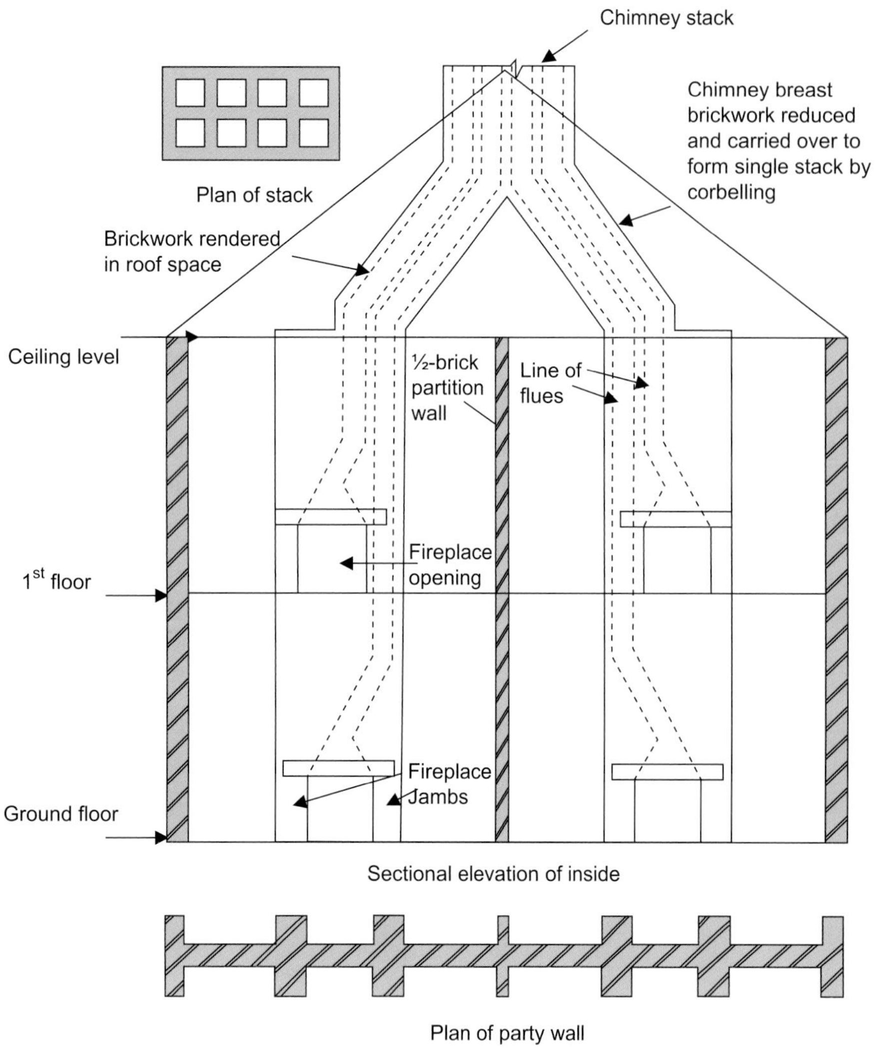

FIGURE 7.15
Back-to-back fireplaces on a party wall

were shown; both have their merits. Figure 7.12 is considered to be more straightforward, while Figure 7.11 gives symmetrical planning. The chimney stack consists of eight flues in a line.

Regulations controlling the construction of chimney breasts and stacks

The construction of chimney breasts and stacks is controlled by Part J of the current Building Regulations.

The Building Regulations are concerned with stability, fire hazards and the discharge of products of combustion to the outside air. For the purpose of relating text and diagrams, points are numbered from 1 to 13, and will differ from references in the Building Regulations.

These cover the construction of flues, chimneys, fireplace recesses and hearths, for the purpose of an ordinary domestic fireplace appliance that

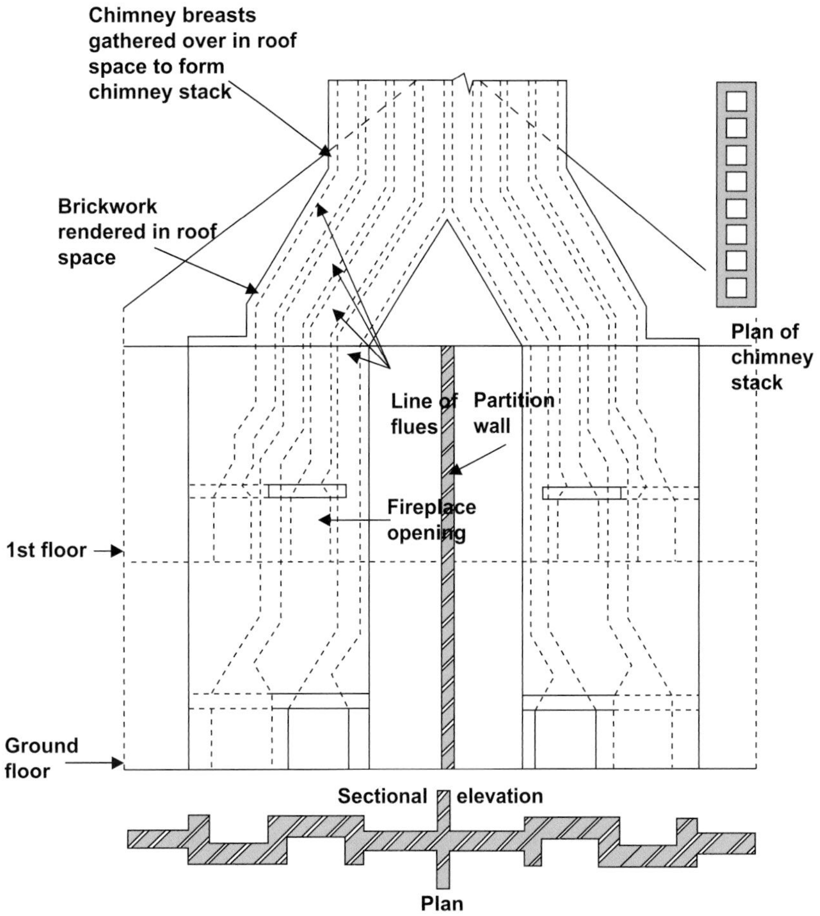

FIGURE 7.16
Interlacing fireplaces on a party wall

will not exceed an output rating of 45 kW and a resulting flue not exceeding approximately 225 × 225 mm.

Other clauses modify construction where gas appliances are used. The following descriptions are therefore an interpretation of the regulations in this sense, and this knowledge is sufficient for the construction of ordinary domestic fireplaces.

A chimney may be defined as the solid material surrounding a flue, while the flue is a duct through which smoke and other products of combustion pass. A chimney is therefore a single flue surrounded by at least 100 mm of brickwork, and a stack consists of two or more flues surrounded by the requisite amount of brickwork. If the 'withes' or 'midfeathers' (dividing walls between flues) of a stack are half a brick in thickness this will meet the requirements.

A flue formed in a brick wall is normally constructed to the dimensions of 225 × 225 mm or one brick by one brick; the cross-sectional area is reduced by the application of purpose-made flue liners. The minimum diameter of the flue to suit a 45 kW closed appliance is 175 mm.

In simple residential buildings the size of a flue used solely for the purpose of discharging the fumes of a gas appliance into the open air must not be less than 12 000 mm^2 in cross-sectional area if circular, and 16 500 mm^2 if rectangular. Proprietary purpose-made flue blocks are available for the construction of gas flues, which must have a minimum dimension on plan of 90 mm.

1. Every flue must be surrounded by at least 100 mm thickness of brickwork, properly bonded and excluding the thickness of the flue lining.
2. Every chimney or stack must be built to a height of at least 1 m above the last point of contact where it emerges from the roof. In the case of a gas flue this height is modified, but in no case must a chimney or stack be built up to a height greater than 4.5 times its least width at the last point of leaving the roof, unless it is adequately supported against overturning (Figures 7.17 and 7.18). The Building Regulations state that the top of a chimney carried up through the ridge of a roof, or within 1000 mm of it, and which has a slope on both sides of not less than 10 degrees to the horizontal must be at least 600 mm above the ridge, but in all other cases at least 1 m measured from the highest point in the line of junction with the roof.
3. Every chimney must be built on suitable foundations approved by the local building control officer. In many traditional type houses chimney breasts have been projected from the main wall at an upper floor level; Figure 7.19 illustrates a case where the chimney breast of the ground floor is discontinued on the upper floor and

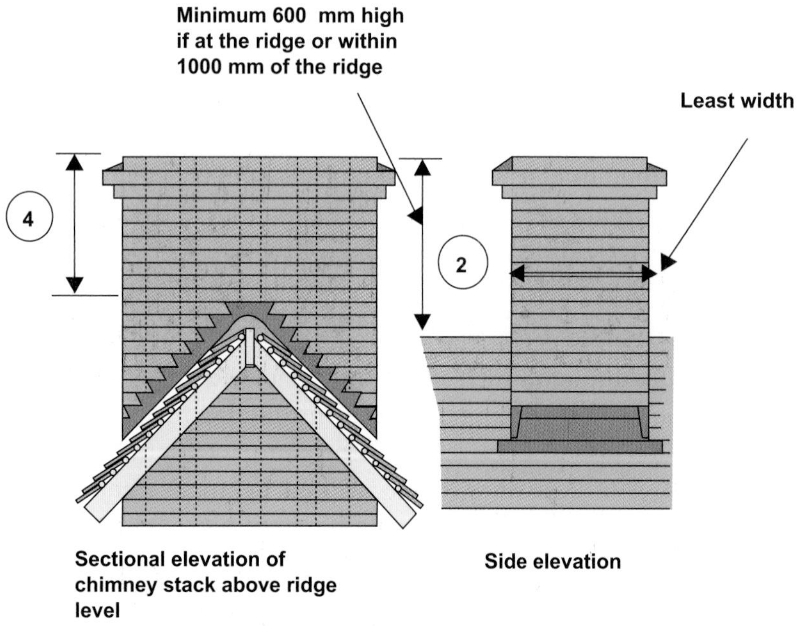

FIGURE 7.17
Lead flashings to weatherproof junction between roof covering and a chimney stack that emerges through the ridge

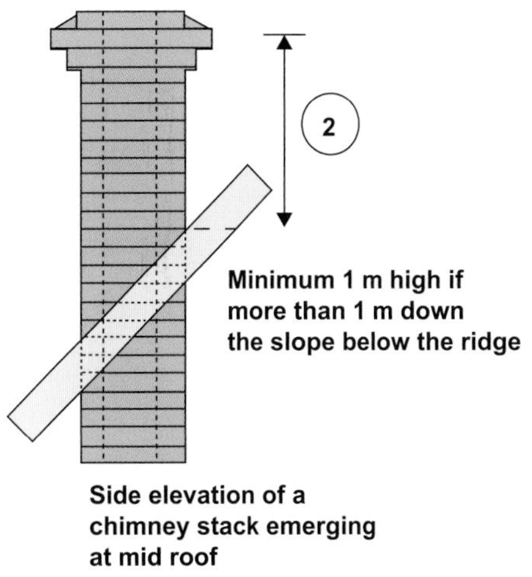

FIGURE 7.18
One metre minimum height where a stack emerges mid-roof

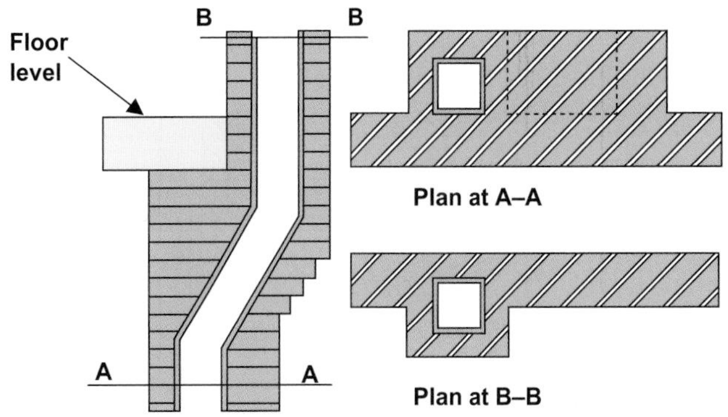

FIGURE 7.19
Corbelling at first floor level to form an external chimney breast

in consequence the chimney has been corbelled out on the external wall.

4. The 12 highest courses of every chimney or stack should be built in cement mortar (Figure 7.17). A chimney stack passing through a roof must be properly protected against downwards penetration of dampness. This is achieved by building in a damp-proof course (DPC) within the brickwork.

5. Where chimney flues are inclined, the angle of travel shall not be less than 60 degrees to the horizontal. Chimney flues must be provided with a means of inspection and cleaning, via a gastight door set in a metal frame (Figure 7.20).

6. The inside of every chimney must be lined with non-combustible materials, e.g. fireclay or terracotta flue liners (Figure 7.21), this work being carried out as the building of the chimney proceeds. The non-combustible linings are purpose-made units, either square or circular in section, to fit a one brick by one brick flue.

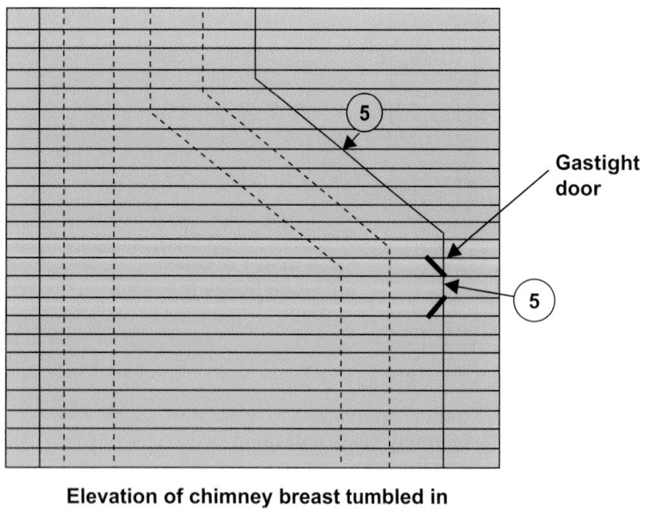

FIGURE 7.20
Travelling a flue in an external chimney

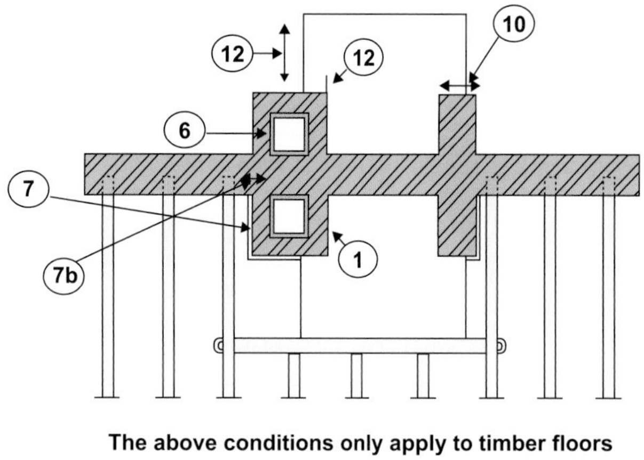

FIGURE 7.21
Plan of back-to-back fireplaces at first floor level. (For numerical references see text)

Flue liners must always be placed with sockets or rebates uppermost, to prevent leakage of any condensation that might form on the inside surfaces. Flue liners must also be jointed with fireproof mortar.

7. It is advisable that the outer surface of every chimney within a building or roof space is rendered up to the level of the outer surface of the roof or gutter. (Plastering to the internal walls of rooms largely takes care of this.) This rendering is for the additional protection of any combustible materials adjacent to the chimney where it passes through floors and roof.

    (a) Woodwork must not be placed under any fireplace opening within 250 mm from the upper surface of the hearth except for the fillets or bearers supporting the hearth, or if there is an air space of not less than 50 mm between the underside of the hearth and any combustible material (see Figures 7.22 and 7.33).

(b) Woodwork must not be built into any wall nearer than 200 mm measured to the inside surface of a flue or to the inside of a fireplace recess (Figure 7.21).

(c) Woodwork must not be placed closer than 40 mm to the surface of a chimney or fireplace recess, excepting floor boarding and skirting.

(d) If the surround of a fireplace opening is constructed of wood it should be at a distance of at least 150 mm measured horizontally and 300 mm measured vertically from the fireplace opening, and it must be backed with solid, non-combustible material.

8. Where a chimney is adjacent to constructional steelwork or reinforced concrete, it is advisable to take precautions to ensure that the steelwork, etc., is not affected by heat, with metal fixings at least 50 mm away.

9. Each fireplace must have its own flue taken to the outside air, with no branching of more than a single fireplace per flue.

10. The jambs of every fireplace opening must be at least 200 mm thick (Figure 7.21).

11. The back of every fireplace opening on a party wall or party structure must be at least 200 mm thick. This 200 mm thickness of brick masonry is also necessary to separate flues on either side of a party wall up to the level of the top floor ceiling. Reduction to 100 mm thickness between flues takes place at this point, where the bulk of the chimney breast is discontinued and the flues are grouped

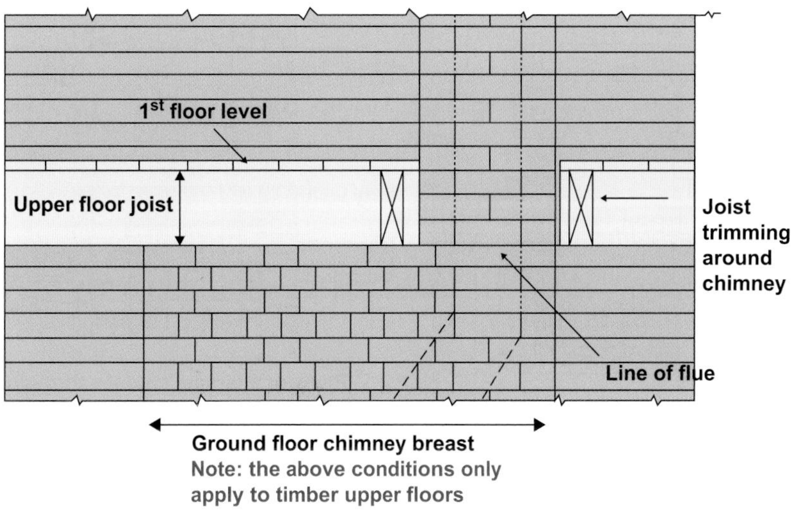

FIGURE 7.22
Construction where a fireplace is not required at first floor level

into one stack for passing through the roof covering (see Figure 7.32).

It is desirable to keep the air in a flue warmer than the outside atmosphere, so that the air in a flue is rising continuously; the warmer the air in a flue, the quicker is the flow of air, and with it the gases of combustion, while the colder the air, the slower will be the flow, a condition which can lead to downdraught and smoky chimneys.

12. The constructional concrete hearth must extend at least 150 mm beyond each side of the fireplace opening and must project at least 500 mm (see Figures 7.21 and 7.26).
13. The hearth must be at least 125 mm thick, formed of combustible materials.

Construction of a chimney through the ground floor, first floor and roof

It should be noted that solid fuel appliances are frequently omitted from upper floor levels and construction is therefore simplified (Figure 7.22). However, the apprentice should be aware of traditional practice and of the problems that may be encountered in the refurbishment of buildings.

A back-to-back fireplace on a party wall has therefore been selected for the purpose of description, as this should cover adequately all types of fireplace construction.

Figures 7.23 and 7.24 illustrate the plan, sectional elevation and section of ground floor fireplace construction.

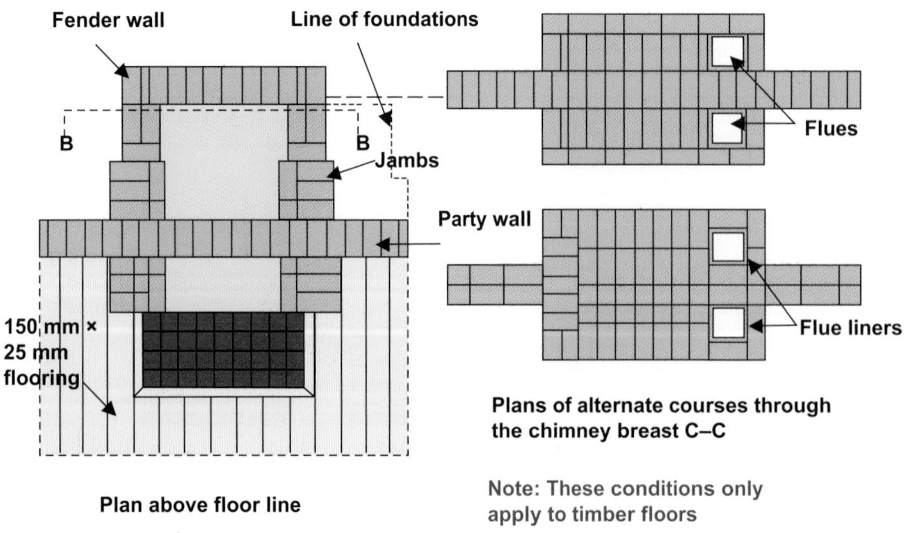

FIGURE 7.23
Plans of the chimney breast

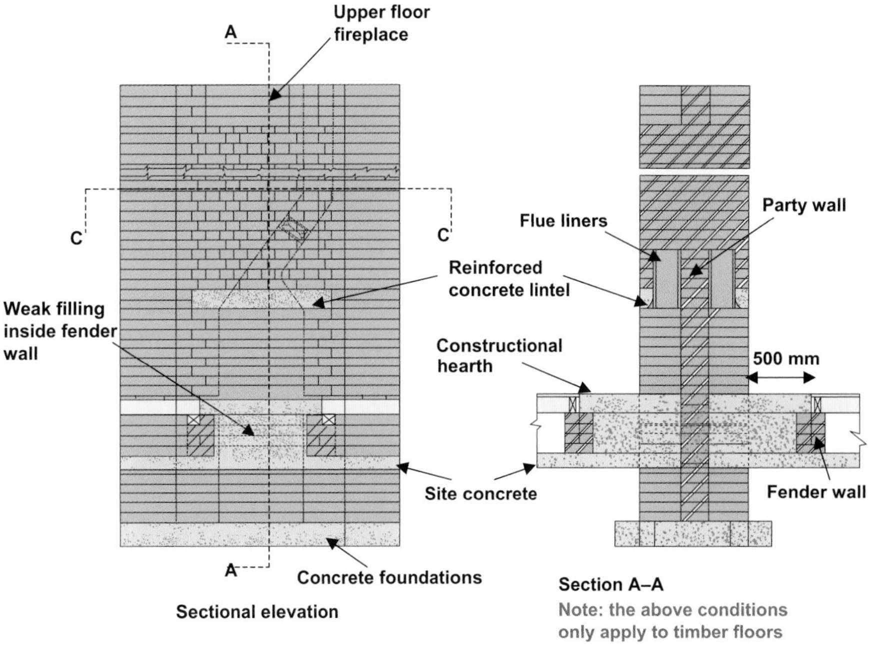

FIGURE 7.24
Ground floor fireplace set in suspended timber flooring

The chimney breast is 1.349 m in width, made up of two 330 mm jambs and a 675 mm fireplace opening, this being sufficiently wide to contain the upper floor fireplace consisting of a 225 mm jamb, a 450 mm chimney jamb containing the flue and a 675 mm fireplace opening.

Fender wall

A one-brick fender wall supports the constructional hearth and is built up on the site concrete. It is advisable to make this wall one brick in thickness, rather than a half brick, as it serves the double purpose of supporting the hearth and part of the timber floor. The void beneath the hearth should be filled with clean hardcore consisting, if possible, of broken brick (Figure 7.25).

The ground floor of many properties has an in situ cast solid concrete floor or suspended flooring of prestressed concrete beams with 440 × 215 × 100 mm blocks between. In such cases, there is no fender wall, and the whole floor is non-combustible, not just the minimum requirement of 500 mm in front of any fireplace.

Constructional hearth

With appliances, it is also necessary to provide a hearth to reduce the fire risk (Figure 7.26). Each appliance shall have a constructional hearth which shall be:

- not less than 125 mm thick
- not lower than the surface of any floor built of combustible material

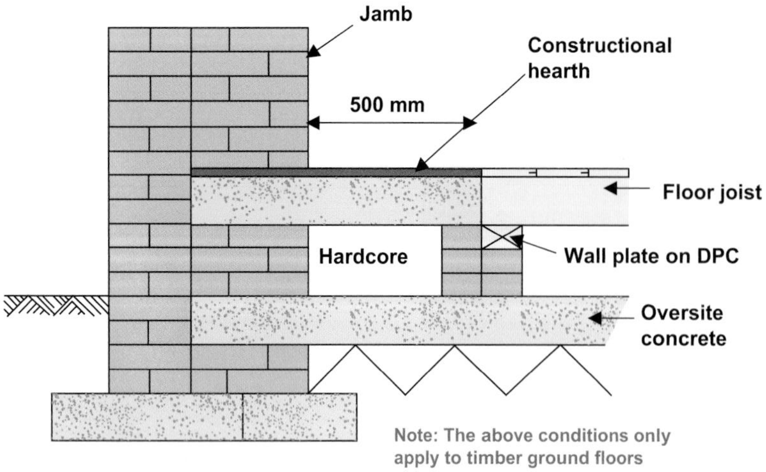

FIGURE 7.25
Section through a ground floor recess

- extended within the recess to the back and jambs of the recess and projected less than 500 mm in front of the jambs and not less than 150 mm each side of the jambs
- not less than 840 mm square if the hearth is not constructed within a recess.

Opening for fireplace

The fire opening is a minimum of 552 mm wide. It has to be reduced at the top of the fire opening to receive a 225 mm flue liner. This can be carried out in two ways:

- traditional throated lintel (Figure 7.27) – the brickwork is then corbelled either side to close the opening to receive the flue liners (Figure 7.28)
- the lintel can be replaced by a precast refractory concrete throat unit which provides access for forming throating when the appliance is fitted and supports the flue liners (Figure 7.29).

Flue liners

Chimneys built before the 1960s were rarely built with flue liners. Instead, they were rendered internally with lime mortar. This operation was called parging.

When modern-day fuels were introduced there was a large increase in the sulphur content discharged into the flues. The sulphur dioxide, when mixed with the condensation in the flue, would seep into the mortar joints. This created sulphuric acid, which would crystallize and expand, causing the chimney stack to lean.

Regulations were changed to make the lining of chimney stacks compulsory.

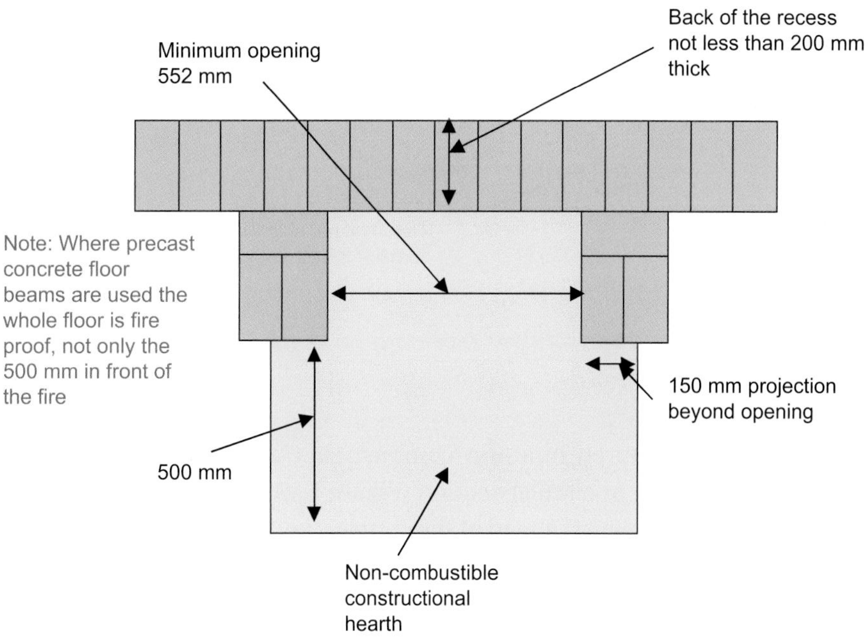

FIGURE 7.26
Plan of constructional hearth

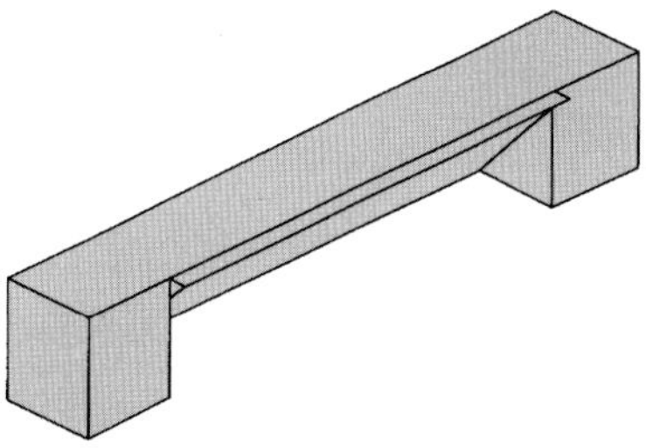

FIGURE 7.27
Throated lintel

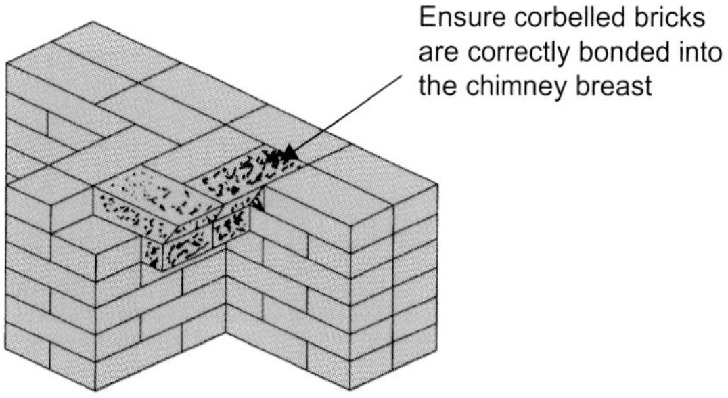

FIGURE 7.28
Corbelling to receive the flue liner

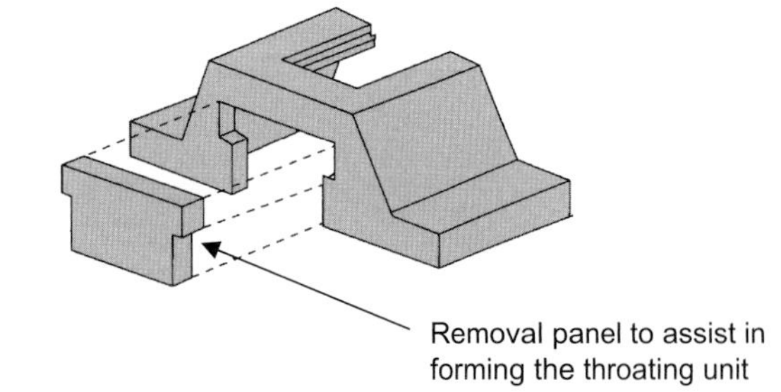

FIGURE 7.29
Throating unit

Liners are manufactured in a non-combustible clay or terracotta and are available in a square or circular section (Figure 7.30). Flue liners are continuously bedded and jointed ahead of the courses, and surrounding brickwork is cut and solidly bedded around them.

Great care must be taken to see that flue liners are always set and bedded with the rebate or socket uppermost. This is to ensure that any acidic condensation which forms inside a cold flue, when a fire is lit, cannot

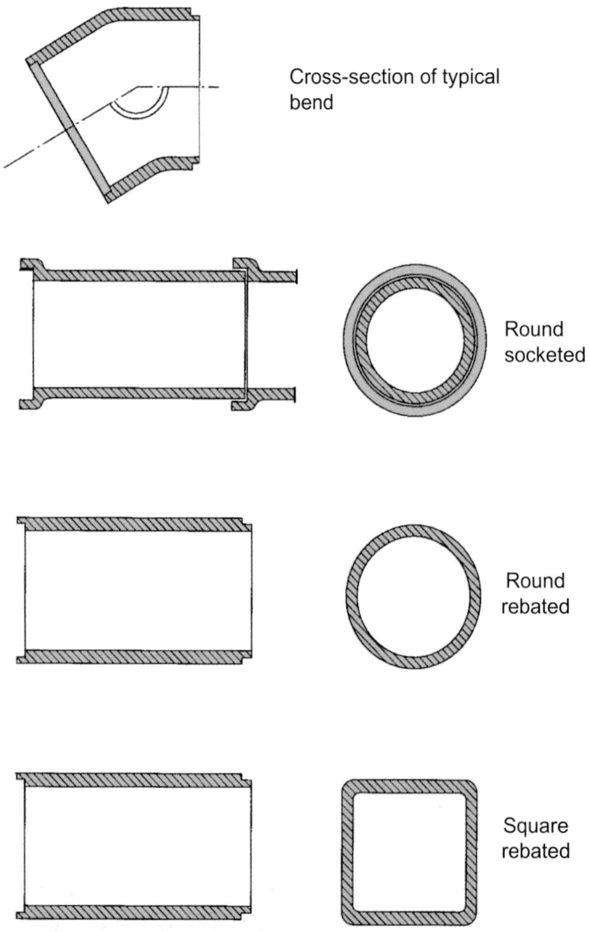

FIGURE 7.30
Square and circular flue liners

seep out of the joints between the flue liners and attack the surrounding brickwork.

Figure 7.31 shows an isometric detail of the gathering over. Note the arrangement of bricks to give quarter bond and a complete tying-in of the cut bricks.

Figure 7.32 illustrates the construction of an upper floor fireplace. The section has been taken through the centre of the fireplace and continued throughout one of the flues. This enables the construction immediately above ceiling level to be clearly illustrated, and it is also amplified in the detail (Figure 7.33). The chimney breast has been set off to the proper stack size and can either be carried straight up or grouped with the adjacent chimney breast to form a single stack as shown in Figure 7.34.

Figure 7.34 illustrates the chimney construction through the roof. The chimney breasts, reducing to proper size, are grouped to form a single stack. This is achieved by the use of temporary timber legs or 'gallopers', supported by a corbel projected from the chimney breast as illustrated. The section has been taken throughout one of the flues. The angled travel of these flues, corbelled out from the party wall, is controlled by the purpose-made flue liner units used at the bends.

Temporary timber gallopers should be erected as soon as the corbel has been sufficiently tailed down by the brickwork immediately above and has the stability to bear the weight of the gallopers.

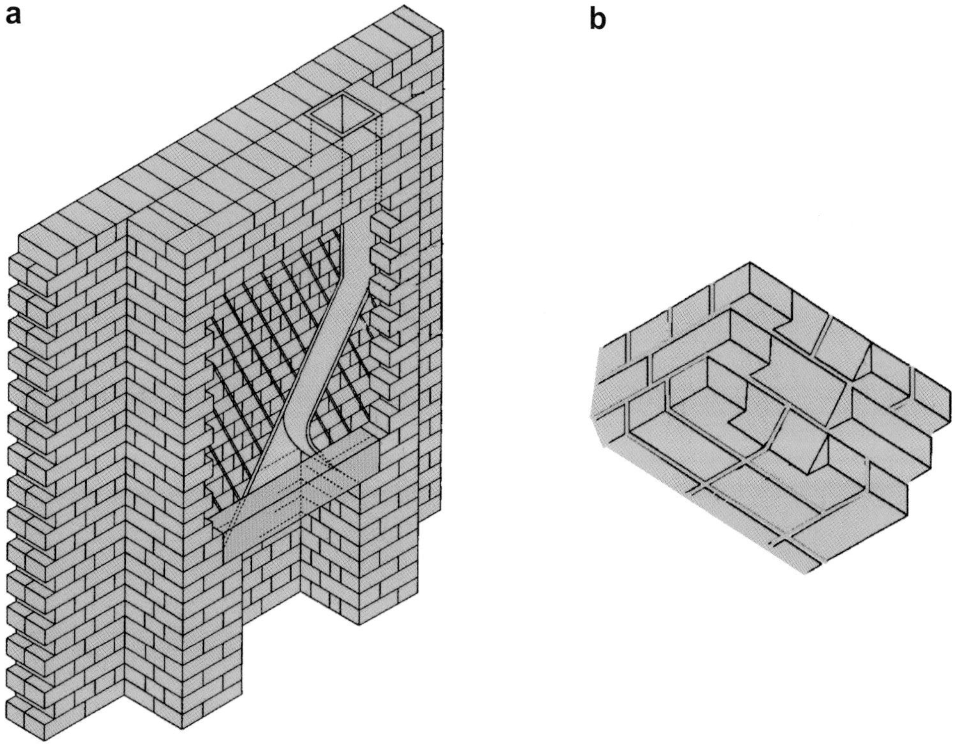

FIGURE 7.31
(a) Typical chimney breast construction; (b) gathering over brickwork around flue linings

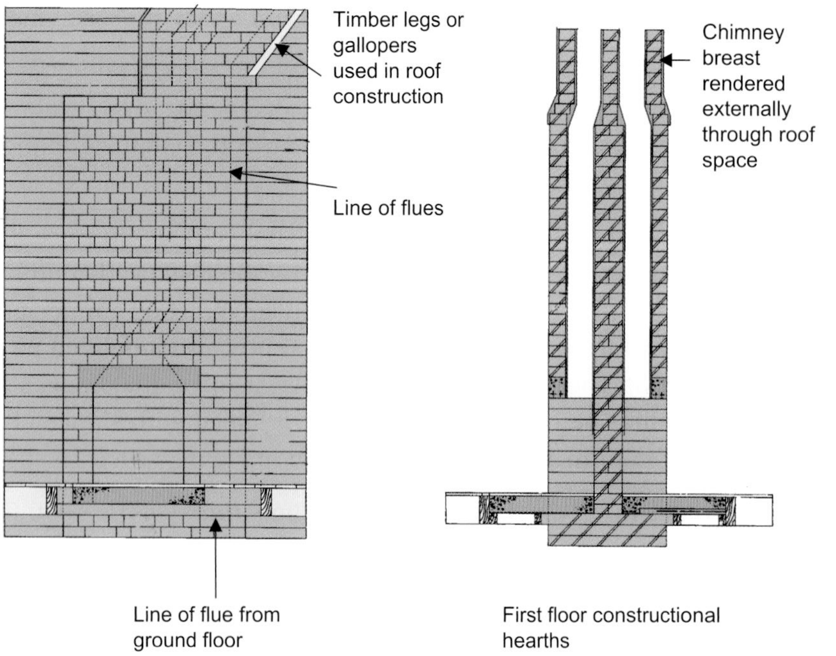

FIGURE 7.32
Upper floor back-to-back fireplace

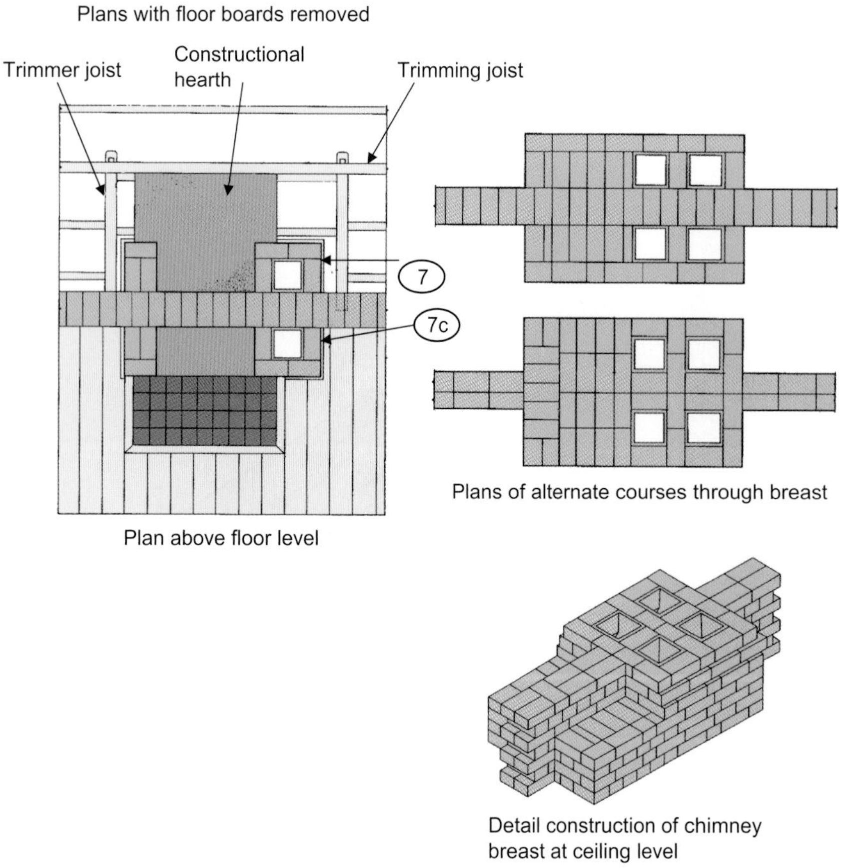

FIGURE 7.33
Upper floor back-to-back fireplace construction

The path of the flue in Figure 7.34 has been shown by dotted lines. Purpose-made bends of fireclay or terracotta are used to provide a continuous smooth inside surface, allowing a flue to be travelled to the left or right within the chimney breast brickwork.

Chimney stacks

Figure 7.35 shows the plans of alternate courses of various sized stacks, and a sectional elevation illustrating the bedding of the chimney pots. It is advisable to place pieces of slate directly under the pots to assist in their correct alignment and to prevent mortar from dropping down the flues where these are not completely covered by the chimney pot (A).

As the pots are usually uneven in shape, it is impracticable to line them in with a straight edge, but they should be lined in by the eye in the direction of the arrows. This may appear superfluous, but as a line of chimney pots can easily be seen from ground level, any lack of alignment presents an unsightly appearance and is indicative of careless workmanship.

B is an alternative method of bonding the withe walls; it is used on the internal withe walls and makes use of the closers that would otherwise be wasted.

Figures 7.36 and 7.37 show two examples of finish to the top of a stack. There are many excellent examples to be seen in all parts of the country where the stacks have been carefully designed by the architect as a special feature of the building.

In Figure 7.37 the outer walls of the stack are built in Flemish bond, one brick thick, with inner walls or withes of half brick. The outer walls are

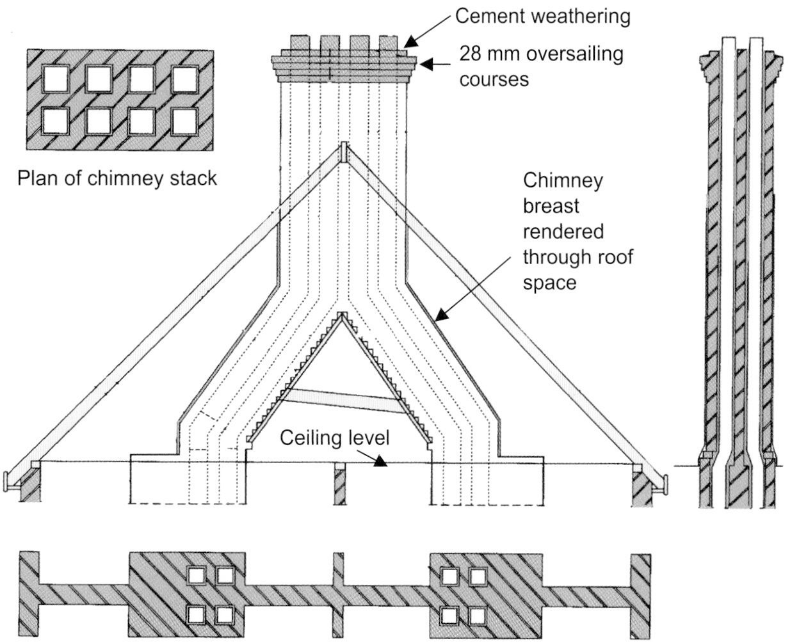

FIGURE 7.34
Gathering of flues

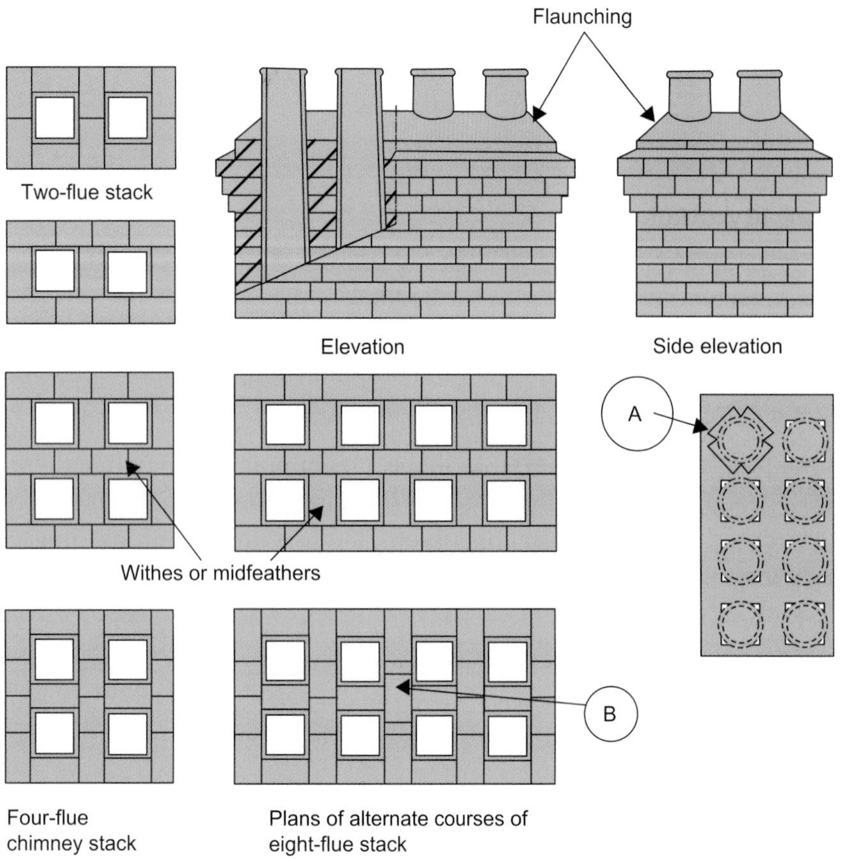

FIGURE 7.35
Bonding of two-, four- and eight-flue chimney stacks

more than half brick in thickness, so that excessively cold flues may to some extent be avoided.

Figure 7.38 shows the plans of the alternate courses of bonding in an eight-flue stack, with the flues in line.

Figure 7.39 illustrates the planning of a diagonal stack; it is usual for these to be built from a rectangular base, which must be of sufficient size. It will be seen that the outer walls and withes of this rectangular base are of one brick thickness. The spaces created on the rectangular base where the diagonal stack commences are filled in with tumbled brick weatherings.

WEATHERING OF CHIMNEY STACKS PASSING THROUGH SLOPING ROOFS

Figure 7.40 shows a horizontal DPC of sheet lead bedded in chimney stack brickwork, and provision of lead flashings between the brickwork and roof tiling.

Fireplaces

There are numerous appliances on the market, making it impossible to cover them all in this chapter. The basic principles, which apply to all appliances, will therefore be covered.

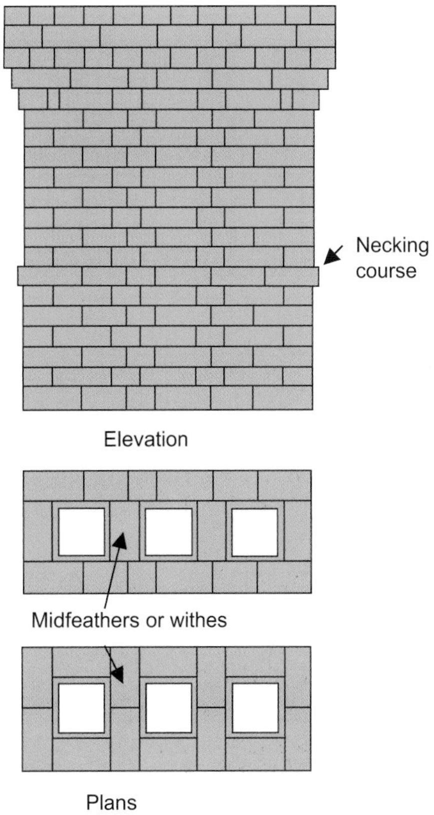

FIGURE 7.36
Bonding of a three-flue chimney stack

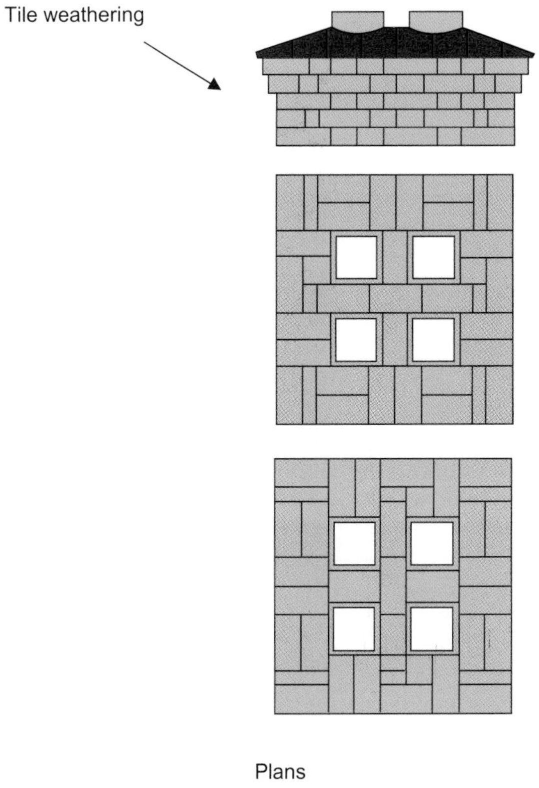

FIGURE 7.37
Bonding of a four-flue chimney stack

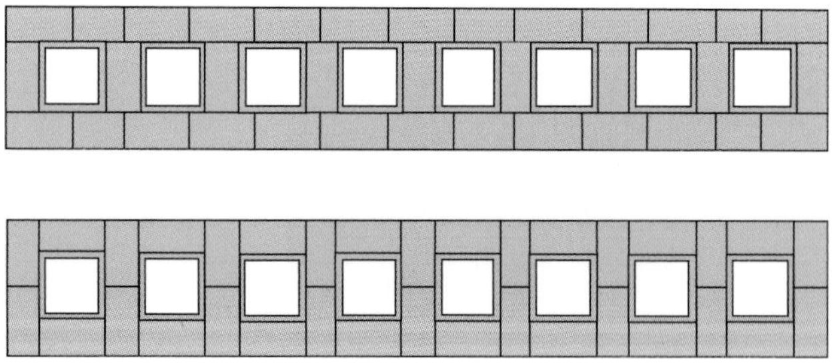

FIGURE 7.38
Bonding of a chimney stack of eight flues in line

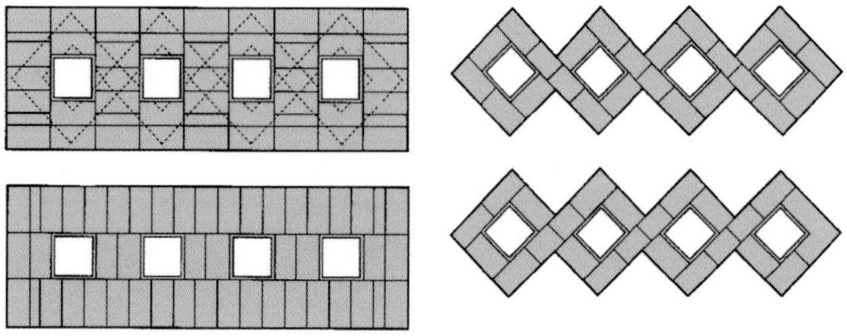

FIGURE 7.39
Bonding of a diagonal chimney stack

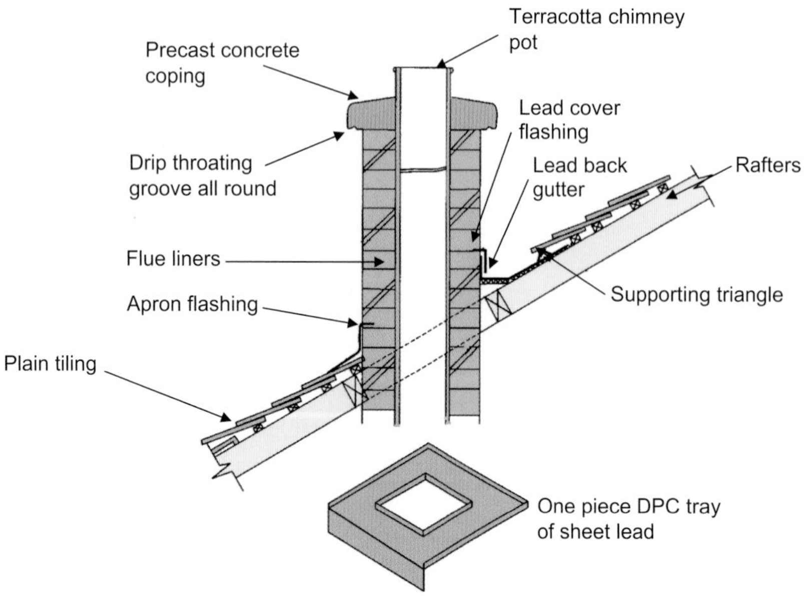

FIGURE 7.40
Typical construction and waterproofing at roof level

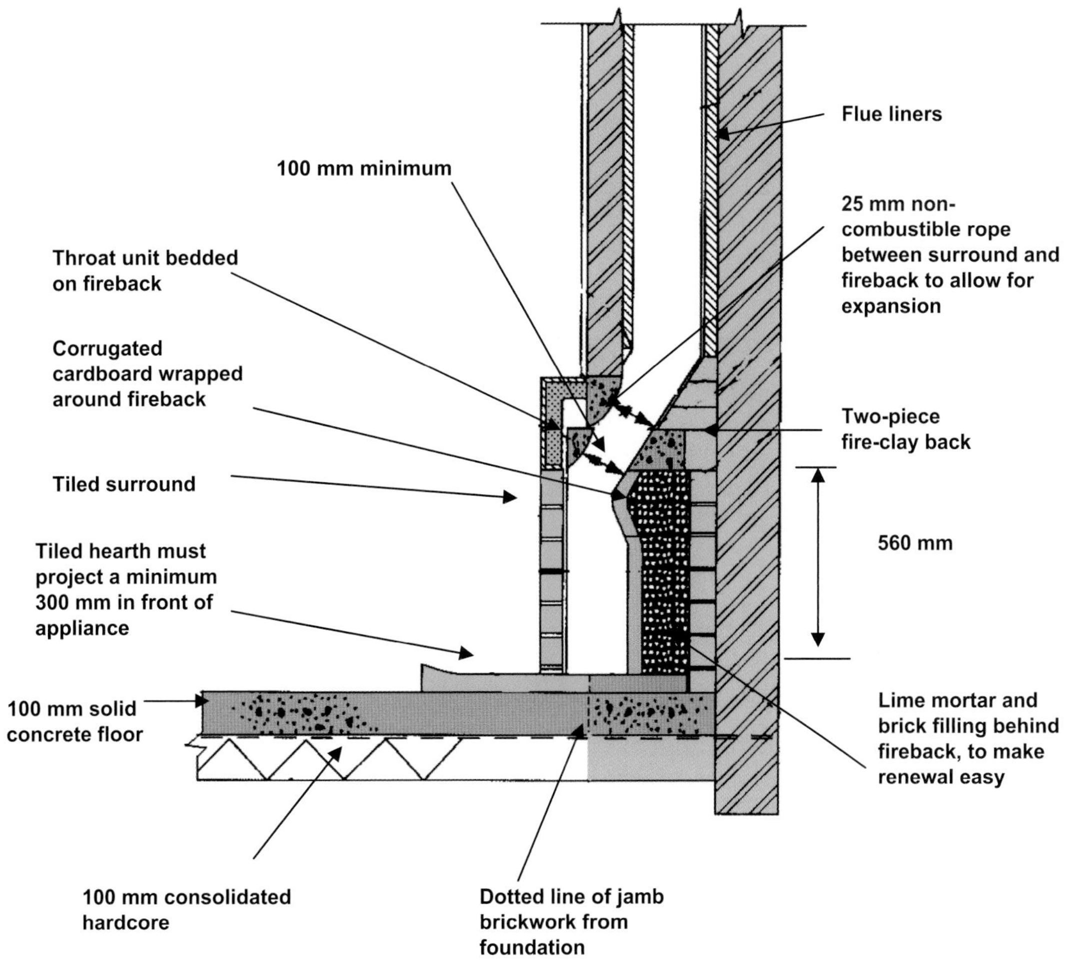

FIGURE 7.41
Section through a ground floor fireplace

IT IS THEREFORE IMPORTANT TO READ THE MANUFACTURER'S INSTRUCTIONS AT ALL TIMES.

Heating appliances may be classified into the following groups:

- open fires
- inset fires
- inset with underfloor primary air supply
- open fires with back boilers
- convector open fires
- free-standing convector open fire
- independent boilers.

This section will deal with open fires only.

The fire surround and hearth should be fixed to a prepared chimney breast. The opening should be 338 mm deep and 572 mm wide.

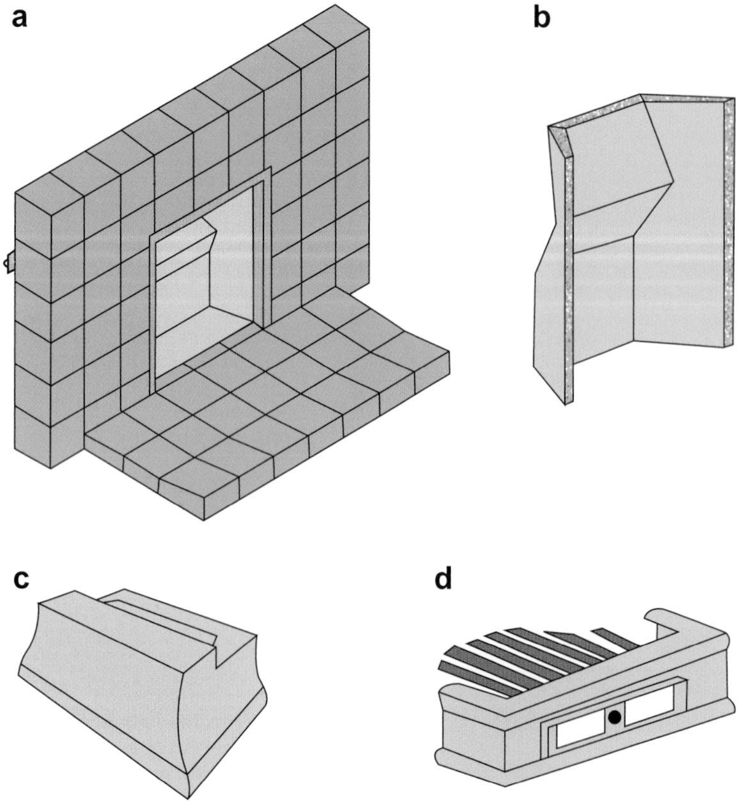

FIGURE 7.42
Main components used in fixing fireplaces: (a) fire surround and hearth; (b) fireback; (c) patent throat unit; (d) fret and basket

Figure 7.41 shows a section through a ground floor fireplace explaining the various constructional details.

The main components include the fire surround and tiled hearth, the fire-back throat unit and fire grate (Figure 7.42).

Firebacks are available in two types: one-piece and two-piece (Figure 7.43).

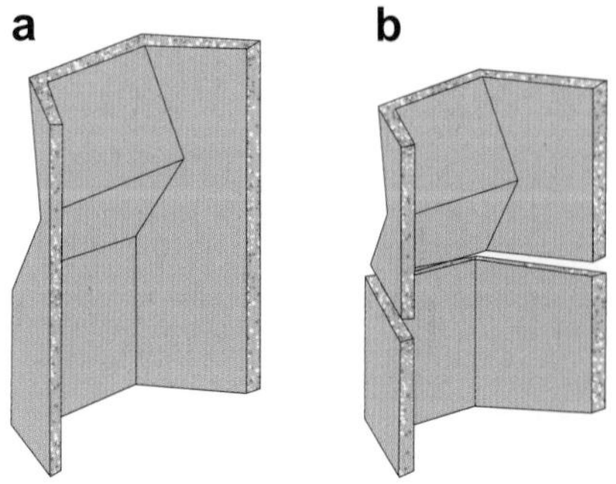

FIGURE 7.43
Types of fireback: (a) one-piece; (b) two-piece

Fixing fireplaces

The first operation is usually to check all the sizes and ensure that you have all the components and materials ready, and then to fix the fireback.

STAGE 1

The back hearth has to be raised to the thickness of the superimposed hearth to allow the fireback to sit at hearth level. Fire bricks should be used, bedded in a lime mortar (Figure 7.44).

STAGE 2

Place the fireback onto the prepared brickwork, centre it and check for plumb and level. You need to place the fire surround in position to check the position of the fireback. The fireback may need to project up to the surround, leaving a small gap for the expansion joint (Figure 7.45).

STAGE 3

Before building in around the fireback corrugated paper should be positioned around the rear of the fireback to allow for expansion. When the paper has burnt with the heat of the fire a small gap will be formed which will allow the fireback to expand when hot.

Build up the gap between the fireback and the chimney breast with bricks and weak mortar.

The void at the back should be filled with vermiculite concrete or brick rubble: 1 part cement, 2 parts lime, 8 parts broken brick. Complete this operation until the top of the fireback is reached (Figure 7.46).

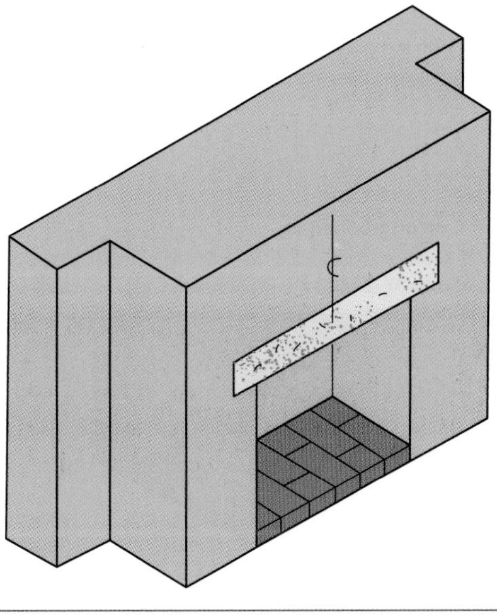

FIGURE 7.44
Stage 1: fixing the back hearth

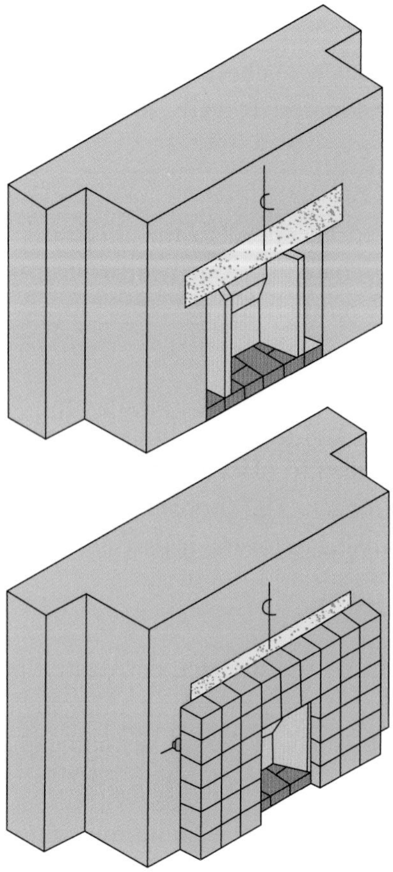

FIGURE 7.45
Stage 2: fixing the fireback and checking for position

STAGE 4

The space from the top of the fireback to the flue liner must be finished with a smooth surface at about 45 degrees.

The gap or throat should be about 100 mm across to give maximum efficiency to the fire and not restrict the flue gases (Figure 7.47).

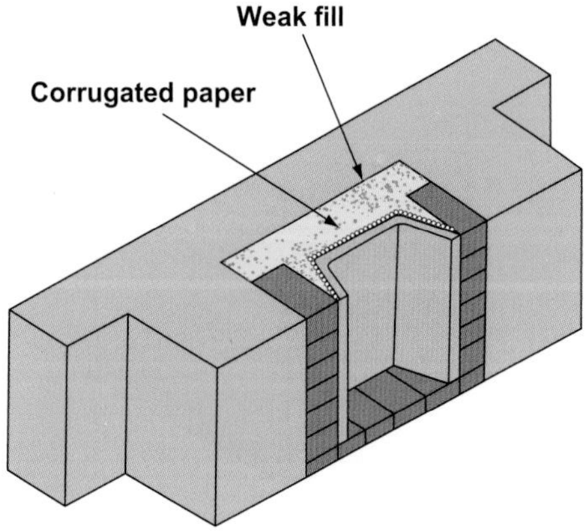

FIGURE 7.46
Stage 3: building in around the fireback

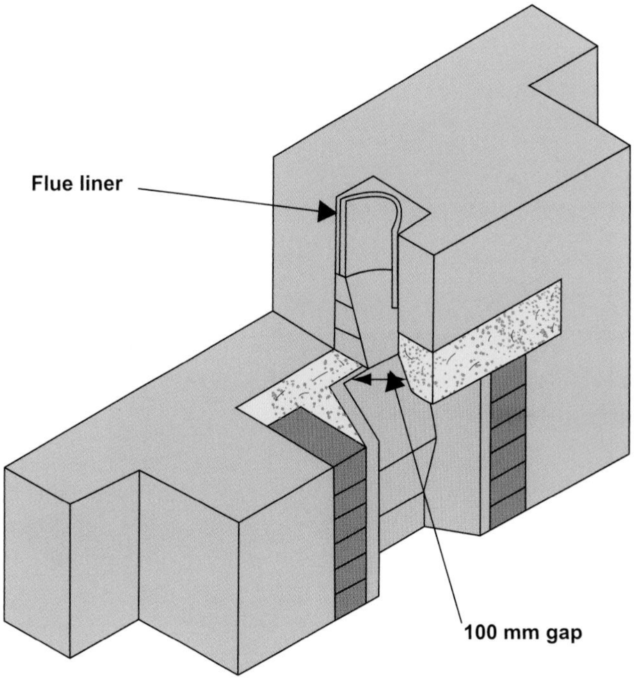

FIGURE 7.47
Stage 4: forming the throat

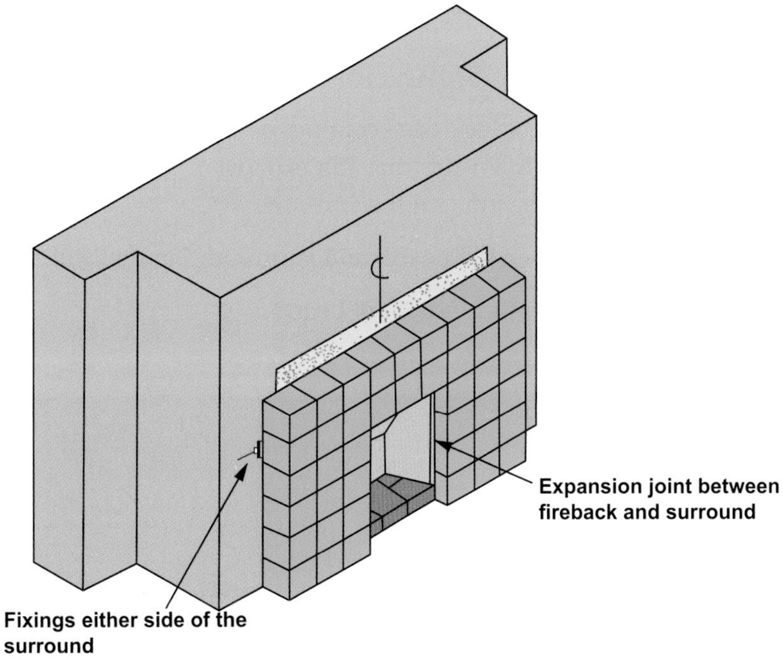

FIGURE 7.48
Stage 5: fixing the tiled surround

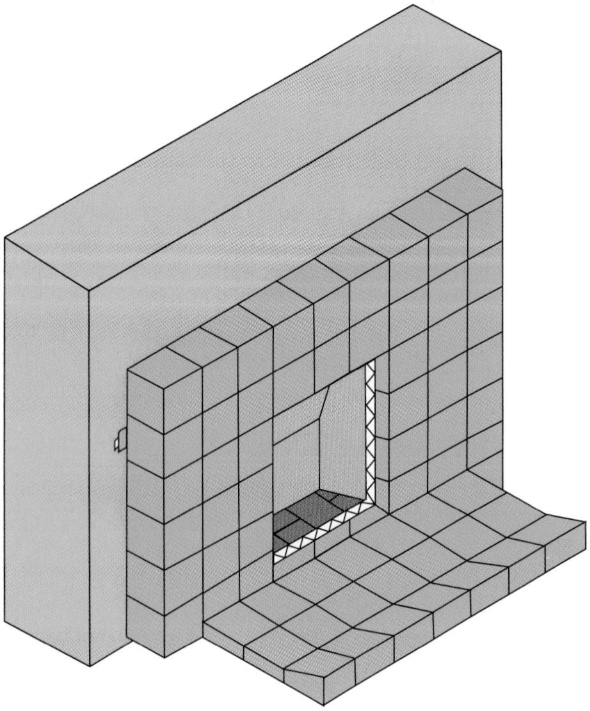

FIGURE 7.49
Stage 6: fixing the tiled hearth

STAGE 5

Lift the surround into position and plumb and level. Make sure the expansion joint is lined up between the fireback and surround as shown in Figure 7.48.

Mark out the position of the lugs and drill the chimney breast using a hammer drill. Screw with brass screws to secure the surround.

STAGE 6

Bed the hearth in position using a weak lime/cement mortar. Check for level along the front and width. Ensure that the expansion joint is in position between the back of the hearth and the surround.

Complete the joint between the hearth and surround (Figure 7.49).

Wipe down all parts of the surround and hearth.

Multiple-choice questions

Self-assessment

This section of the book is designed to allow you to check your level of knowledge. The section consists of revision questions for this chapter. The questions are all multiple choice and have four possible answers. The answers are to be found at the end of the book.

The main type of multiple-choice question will be the four-option multiple-choice question. This will consist of a question or statement, known as the stem, followed by a choice of four different answers, called the responses. Only one of these responses is the correct answer; the others are incorrect and are known as distracters.

You should attempt to answer the questions by choosing either (a), (b), (c) or (d).

Example

The person employed by the local authority to ensure that the Building Regulations are observed is called the:

(a) clerk of works
(b) building control officer
(c) council inspector
(d) safety officer

The correct answer is the building control officer, and therefore (b) would be the correct response.

Chimneys, flues and fireplaces

Question 1 What are the attached piers called when forming a fireplace opening?

(a) jambs
(b) walls
(c) abutments
(d) withes

Question 2 What is the minimum height a stack should project above the ridge?

(a) 1500 mm
(b) 1000 mm
(c) 750 mm
(d) 600 mm

Question 3 Identify the following component used when constructing a fireplace opening:

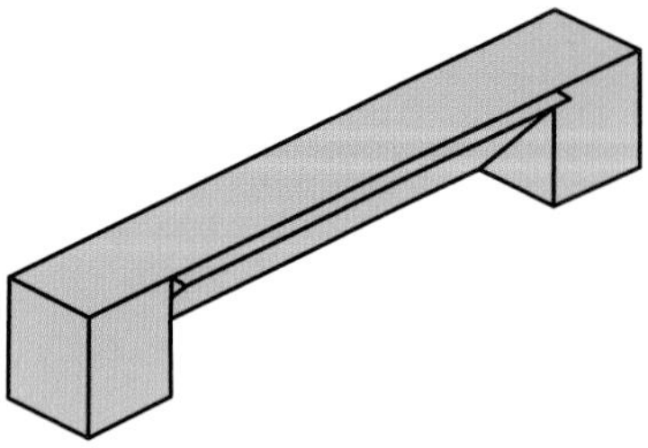

(a) throated lintel
(b) lintel
(c) beam
(d) concrete slab

Question 4 Name the wall built off the oversite concrete to support the hearth at ground level:
(a) foundation wall
(b) honeycombed wall
(c) fender wall
(d) party wall

Question 5 Where on a chimney stack would you find an oversailing course?
(a) at the start of the capping
(b) as the stack leaves the roof
(c) just above the roof line
(d) anywhere

Question 6 When fitting flue rebated flue liners which is the correct way up?
(a) rebate downwards
(b) rebate uppermost
(c) either way
(d) remove rebate before fitting

Question 7 What is the minimum distance the constructional hearth should project into the room?
(a) 200 mm
(b) 300 mm
(c) 400 mm
(d) 500 mm

Question 8 Why is corrugated paper placed all around the rear of the fireback before filling with weak infill material?

(a) to stop the infill material setting too quickly

(b) to allow for expansion

(c) to prevent the infill material falling out

(d) to seal the back of the fireback

CHAPTER 8

Arches

This chapter will cover the following NVQ and Diploma units:

- NVQ VR49
- CC 3028

This chapter is about:

- Interpreting building information
- Adopting safe and healthy working practices
- Selecting materials, components and equipment
- Preparing and erecting brickwork arches.

The following NVQ performance criteria will be covered:

- Interpretation of information
- Safe work practices
- Selection of resources
- Preparing and erecting brick arches

The following Diploma outcome will be covered:

- Know how to set out and build arches

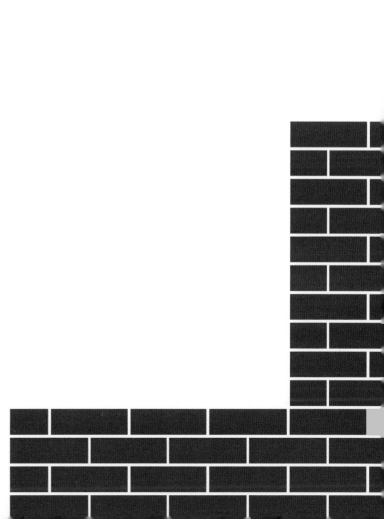

Building information

This chapter will deal with the construction of arches. Basic rough arches have been dealt with in Chapter 10 in Level 2. Please refer back to this chapter for more information on bridging openings and rough arches.

Information can be gathered from numerous sources such as drawings, specifications and manufacturers' information sheets.

Information from drawings

The student should be able to extract information required for setting out from drawings. Drawings have been dealt with in Chapter 3. Please refer back if necessary.

Information from specifications

The specification is a precise description of all the essential information and job requirements that will affect the price of the work but cannot be shown on the drawings.

Manufacturers' information

Technical information can be produced in several formats. When items of equipment or components are purchased there will always be manufacturers' information sheets with them.

Safe work practices

As stated before, there are several hidden dangers when working on the building site, especially when working at height. It is therefore essential to be aware and to ensure that correct personal protective equipment is worn, especially safety boots.

The scaffold should have been erected by a trained scaffolder and all ladders fixed in position.

It is vital that the highest standard of workmanship and the correct interpretation of the current Building Regulations are maintained throughout the construction of all types of arches. Failure to do so can have devastating results.

Selection of resources

The most important resources when constructing arches are the bricks and mortar.

Normally the bricks used match the surrounding brickwork, but occasionally arches can be built with contrasting bricks.

When preparing to construct an axed or gauged arch special bricks may be required.

Most arches require temporary centres to support the brick arch until it has set.

Brick arches

Arch construction is a decorative means of spanning openings. True arches are made to curve upwards when looked at in elevation, so that they are always in a state of compression, wedged between abutments. As bending under load cannot take place, there will be no tensile stress in the arch.

All arches are formed on a temporary support called an arch centre, which must not be removed until the jointing mortar has been allowed to harden for at least a week. Only when the arch centre has been removed does the arch become self-supporting.

Terminology

Arches comprise small units bonded together around a curve.

If an arch is constructed correctly it will not require any additional reinforcement, as brick lintels do, because the units are wedge shaped and any load applied on to the arch will tighten the units together.

The curved shape of an arch distributes the load of the walling above an opening down through the abutments on either side of the opening.

Arches are classified generally into three groups according to their shape, the number of centres and the method of cutting and preparing the arch.

The names of the various parts of an arch are shown in Figure 8.1.

Arch types

The three types of arches are:

- rough arches
- axed arches
- gauged arches.

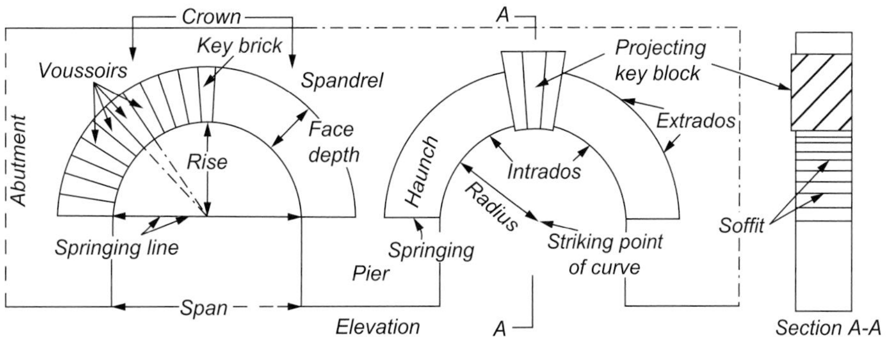

FIGURE 8.1
Craft terms in arch construction

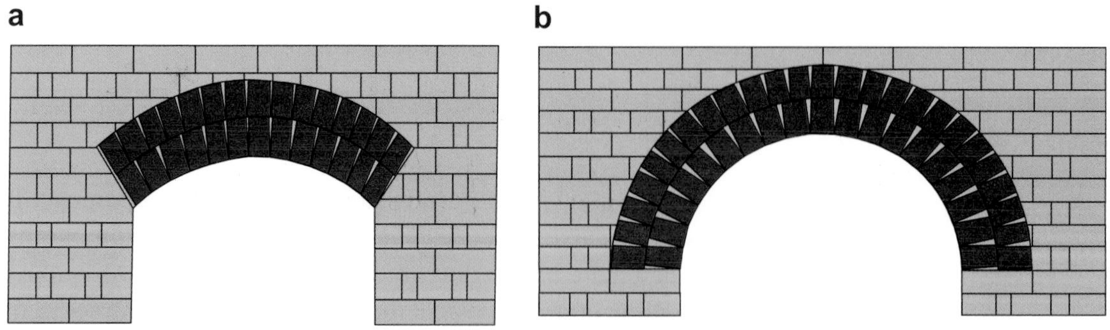

FIGURE 8.2
Rough ring arches; (a) segmental arch; (b) semi-circular arch

Rough ringed arch

This type of arch was dealt with in Level 2.

It is the simplest type of arch to construct. It is formed of uncut bricks and its shape is controlled by the type of turning piece or centre adopted.

The joints are wedge shaped and thus play an important part in the stability of the arch.

Their use is usually restricted to semi-circular and segmental arches (Figure 8.2).

The arch is constructed of a number of half-brick rings, hence the name ringed arch, the number of half-bricks varying according to the size of the opening. This method is adopted to reduce the size of the mortar joint at the extrados of each separate ring.

Axed arches

This type of arch is formed by bricks cut to a wedge shape using an ordinary facing brick (Figure 8.3). It can be of the same colour and texture as the wall facings, or of some contrasting colour and probably of finer texture.

The tools used for cutting and trimming are the hammer, bolster and scutch, with a carborundum block.

The thickness of the mortar joint varies from 4 to 10 mm.

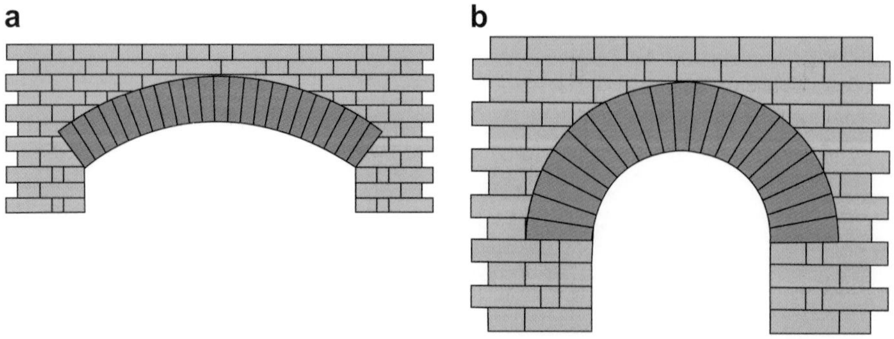

FIGURE 8.3
Axed arches

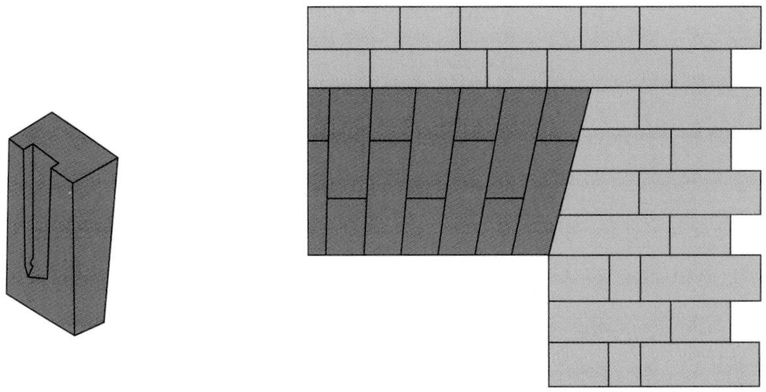

FIGURE 8.4
Gauged arch

The operations of cutting and setting will be explained later in this chapter. It is also possible to bond the face of an axed arch.

Gauged arches

These arches are sometimes known as rubbed arches and have a much finer joint between the bricks. It is usually 2 mm and made from lime putty (Figure 8.4).

This work requires specially made clay bricks called red rubbers. These bricks contain approximately 30 per cent fine sand mixed with the clay, and are carefully fired so that the same even orange–red colour and texture is present throughout the body of each brick.

The high content of sand allows these bricks to be hand sawn and rubbed, on a block of York stone, to permit joints of 2 mm thickness.

Gauged work demands the very highest level of skills from the bricklayer in setting out, cutting, rubbing and bedding these bricks, which are 'white-line' jointed, using a putty made only from freshly slaked lime and water.

Purpose-made arches

A further addition can be made to this section as arches can now be purchased purpose made.

Several brick manufacturers produce purpose-made arches to order. The complete arch arrives on site in kit form. All the voussoirs have been specially made to the required dimensions. Some even provide the centre.

The whole arch is simply built in situ along with the rest of the wall.

Semi-circular axed arches

As the name implies, the arch forms a semi-circle.

Any load placed on the walling above the arch is transferred vertically down through the arch onto the abutments.

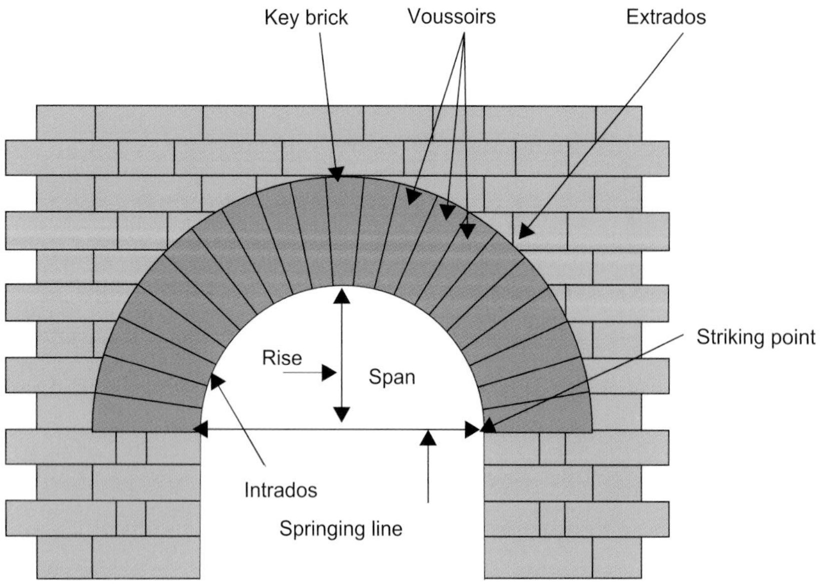

FIGURE 8.5
Semi-circular axed arch details

Figure 8.5 shows the various parts of an axed semi-circular arch.

Geometry of arches

This section will deal with the shape of arches and their geometrical setting out. These arches will all be shown straight in plan.

Various simple geometrical constructions are illustrated and explained to assist in the understanding of arch setting out and correct construction on site (Figures 8.6 and 8.7).

1. To bisect a given line AB – With the compasses set at a radius greater than half the length of the line, and used successively at points A and B, describe the intersecting arcs AC, AC' and BC, BC'. Then C'C is the perpendicular bisector.
2. To bisect the angle ABC – Set the compasses at any radius. From point B describe arc DD'. With centres at D, D' alternately, and with the same radius or a radius greater than a half DD', describe the intersecting arcs E. Then BE is the bisector.
3. To erect a perpendicular on line AB from point A – From any centre C and with the distance CA as the radius, describe a circle cutting AB at D. Draw the line DCE. Then the line AE is the perpendicular.
4. To divide a given line into a given number of equal parts – From one end of the line AB, draw a straight line at any angle. From point A on this line mark off the given number of equal lengths, say 5. Join B5. Through the other points draw parallels to the line B5. Then the line has been divided into five equal parts.

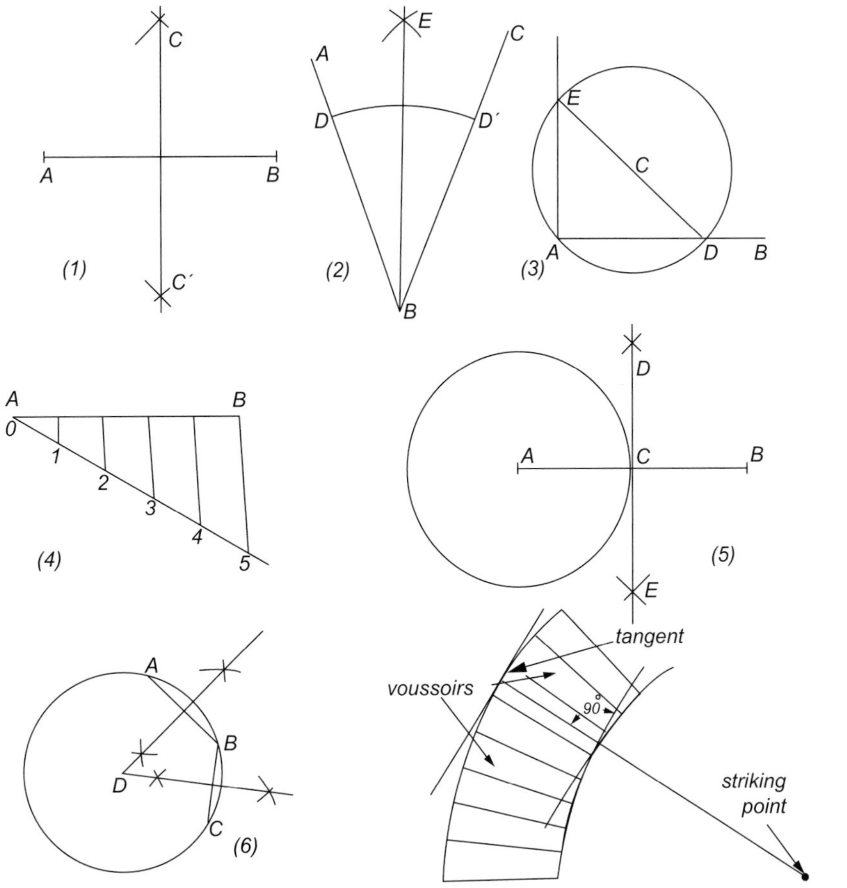

FIGURE 8.6
Geometry of arches

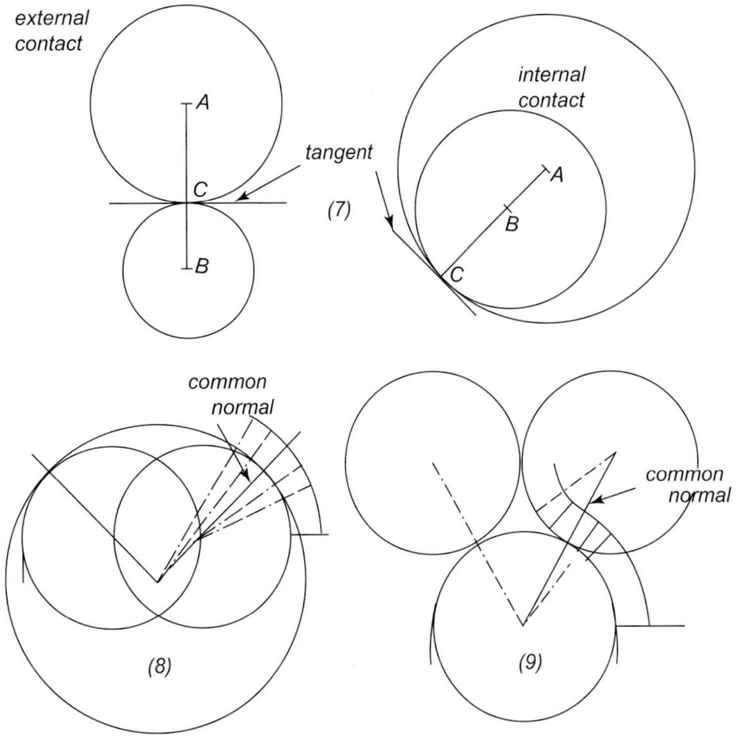

FIGURE 8.7
Geometry of arches

5. To draw a tangent through a given point C on the circumference of a circle – Draw the line AB from the centre of circle to cut the circumference at C, the given point. Make CB equal to AC and bisect AB by drawing arcs D and E as in Method 1. Then line DE is at a tangent to the circle.

   Note that the intrados and extrados of an arch are made up of a series of tangents through its voussoirs. A line taken through the centre of a voussoir from the striking point is at right angles to this tangent.

6. To draw a circle to pass through three given points A, B and C – Join AB, BC. Bisect the lines or chords AB, BC. Then the centre of the required circle, point D, is at the point where the perpendicular bisectors intersect.

7. Figure 8.7 illustrates circles in external and internal contact. They touch one another and are said to be at a tangent. Notice that their centres are on the same line AB, which passes through the point of contact, and that a tangent drawn at C is common to both circles. Line AB is called the common normal.

8. An application of 7 is that circles in internal contact form the curve of a three-centred or semi-elliptical arch. Note that a three-centred arch is not truly elliptical; it follows a similar shape and is therefore called elliptical, but if a true ellipse were formed, each brick would need a separate template.

   By forming an approximate ellipse of three or five centres only, two or three templates, respectively, are needed. This will be further explained in a later section.

9. This is a further application of 7, showing circles in external contact forming an ogee curve.

Geometry of semi-circular axed arches

Remember when setting out for semi-circular or segmental axed arches the voussoirs must be set out on the extrados.

This means that it cannot be done on the arch centre, as for rough ring arches. The arches need to be set out on drawing paper and a template for the voussoirs produced. If an axed arch is to look correct, each voussoir must have exactly the same tapering shape, so that the mortar joints between them are parallel.

Information required:

- the span – assume 150 mm (using a scale of 1:10)
- depth of face – assume a one-brick ring.

Construct a 150 mm span and bisect to find the centre (Figure 8.8).

Place a compass point on the centre point and set the radius to half span. Scribe the arch. This curve is the intrados of the arch.

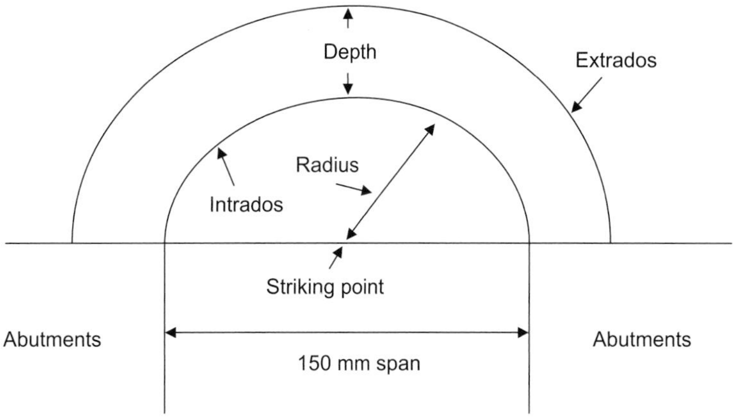

FIGURE 8.8
Geometry of a semi-circular arch

The depth can be set out, say 22.5 mm, and the extrados drawn.

Use dividers to set out the key brick on the extrados (Figure 8.9). Then use the dividers to set out the remaining voussoirs of the arch from the key brick down to the springing line either side.

The dividers should be set to the width of a brick plus a 10 mm joint. This should always give an odd number of voussoirs.

Note

If the spacing does not work exactly first time, reduce the dividers to produce a tighter joint.

NEVER widen the dividers.

When drawing the voussoirs the key brick should be marked out first in each ring.

Using a ruler, draw a line from the striking point to the voussoir points.

Traditionally, axed brick arches have been produced by cutting the bricks using a club hammer and bolster chisel, after making the template of the voussoir shape required. This template can be made from a piece of timber 6–12 mm thick, once the arch has been drawn.

Cutting the voussoirs

Hopefully consideration has been given to the type of brick for the axed arch. Bricks should only be selected that can be easily cut with a hammer and bolster without excessive damage or waste. Mechanical masonry bench saws are available and make the work easier.

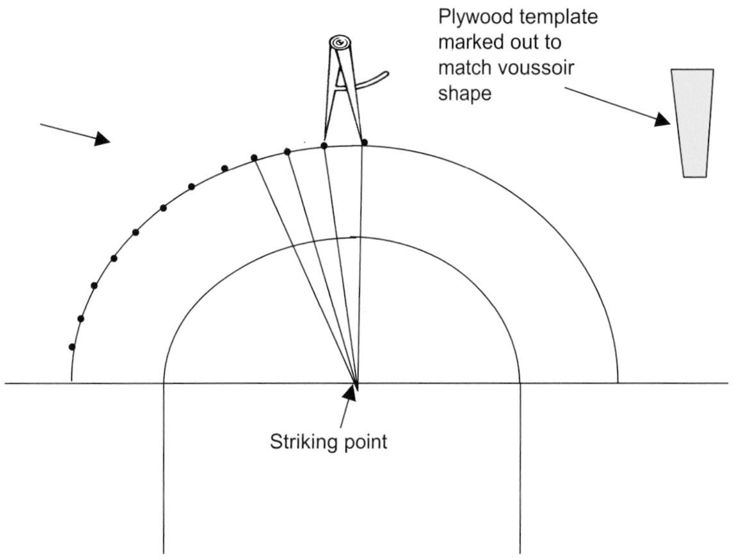

FIGURE 8.9
Marking out the voussoirs

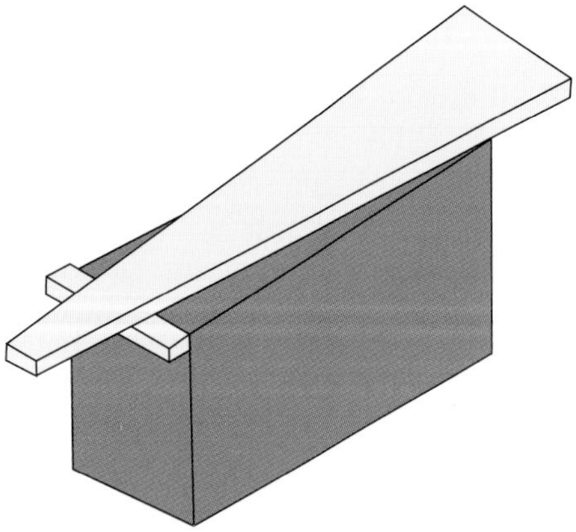

FIGURE 8.10
Using the template

All the voussoirs can now be cut to the shape of the template.

Apply the template and scribe with a pencil (Figure 8.10) and proceed to cut the voussoir.

When finished to the required shape, cut a joggle which allows the arch to be grouted when set in position, giving added security (Figure 8.11).

Stack the bricks carefully when cut. A well-cut arch should stack evenly, each brick fitting close to the one above and the one below it.

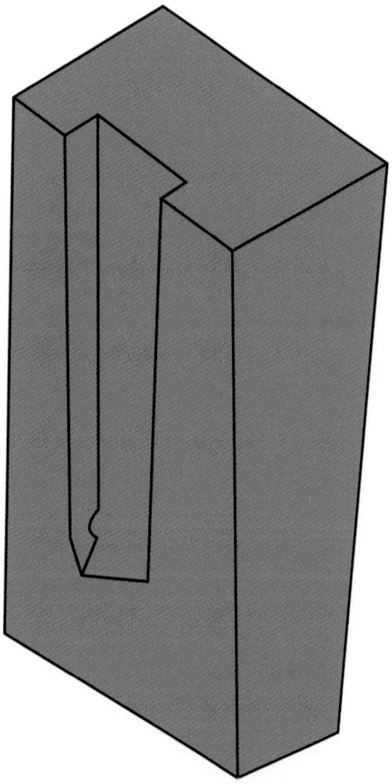

FIGURE 8.11
Cut voussoir with joggle

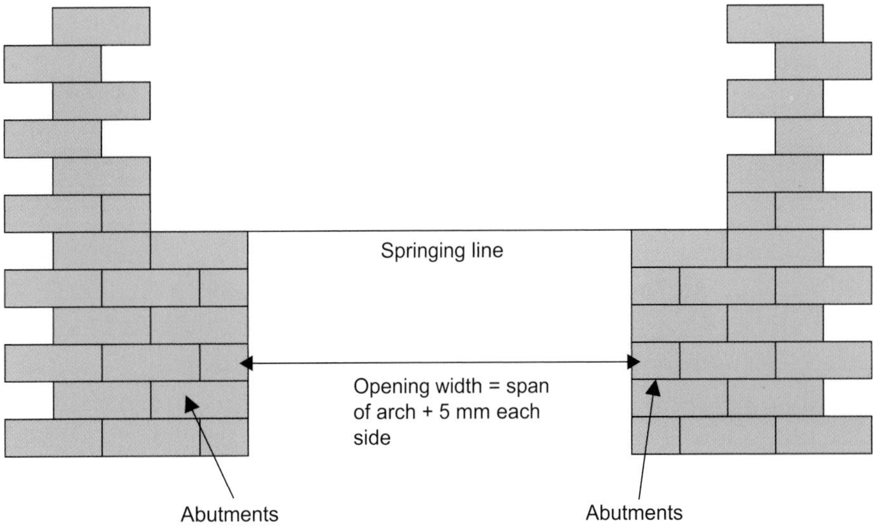

FIGURE 8.12
Building up the abutments to receive the arch centre

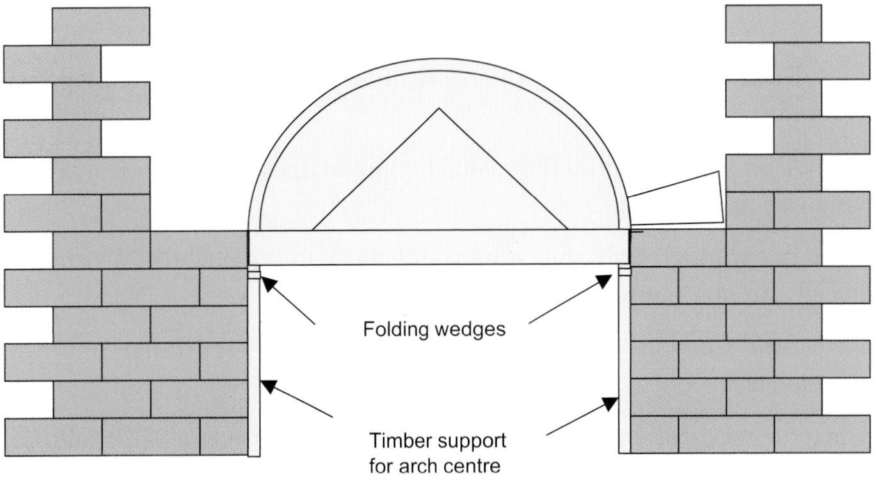

FIGURE 8.13
Setting the arch centre

Constructing a semi-circular arch

The method of constructing a semi-circular arch is similar for all types (Figures 8.12–8.14).

Set out the opening with a tolerance of 5 mm.

Use the arch centre as a guide or produce a pinch rod, to check the abutments for plumb as the work proceeds.

Build the abutments up to the springing line. Position the centre supported on timber props and folding wedges. These will be dealt with later in this chapter.

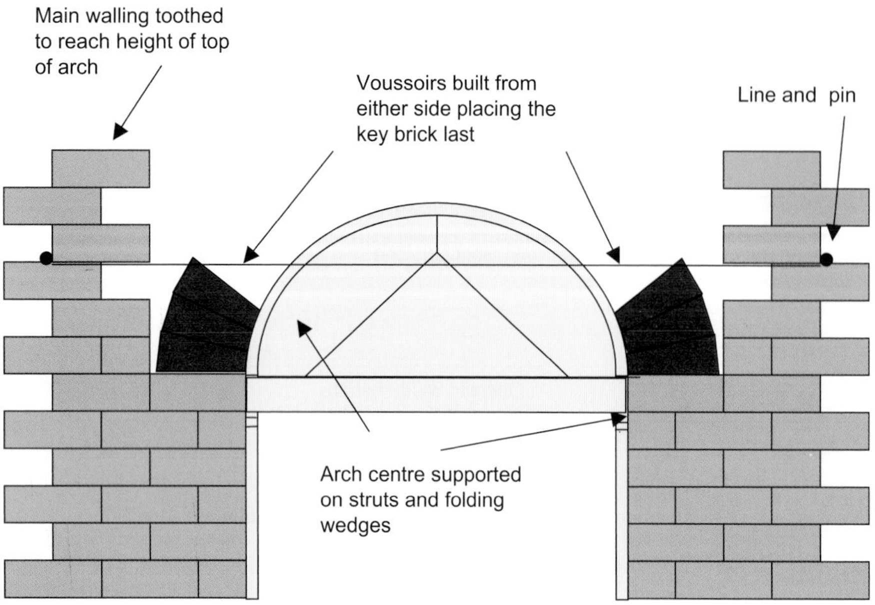

FIGURE 8.14
Laying the voussoirs

It is advisable to build the abutments to the height of the crown of the arch as support for the lines.

Position and fix the arch centre using folding wedges for easy removal after the arch has been built.

Pull a line through from the corners and check that the centre is correctly placed and not interfering with the line.

Lay a few voussoirs on either side of the arch; never build one side completely.

It may be necessary to raise the line while the voussoirs are being laid.

Raise the line course by course to complete the cutting around the extrados of the arch.

Always ensure that the joints are full to provide maximum strength to the arch. Strike the centre as soon as the arch is complete to allow the joints to contract uniformly when setting.

Always remember to joint the soffit of the arch immediately the centre is dropped.

Segmental arches

As the name implies, a segmental arch forms part of a circle.

Any load placed on the walling above the arch is transferred vertically down through the arch onto the abutments.

Figure 8.15 shows the various parts of a segmental arch. Recap on previous information if necessary.

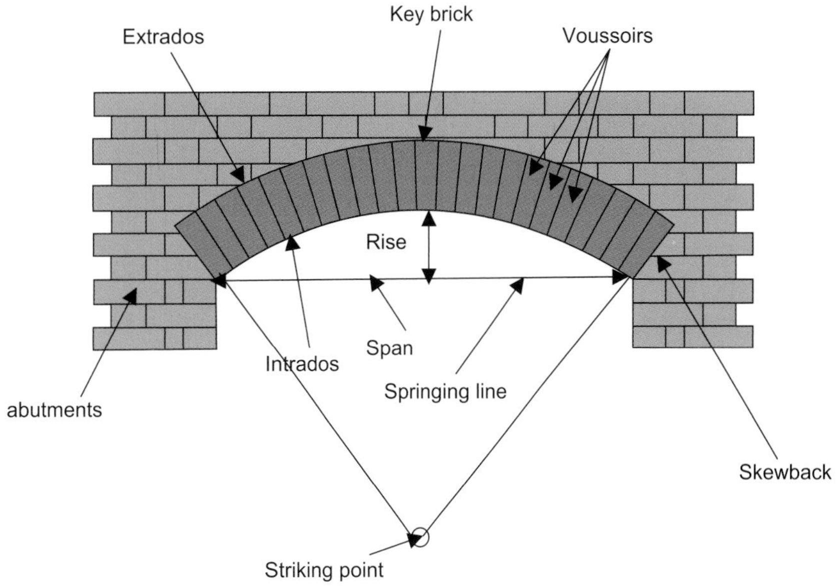

FIGURE 8.15
A two-ring, rough, segmental arch

GEOMETRY OF AXED SEGMENTAL ARCHES

Again, the span must be known, but in the case of the segmental arch we also need to know the rise.

The rise to any segmental arch is normally one-sixth of the span.

First draw the span – assume it to be 900 mm – and set up a perpendicular bisector.

Mark off the rise: 150 mm. Join AB and bisect.

The point where this bisector intersects the centre line, C, is the striking point of the required arch.

A face depth of 225 mm is shown in Figure 8.16.

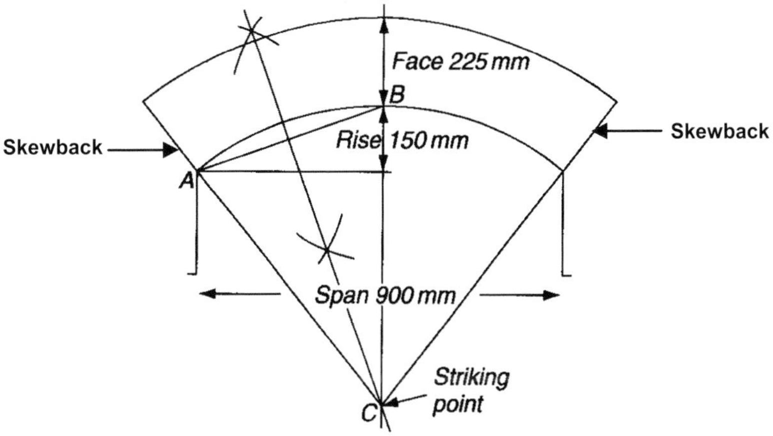

FIGURE 8.16
Geometrical setting out of a segmental arch

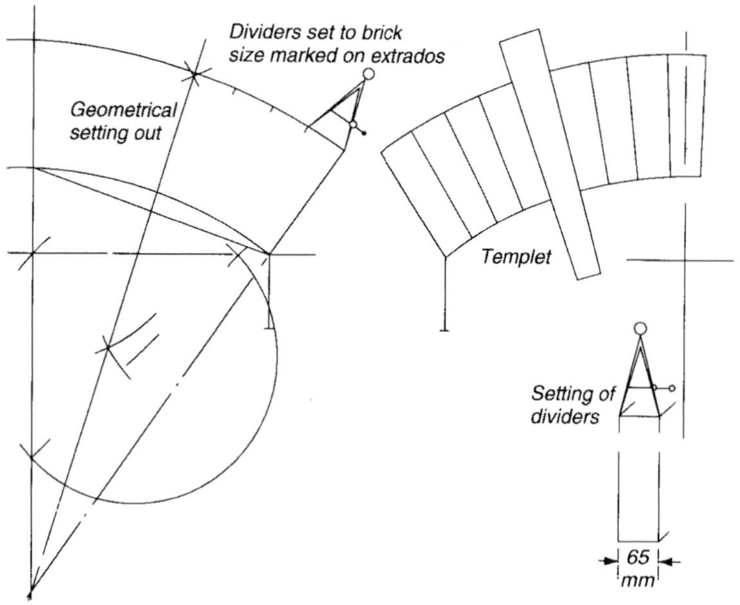

FIGURE 8.17
Geometrical setting out of the voussoirs and template

The angle produced by the line from the striking point, passing through the springing point to the extrados, is known as the skewback.

To ascertain the number of voussoirs and the position of the joint lines in the axed arch, set the dividers to 65 mm. This is accepted in general drawing practice for both axed and gauged arches. In practice, the voussoir size should be taken from the actual brick being used.

Place the points equidistant on each side of the centre line, thus forming the key brick and step round the extrados (Figure 8.17).

If the last step fails to connect with the springing point at the first attempt, make a further attempt by slightly closing the dividers.

The angle of skewback can now be set using one of two methods.

Place a brick on the centre adjacent to the springing point and set a sliding bevel to the angle produced.

Pull a line from the striking point through the springing line at the point of support and set the bevel to this angle (Figure 8.18).

CUTTING SKEWBACKS

For cutting the skewbacks in preparation for the building of a segmental arch, always use a gun template.

Its function is to keep the angle of the skewback constant, especially when there are several arches to be constructed (Figure 8.19).

CONSTRUCTING AN AXED SEGMENTAL ARCH

The method of constructing a segmental arch is similar for all types.

Set out the opening with a tolerance of 5 mm.

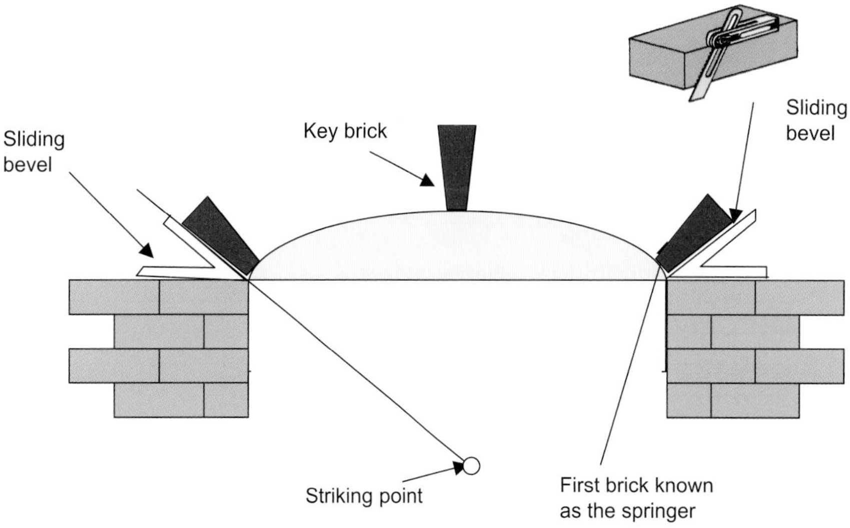

FIGURE 8.18
Setting out skewback

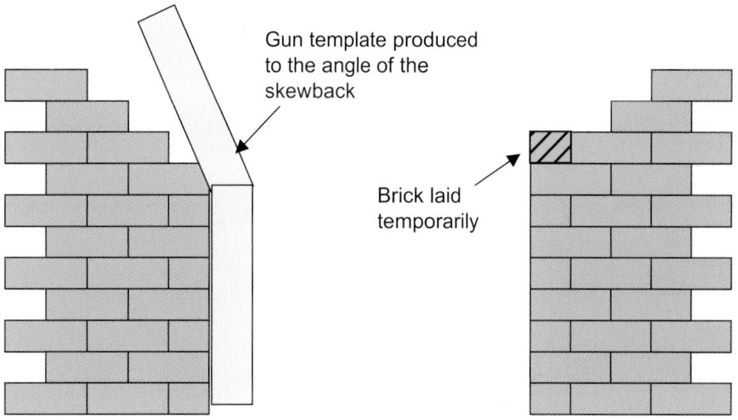

FIGURE 8.19
Use of the gun template

Use the arch centre as a guide or produce a pinch rod, to check the abutments for plumb as the work proceeds.

Build the wall up to the springing line, and position the centre supported on timber props and folding wedges. Extend the abutments to a height above the top of the arch to support the lines (Figure 8.20).

Skewbacks must be marked and cut as shown in Figure 8.21.

Pull a line from the striking point through the springing line at the point of support and set the bevel to this angle (Figure 8.18) or use the gun template.

Pull a line through from the corners and lay the voussoirs to the line.

Raise the line course by course to complete the cutting around the extrados of the arch. Strike the centre as soon as the arch is complete to allow the joints to contract uniformly when setting.

Always remember to joint the soffit of the arch.

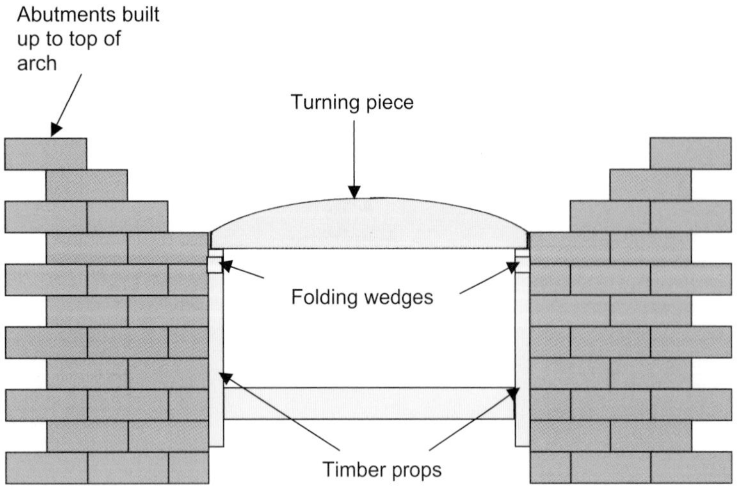

FIGURE 8.20
Setting the turning piece for a segmental arch

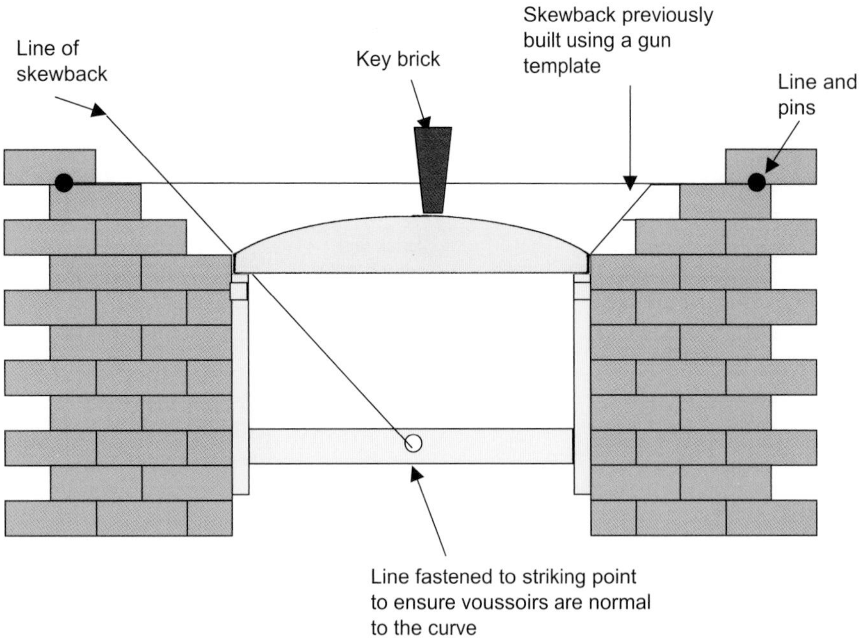

FIGURE 8.21
Setting the skewbacks and voussoirs to a segmental arch

Types of support

No attempt will be made here to describe temporary arch supports over large spans. Only those met with by the bricklayer in the course of everyday work will be dealt with.

Turning piece

A turning piece is made from solid timber for arches of limited span.

For an arch with a 225 mm soffit, two 75 mm timbers can be placed side by side. Alternatively, a turning piece could be constructed from plywood and packed out with timbers. See examples in Figure 8.22.

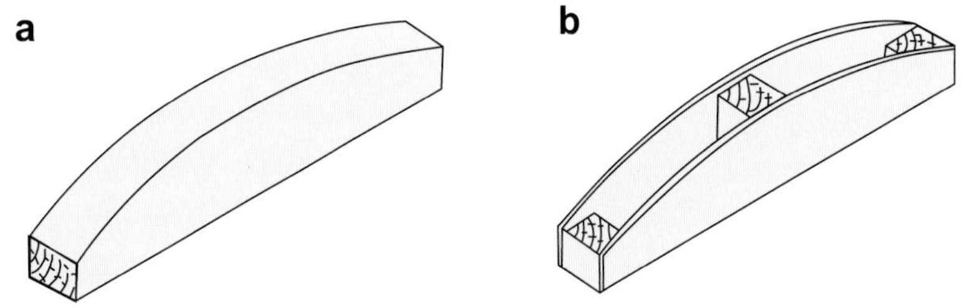

FIGURE 8.22
Arch turning pieces: (a) solid turning piece; (b) built-up turning piece

Arch centres

Arch centres are usually produced for larger spanning arches. They can either be open or closed lagged:

- An open-lagged centre is framed from light timber and used for the turning of ringed arches (Figure 8.23).
- A close-lagged centre is suitable for the turning of an axed or gauged arch. Its use facilitates the marking of the voussoir positions (Figure 8.24).

When using temporary arch supports, folding wedges are placed directly under the centre (Figure 8.25). This facilitates the removal of the centre or turning piece and, in particular, avoids the chipping of the arch face on the intrados arris, which invariably occurs if wedges are not used.

Modern methods

There are also new methods of supporting arches in both solid and cavity walls. These arch formers are available in plastic and steel.

The centres are built in and remain as part of the structure.

A semi-circular centre is shown in Figure 8.26, but segmental centres are also available. A version that sits on a steel lintel is shown in Figure 8.27.

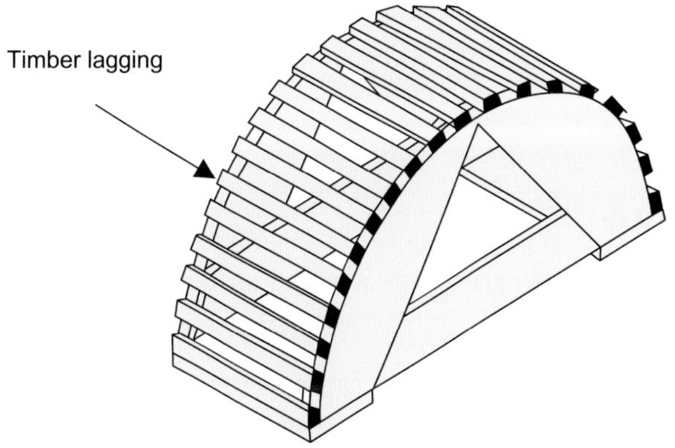

FIGURE 8.23
Open lagged centre for semi-circular arch

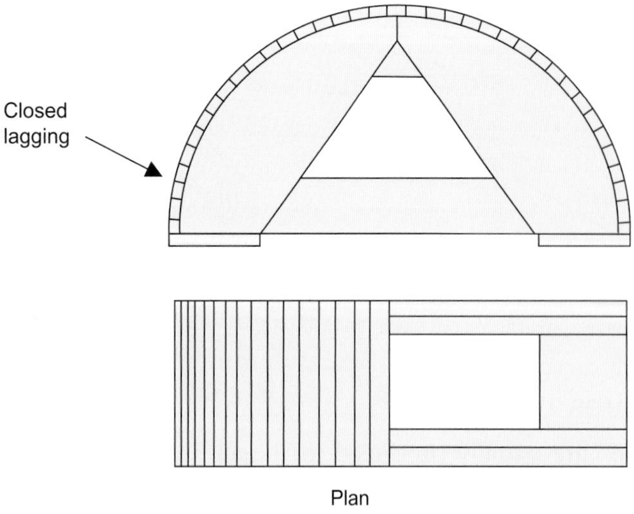

FIGURE 8.24
Closed lagged centre for semi-circular arch

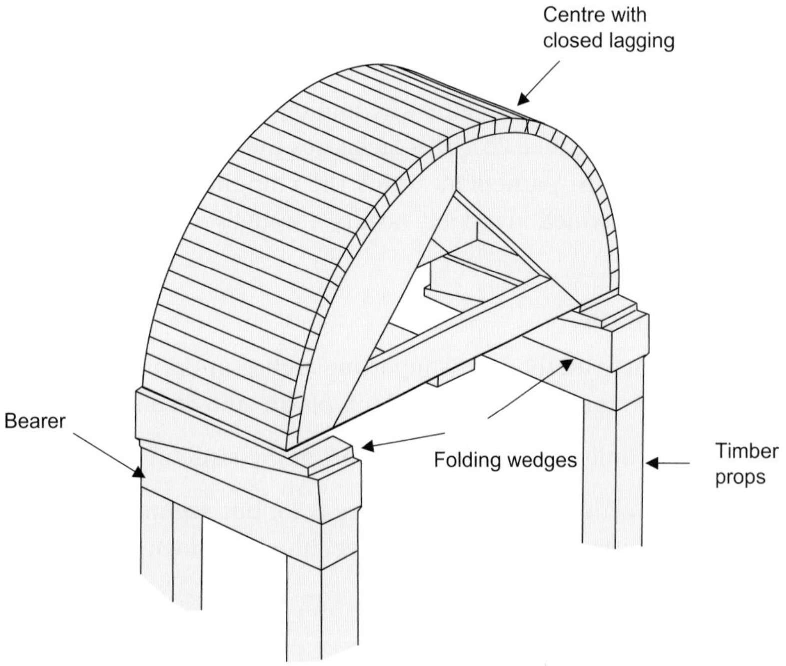

FIGURE 8.25
Centre supported on timber props and folding wedges

Bull's eye

This type of arch is also known as a wheel arch and consists of a complete circle.

The setting out and cutting are the same as for a semi-circular arch, except that it is common practice to arrange the voussoirs so that the key brick is placed at both horizontal and vertical lines.

The setting out of the arch is carried out in two main stages.

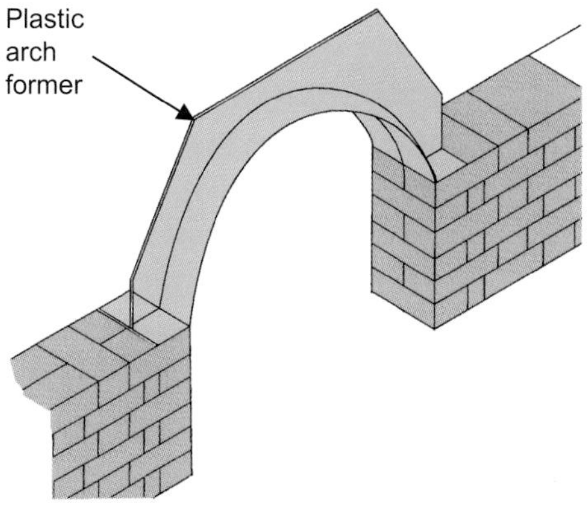

FIGURE 8.26
Plastic arch former for solid walls

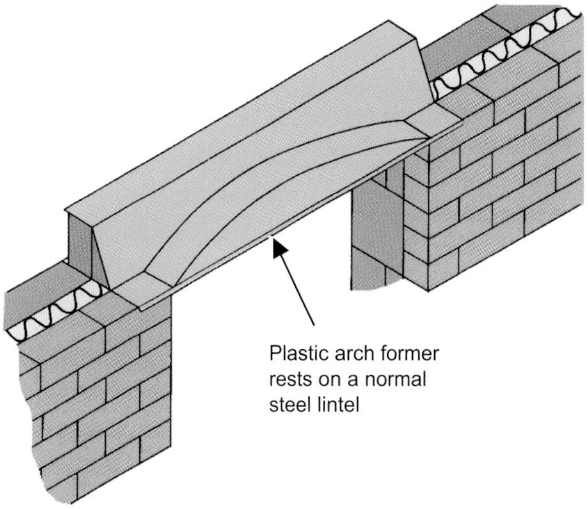

FIGURE 8.27
Plastic arch former sitting on a boxed steel lintel on cavity walls

The lower half of the bull's eye is similar to the method used for curved arches (Figure 8.28).

After the brickwork has been built up on either side of the bull's eye a timber beam is placed over it so that the horizontal centre line of the arch occurs in the middle of the timber beam. Weight is placed on the timber beam to prevent any movement during building operations.

A trammel is cut. The radius of the bull's eye arch is marked from one end and a small hole is drilled at this point. A nail is inserted in this hole and nailed to the centre of the timber beam.

The trammel should be able to swing freely around the brickwork, marking with pencil the bricks to be cut.

Once the bottom of the bull's eye has been prepared the previously prepared voussoirs can be laid (Figure 8.29).

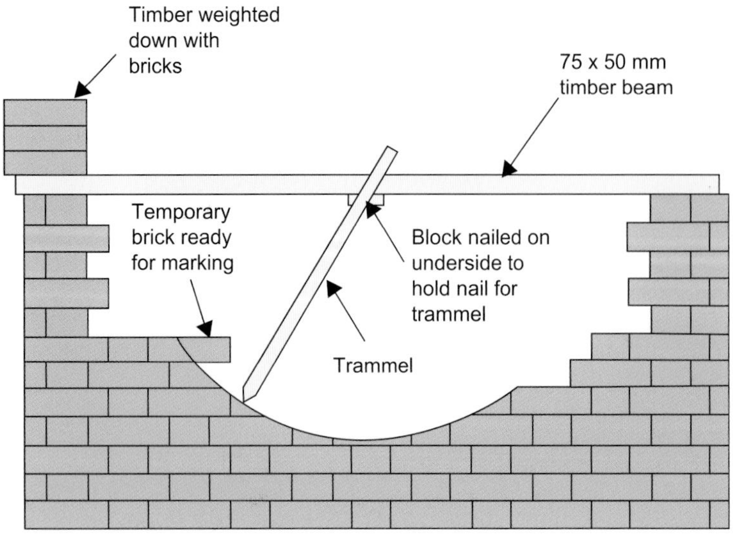

FIGURE 8.28
Setting out the trammel for a bull's eye

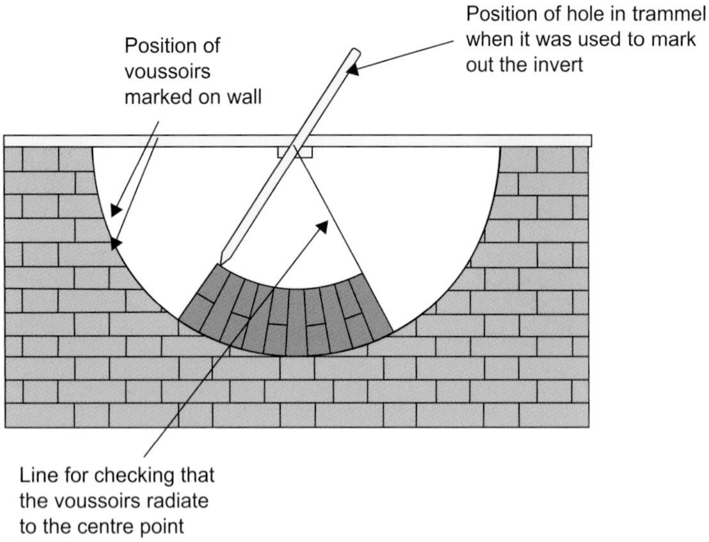

FIGURE 8.29
Constructing the bottom half of the bull's eye

The trammel is removed and a fresh hole drilled in it at a distance equal to the depth of the arch.

The voussoirs are bedded, starting from the lower key brick and building up each side to the horizontal centre line.

The upper part of the bull's eye is similar to the semi-circular arch (Figure 8.30). The centre is fixed in position and the voussoirs are marked off.

A completed bull's eye is shown in Figure 8.31.

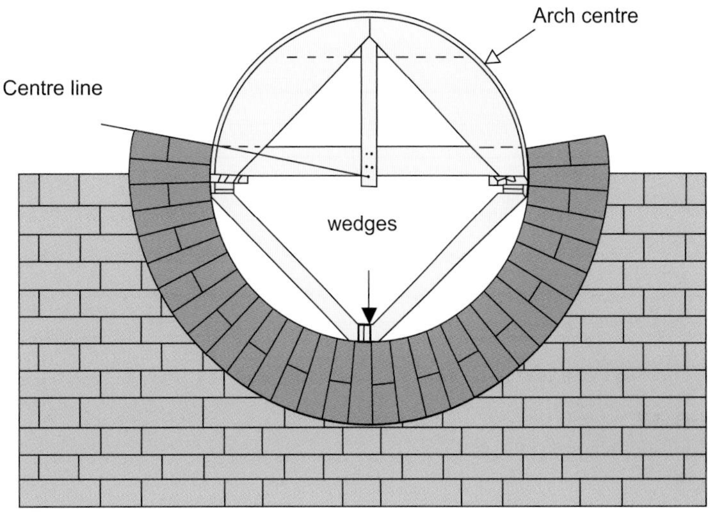

FIGURE 8.30
Setting the arch centre

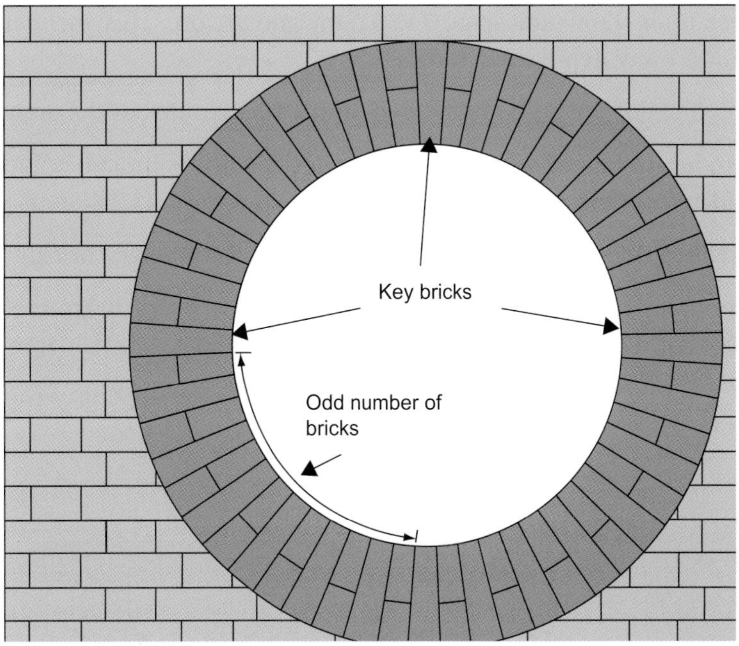

FIGURE 8.31
Completed bull's eye

Flat camber arch

This is basically a flat arch, although the voussoirs are set out on an arc.

A common method for setting out this type of arch is shown in Figure 8.32.

Draw the springing line, and parallel to this and 300 mm above it, draw the setting-out line.

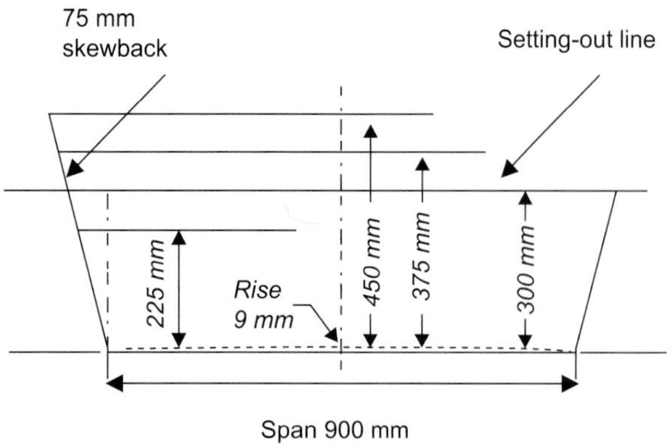

FIGURE 8.32
Setting out a flat camber arch

Erect a perpendicular at the springing point to intersect the setting-out line. From this intersection, mark off the skewback, allowing 25 mm of skewback to every 300 mm of span. Therefore, for a 900 mm span, a 75 mm allowance is given; for a 1350 mm span, 112.5 mm; and so on. This inclination is constant for any depth of arch face.

The voussoirs of the arch are marked off on the extrados of the arch.

If the arch is to be bonded, an even number of courses must be plotted on each side of the key brick.

This ensures that the springing bricks correspond to the key brick.

The amount of camber given to the soffit is 3 mm to every 300 mm of span, therefore for a 900 mm span the rise is 9 mm, and for a 1350 mm span, 14 mm.

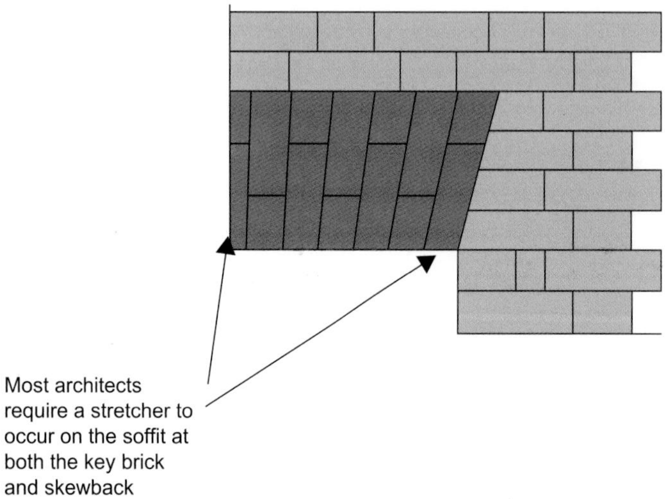

FIGURE 8.33
Half-elevation of a flat camber arch

Template

The cutting of the flat camber arch is different from any other arch in that the bevels are not at a tangent to the same curve and are therefore all different (Figure 8.33).

Some bricklayers consider this arch to be the most difficult to prepare and cut, but if the operations are carried out systematically, no problems should arise.

Half of the arch can be drawn out on plywood and each individual voussoir can be numbered and cut out.

Multiple-choice questions

Self-assessment

This section of the book is designed to allow you to check your level of knowledge. The section consists of revision questions for this chapter. The questions are all multiple choice and have four possible answers. The answers are to be found at the end of the book.

The main type of multiple-choice question will be the four-option multiple-choice question. This will consist of a question or statement, known as the stem, followed by a choice of four different answers, called the responses. Only one of these responses is the correct answer; the others are incorrect and are known as distracters.

You should attempt to answer the questions by choosing either (a), (b), (c) or (d).

Example

The person employed by the local authority to ensure that the Building Regulations are observed is called the:

(a) clerk of works
(b) building control officer
(c) council inspector
(d) safety officer

The correct answer is the building control officer, and therefore (b) would be the correct response.

Arches

Question 1 Identify the item of equipment being used:

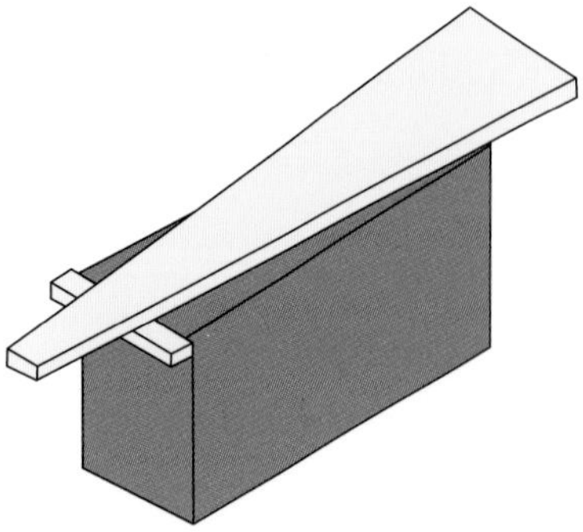

(a) template
(b) wedge
(c) gauge
(d) datum

Question 2 Which of the following is the correct name for the bricks in an arch?

(a) key bricks
(b) voussoirs
(c) soldiers
(d) soffit bricks

Question 3 What is the timber support to a semi-circular arch known as?

(a) formwork
(b) springer
(c) centre
(d) turning piece

Question 4 What is the timber support to a segmental arch known as?

(a) formwork
(b) springer
(c) centre
(d) turning piece

Question 5 Which of the following items of equipment can be used for setting out arch curves?

(a) gauge
(b) turning piece
(c) trammel
(d) datum

Question 6 What is the name given to the slope of brickwork cut to receive an arch?

(a) skewback
(b) springer
(c) closer
(d) voussoir

Question 7 An arch built with parallel joints is known as:

(a) a cut arch
(b) an axed arch
(c) a rough arch
(d) a soldier arch

Question 8 Where should the voussoirs be marked on an axed segmental arch?

(a) soffit
(b) crown
(c) extrados
(d) intrados

CHAPTER 9

Curved Walls on Plan

This chapter will cover the following NVQ and Diploma units:

- NVQ VR49
- CC 3029

This chapter is about:

- Interpreting building information
- Determining quantities
- Relaying information
- Preparing and erecting curved walls on plan

The following NVQ performance criteria will be covered:

- Interpretation of information
- Safe work practices
- Selection of resources
- Preparing and erecting curved walls on plan

The following Diploma outcomes will be covered:

- Know how to set out brickwork curved on plan
- Know how to build brickwork curved on plan

Building information

This chapter will deal with the construction of curved walls on plan.

Information can be gathered from numerous sources such as drawings, specifications and manufacturers' information sheets.

Information from drawings

The student should be able to extract information required for setting out from drawings. Drawings have been dealt with in Chapter 3. Please refer back if necessary.

Information from specifications

The specification is a precise description of all the essential information and job requirements that will affect the price of the work but cannot be shown on the drawings.

Manufacturers' information

Technical information can be produced in several formats. When items of equipment or components are purchased there will always be manufacturers' information sheets with them.

Safe work practices

As stated before, there are several hidden dangers when working on the building site, especially when working at height. It is therefore essential to be aware and to ensure that correct personal protective equipment is worn, especially safety boots.

The scaffold should have been erected by a trained scaffolder and all ladders fixed in position.

It is vital that the highest standard of workmanship and the correct interpretation of the current Building Regulations are maintained throughout the construction of all types of curved walls. Failure to do so can have devastating results.

Selection of resources

Various items of equipment will be required when setting out curved walls on plan, such as tapes, builder's squares and a selection of pegs and boards, along with extra items of equipment such as trammels and templates.

Along with the normal bricks and mortar there may be a need for special header and stretcher curved bricks.

Geometry of curved walls

Setting out curved work when the centre is accessible

Setting out circular, segmental and elliptical curves is required for bay windows, paths, kerbs, roads, and retaining and boundary walls.

Three basic methods may be used without the aid of surveying instruments, when the centre of the curve is accessible.

- tape measure
- trammel
- template.

TAPE MEASURE METHOD

To set out a shape that is circular on plan, you must first determine the centre or striking point of the curve.

Please refer back to Chapter 8 on arches to revise setting out segmental and semi-circular arches.

If the curve is semi-circular then dividing the span in half will give the centre or striking point (Figure 9.1).

If the curve is segmental then a similar approach to arches should be taken (Figure 9.2).

Once the centre or striking point has been established the curve can be marked out by holding the tape on the striking point and extending the

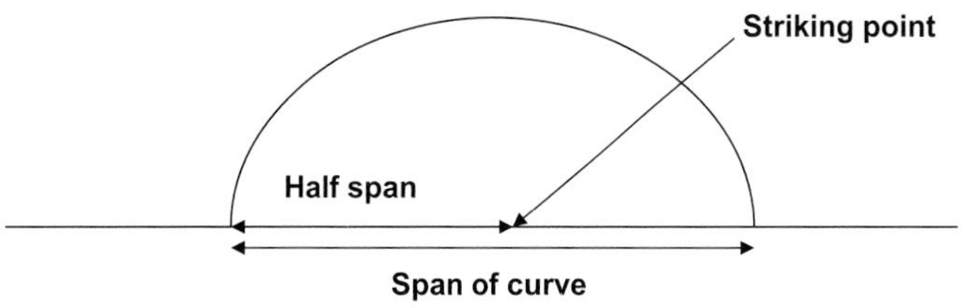

FIGURE 9.1
Finding the striking point of a semi-circular curve

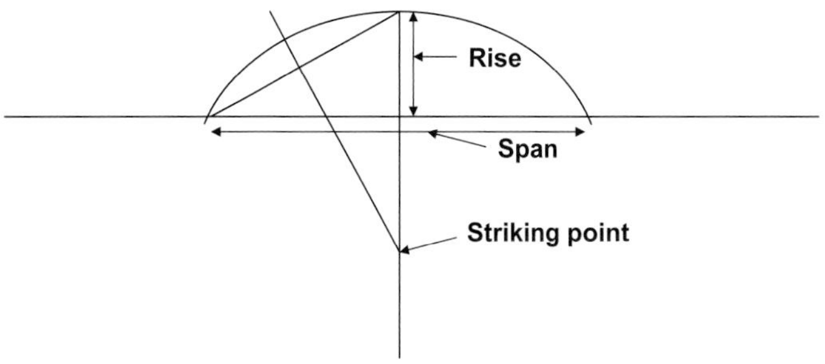

FIGURE 9.2
Finding the striking point of a segmental curve

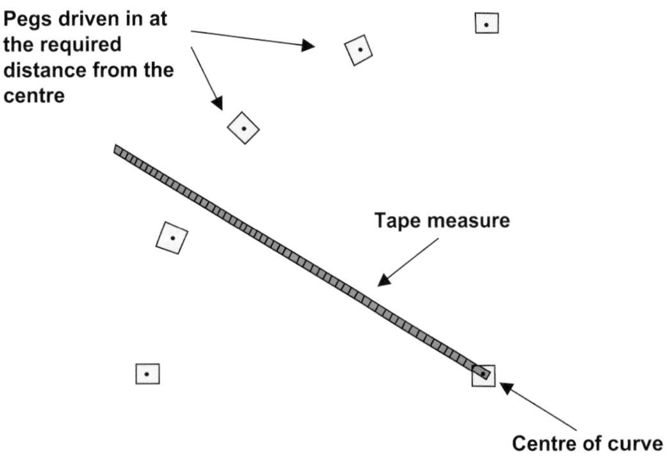

FIGURE 9.3
Marking out a curve with a tape measure

tape to the radius of the curve. The outline of the curve can be drawn by swinging the tape round in an arc (Figure 9.3).

TRAMMEL METHOD

The trammel consists of a piece of timber with two trammel heads, one at either end (Figure 9.4).

The span of the curve can be set between the two trammel heads. Holding one point on the centre of the curve, the other point should scribe out the path of the required curve (Figure 9.5).

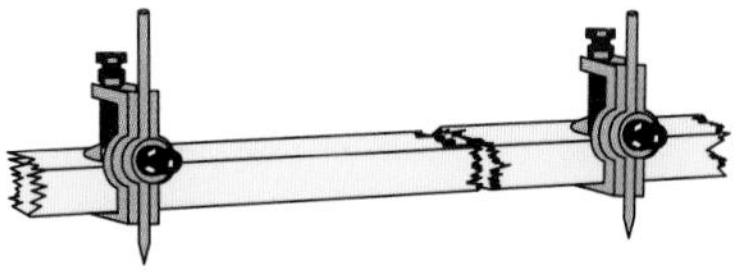

FIGURE 9.4
Special trammel heads

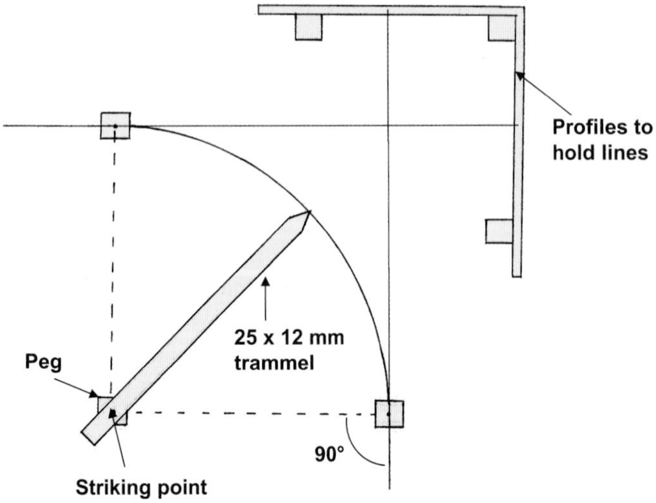

FIGURE 9.5
Using a trammel

TEMPLATE METHOD

For small curves, prepared wooden templates (Figure 9.6) may be used, similarly to the application of a builder's square.

The span AB and rise CD are marked out and, if possible, nails are fixed at points A, B and C.

Small-section battens are nailed together so that the first batten touches points A and C and the second batten touches points B and C.

A batten known as a spreader is then nailed across these two battens to secure them in shape.

The required curve can then be set out by traversing this frame, so that it keeps in contact with points A and B.

A pencil should be held at the intersection of the battens and will scribe out the required curve.

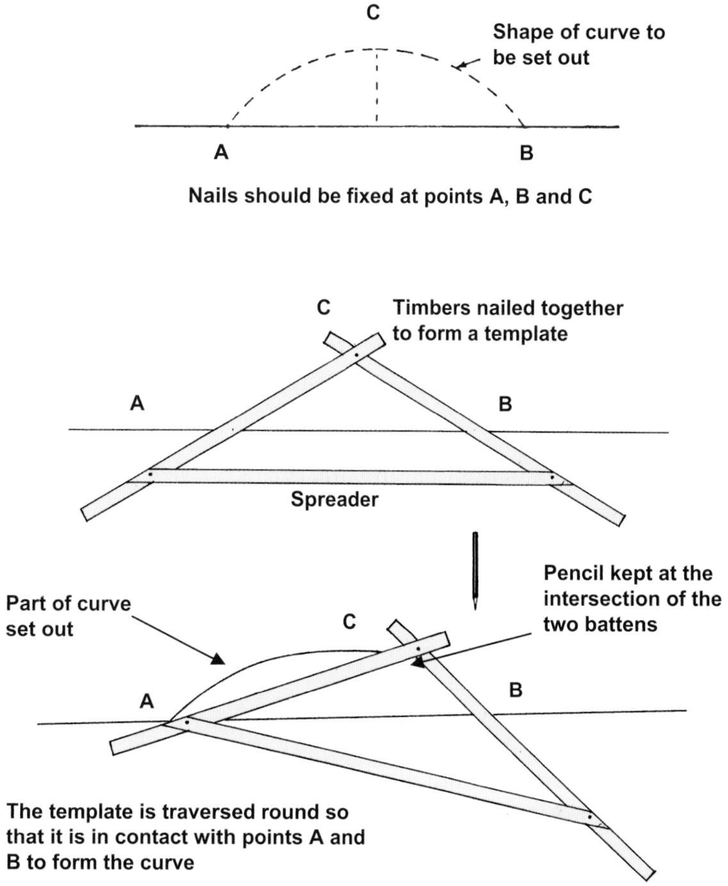

FIGURE 9.6
Using a template to scribe the path of a curve

Note

The tape measure is the preferred method for setting out larger curves.

If, however, the striking point is inaccessible because buildings or materials are in the way, it will be impossible to use any of the above methods. The next part deals with a method of setting out curves when the striking point cannot be reached.

Setting out a curve when the centre is inaccessible

The curve is set out by driving pegs in at regular intervals.

To calculate the lengths of the offsets the following formula is used:

$$x = \frac{L^2}{2R}$$

where x = the first offset, L = the distance between the pegs around the curve, and R = the radius of the curve.

Example 1

If the radius of the curve is 5 m and the distance between the pegs will be 300 mm, then the calculation will be:

$$x = \frac{300^2}{2 \times 5000} = \frac{90\,000}{10\,000} = 9 \text{ mm}$$

This calculation gives the length of the first offset. *All* other offsets are twice this length. In this example all other offsets would be 18 mm.

It is important when using this method to ensure that the dimension is correctly marked on each peg. The best method is to drive a nail into the top of the peg at the correct position.

A curve may be set out by stretching a line along the base line. From the starting point of the curve that is the tangential point, peg 1, a distance equal to length L is marked off along this line at B.

Using the dimensions in the example, extend the line 300 mm from peg 1 to point B.

Using a builder's square, produce a right angle from point B to peg 2.

Mark off the offset of 9 mm from point B to peg 2

Extend the line from peg 1 to peg 2 and a further distance of 300 mm from peg 2 to a point C.

Repeat the process again, but this time the offset is twice the distance of x = 18 mm. This will give peg 3.

This procedure is continued around the curve until the required length of the curve is obtained.

See the example shown in Figure 9.7.

Setting out an ellipse

The student may have to set out an elliptical curve, but it is over and above the requirements for both Diploma and NVQ programmes.

In a true ellipse no portion of the curve is a portion of a circle. If a line is drawn parallel to an ellipse the curve may not be a true ellipse.

The more striking points an ellipse has the more true it becomes. Most elliptical curves are not true ellipses.

An ellipse has two axes which pass through the centre point and are perpendicular to each other. The longest line is known as the major axis and the shortest line is known as the minor axis (Figure 9.8).

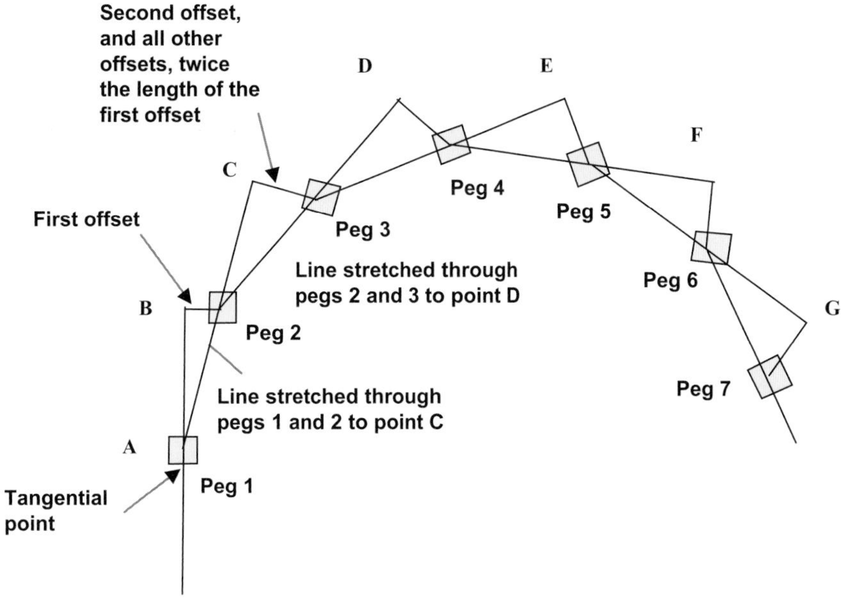

FIGURE 9.7
Setting out a curve when the centre is inaccessible.

An ellipse has two fixed points called focal points.

To find the focal points of an ellipse, swing arcs from one end of the minor axis using a radius of half the major axis, and cutting the major axis each side of the centre point.

These two points, F1 and F2, are known as the focal points of the ellipse.

The two most common methods for setting out an ellipse on site are the trammel and string methods.

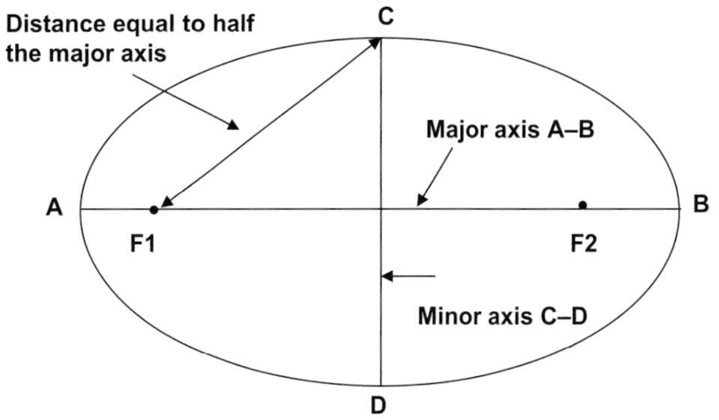

FIGURE 9.8
Setting out an elliptical curve

TRAMMEL METHOD

The trammel is constructed from a piece of batten, possibly 50 × 25 mm in section and longer then the span of the ellipse.

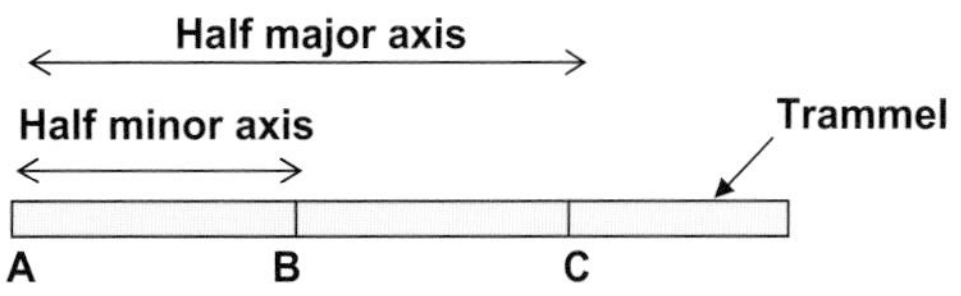

FIGURE 9.9
Constructing the trammel

Owing to the length of timber available it is advisable to use this method when setting out an ellipse with a major axis of no more than 3 m.

Mark off two distances, with saw cuts, one equal to the length of half the major axis (C) and another equal to half the minor axis (B) (Figures 9.9 and 9.10).

Setting out the curve can be done directly on the ground by marking out the major axis and driving pegs in at either end. Then mark out the minor axis with pegs at either end. If possible produce a line, in chalk or pencil, to show the major and minor axes.

Place the trammel across the major and minor axis so that mark B is kept in line with the major axis and mark C is kept in line with the minor axis. A true elliptical curve can be obtained by setting pegs or marking out with a pencil or chalk at point A.

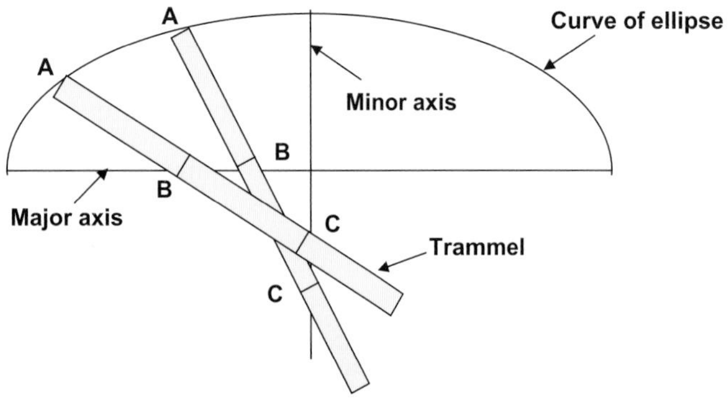

FIGURE 9.10
Setting out an elliptical curve with a trammel

STRING METHOD

This method can be used for larger spans.

The major and minor axes should be set out at right angles to each other and bisecting each other (Figure 9.11).

Half the length of the major axis is struck from one end of the minor axis to cut the former at points F1 and F2, the focal points. Nails or pegs can be driven into the ground at the focal points.

String is attached to one peg, F1, and after passing round a marking pencil held at one end of the minor axis, is then attached to the second focal point, F2, and the tension held.

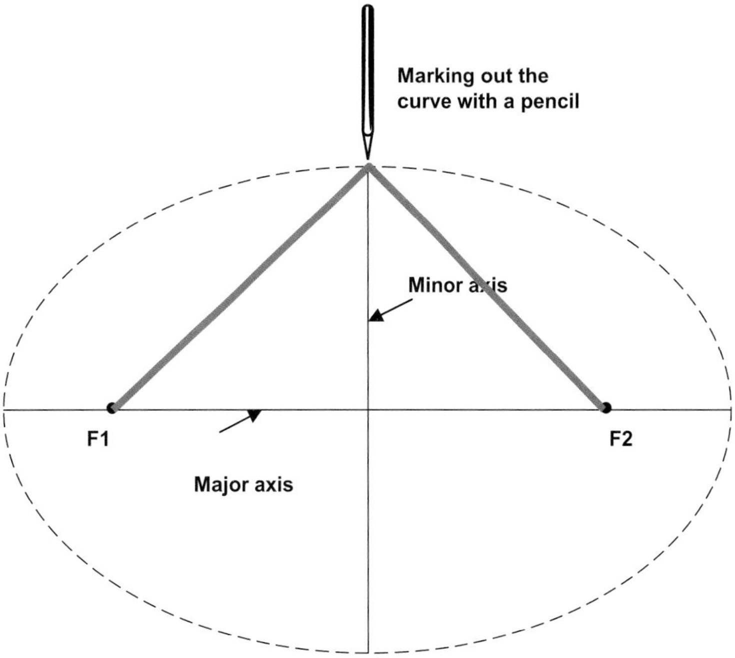

FIGURE 9.11
Setting out an elliptical curve with a string

The marking pencil is then moved to scribe one half of the ellipse.

The loose loop is then transferred to the opposite half of the ellipse and the operation repeated to complete the shape.

Building curved work on plan

Curved brickwork is constructed in modern buildings to provide semi-circular, segmental and elliptical curves.

To build this type of work vertical plumbing and horizontal levelling of the courses are very important as it is not possible to use line and pins on curved work.

Three methods are available to construct curved brickwork:

- full templates or moulds
- small template and plumbing points around the curve
- trammels.

Full templates

Templates can be produced when constructing bay windows and are therefore known as bay moulds in some parts of the country.

They are constructed to give the line of the main wall and the full curvature of the outer face of the bay brickwork.

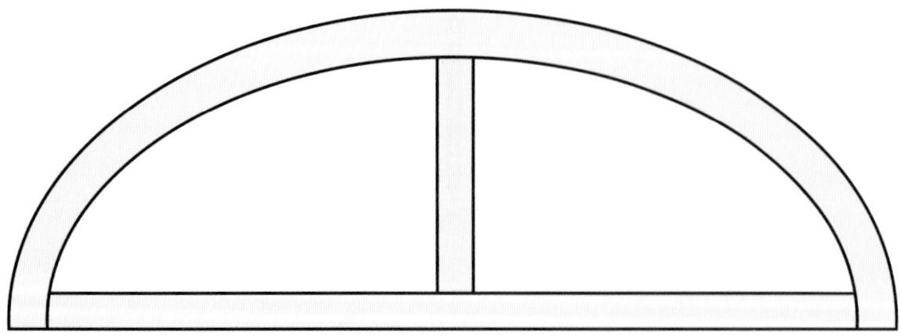

FIGURE 9.12
Typical template for a segmental bay window

The outer face of the timber is carefully shaped to the required curve.

Such templates are used for semi-circular, segmental and other bay windows incorporating curved brickwork (Figure 9.12).

Small template and plumbing points around the curve

The complete template can be used with small radii.

When working with larger radii a small template can be produced, approximately 600–1200 mm long, and one face shaped to give the exact curvature of the brickwork (Figure 9.13).

The first course of brickwork is laid from the large template or trammel.

Several plumbing points are marked around the curve.

As each course of brickwork is begun a brick is bedded in bond at each of these points. These bricks are levelled horizontally from the main wall.

Each marked point is plumbed vertically from the first course.

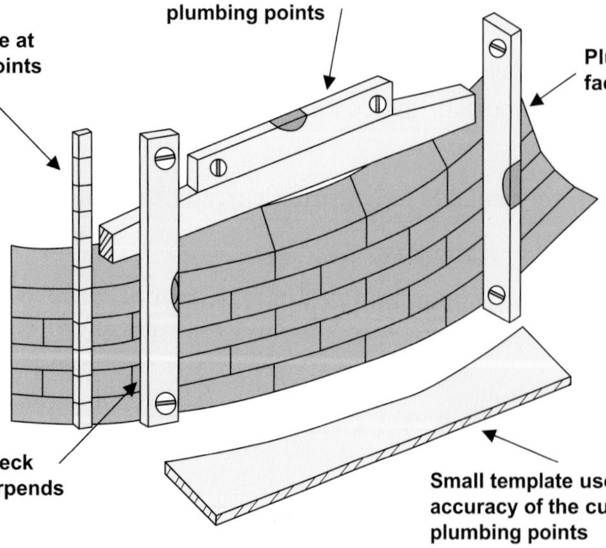

FIGURE 9.13
Method of checking the accuracy of curved brickwork

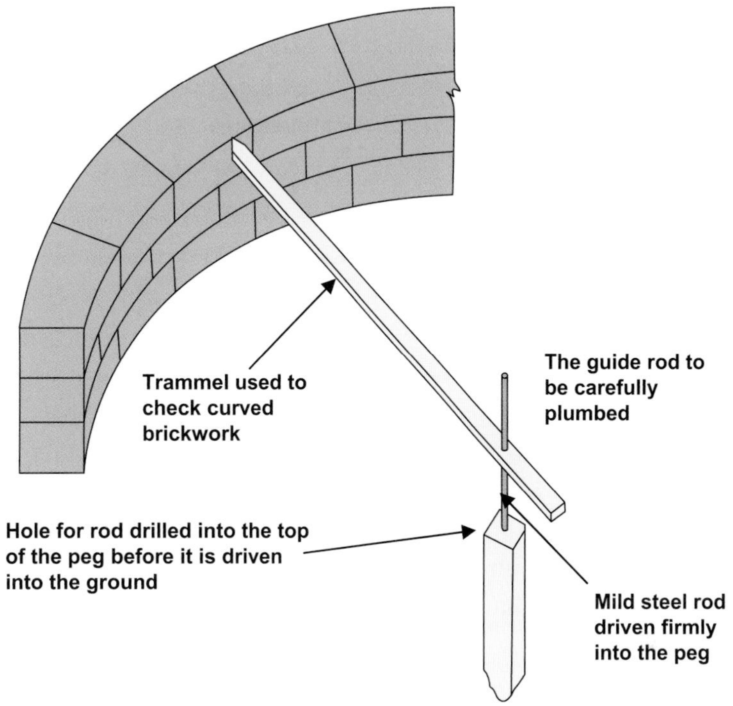

FIGURE 9.14
Trammel being used to check curved brickwork

Trammels

Curved work can also be set out using trammels (Figure 9.14).

A trammel point must be set up in the correct position.

This is a rod of mild steel, 20–25 mm diameter, set in a wooden peg or concrete block. This pivot must be vertically plumb when fixed.

The trammel is constructed from a piece of timber measuring approximately 20 × 125 mm. A hole is drilled in one end of the trammel so that it can be placed over the mild steel rod, and a point is cut at the other end.

The courses of brickwork are laid horizontally level to the curve given by the swing of the trammel.

The trammel can be used for either face of the wall.

Bonding

Walls that are curved on plan can be built from straight bricks by forming wedge-shaped cross-joints (Figure 9.15).

Stretcher bond may be used in half-brick walls to as little as 3 m radius without the need to cut back corners and to give an acceptable cross-joint on the face.

A certain amount of cutting may be required on the inside of the wall.

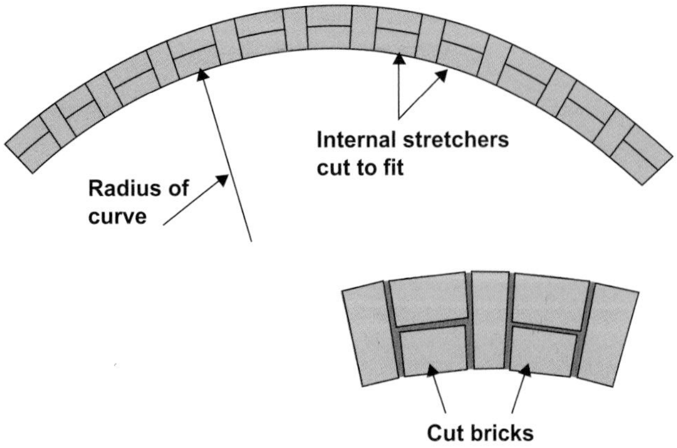

FIGURE 9.15
Bonding of curved work

If neat facework is required on both faces it may be necessary to use purpose-made bricks (Figure 9.16).

Header bond

One type of bond that is often used on curved walling is header bond (Figure 9.17). This bond consists entirely of headers on the face.

When using header bond the smaller the radius the more cutting is required. If the radius is small then alternative methods have to be adopted to avoid large wedge-shaped joints.

Figure 9.18 shows a bonding method where only one in three headers traverses the wall; the remainder are snap headers with cut bricks built into the rear of the wall.

Serpentine walling

Serpentine walling of either 102.5 mm stretcher bonded brickwork or 225 mm English or Flemish bonded brickwork is a very effective shape to resist overturning, and is normally found in boundary walls.

This type of walling curves in and out along its length and creates a pleasant and interesting effect.

It also may be seen on some large housing estates where the roads are deliberately constructed with curves in them to control the speed of the traffic.

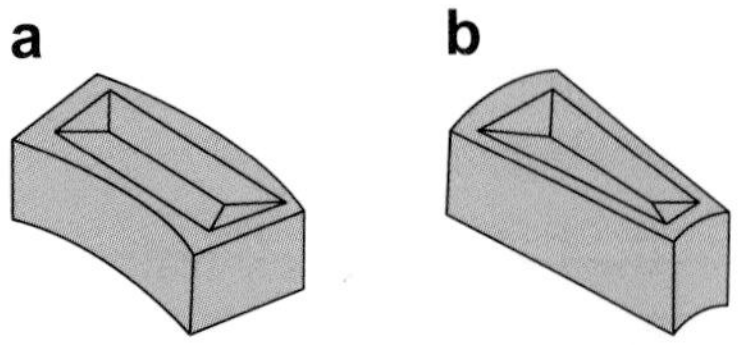

FIGURE 9.16
Purpose-made bricks: (a) stretcher; (b) header

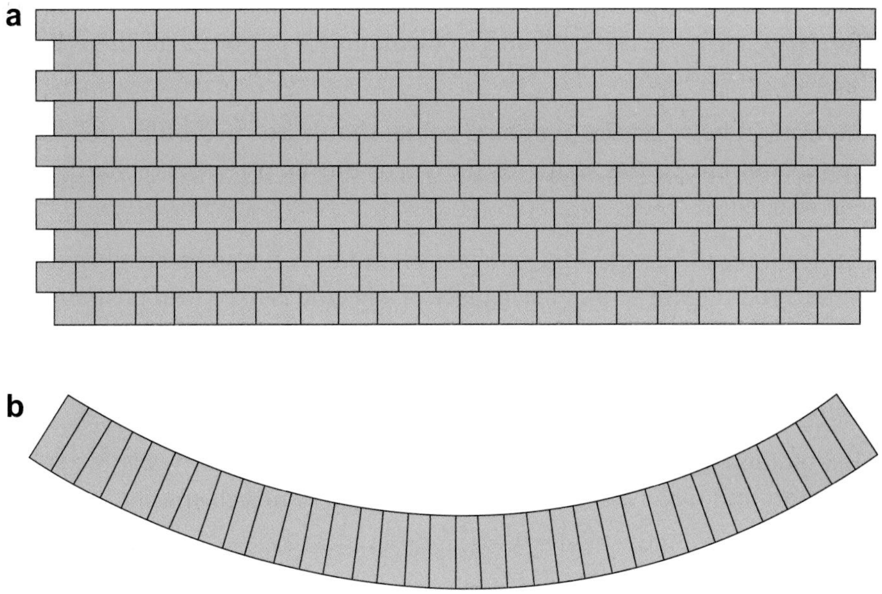

FIGURE 9.17
Header bond: (a) stretcher; (b) header

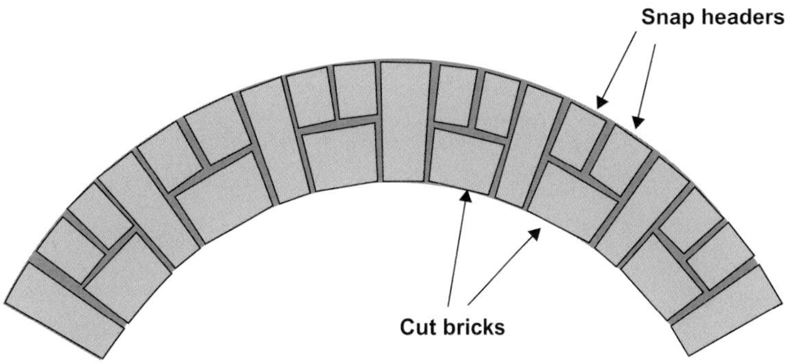

FIGURE 9.18
Bonding arrangement for small radius curves

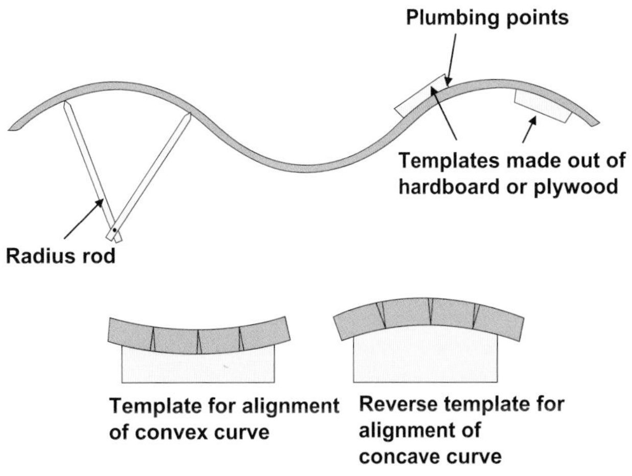

FIGURE 9.19
Constructing serpentine walling

When building walls that are curved on plan, it is important to set out the plumbing points at the base and to maintain these points all the way up the wall.

The work in between the plumbing points should be checked by the use of a template, cut to the shape of the curve out of plywood or hardboard (Figure 9.19).

The accuracy of curved work with a small radius can also be checked using a radius rod (Figure 9.19). First, a piece of steel rod is fixed into position and plumbed. A batten is then drilled so that it fits easily over the rod and is cut to the length of the radius. The batten is threaded over the rod and the wall can then be built to the batten.

This method is particularly useful in the construction of walls for spiral staircases and other similar work carried out in confined spaces.

Multiple-choice questions

Self-assessment

This section of the book is designed to allow you to check your level of knowledge. The section consists of revision questions for this chapter. The questions are all multiple choice and have four possible answers. The answers are to be found at the end of the book.

The main type of multiple-choice question will be the four-option multiple-choice question. This will consist of a question or statement, known as the stem, followed by a choice of four different answers, called the responses. Only one of these responses is the correct answer; the others are incorrect and are known as distracters.

You should attempt to answer the questions by choosing either (a), (b), (c) or (d).

Example

The person employed by the local authority to ensure that the Building Regulations are observed is called the:

(a) clerk of works
(b) building control officer
(c) council inspector
(d) safety officer

The correct answer is the building control officer, and therefore (b) would be the correct response.

Curved walls on plan

Question 1 Which of the following is the correct use of a trammel?

(a) to mark out the gauge of brickwork
(b) to support the overhang of bricks
(c) to mark out the path of a curve
(d) to support the brick corbels

Question 2 Identify the type of brick shown:

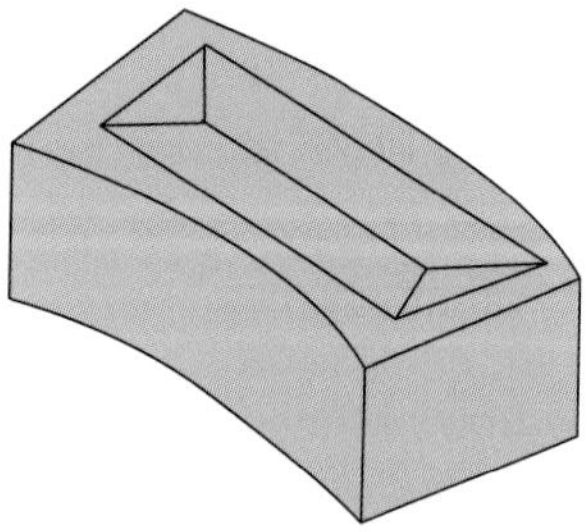

(a) purpose-made header brick
(b) purpose-made plinth brick
(c) purpose-made stretcher brick
(d) purpose-made squint brick

Question 3 Identify the item of equipment shown:

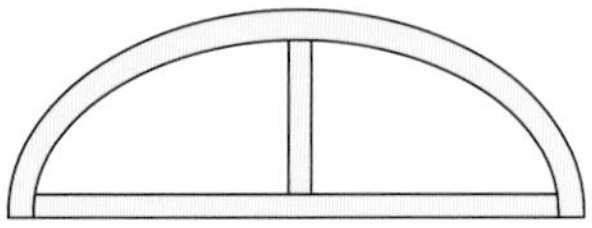

(a) template
(b) trammel
(c) centre
(d) protractor

Question 4 If the larger dimension of the ellipse is known as the major axis, what is the smaller dimension known as?

(a) rise
(b) minor axis
(c) span
(d) centre

Question 5 Identify the item of equipment shown:

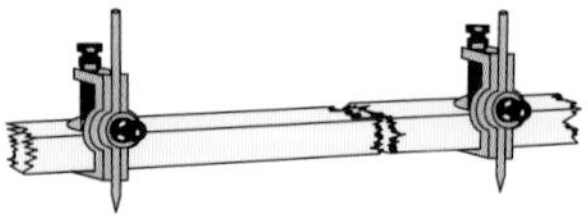

(a) trammel heads
(b) dividers
(c) striking points
(d) compass

Question 6 What is the name for a wall that curves in and out along its length?

(a) curved wall

(b) snake wall

(c) serpentine wall

(d) elliptical wall

Question 7 When building a curved wall, what is used to check the accuracy of the curve between plumbing points?

(a) gauge rod

(b) small template

(c) trammel

(d) radius rod

Question 8 Define the term 'header bond':

(a) the first brick on every course is a header

(b) every course consists entirely of headers.

(c) every other course consists of headers.

(d) every third brick in every course is a header

CHAPTER 10

Ramped Brickwork

This chapter will cover the following NVQ and Diploma units:

- NVQ VR49
- CC 3029

This chapter is about:

- Interpreting building information
- Determining quantities
- Relaying information
- Preparing and erecting decorative brickwork

The following NVQ performance criteria will be covered:

- Interpretation of information
- Safe work practices
- Selection of resources
- Preparing and erecting decorative brickwork

The following Diploma outcomes will be covered:

- Know how to set out ramped brickwork
- Know how to build ramped brickwork

Building information

This chapter will deal with the construction of ramped brickwork.

Ramped brickwork can be either sloping or curved and is usually found as a decorative finish to boundary walls.

Information can be gathered from numerous sources such as drawings, specifications and manufacturers' information sheets.

Information from drawings

The student should be able to extract information required for setting out from drawings. Drawings have been dealt with in Chapter 3. Please refer back if necessary.

Safe work practices

As stated before, there are several hidden dangers when working on the building site, especially when working at height. It is therefore essential to be aware and to ensure that correct personal protective equipment is worn, especially safety boots.

The scaffold should have been erected by a trained scaffolder and all ladders fixed in position.

It is vital that the highest standard of workmanship and the correct interpretation of the current Building Regulations are maintained throughout the construction of all types of ramped walls. Failure to do so can have devastating results.

Selection of resources

Various items of equipment will be required when setting out ramped walls, such as tapes, trammels and supporting timber.

Along with the normal bricks and mortar there may be a need for special bricks.

Setting out ramped brickwork

Setting out for ramped brickwork, either sloping or curved, is required for retaining and boundary walls.

Curved ramps

The setting out for curved ramps is similar to that for arch construction.

To set out a shape that is circular on elevation, you must first determine the centre or striking point of the curve or slope.

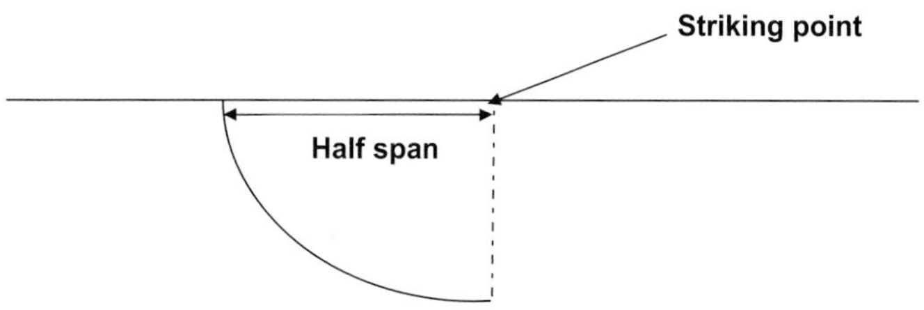

FIGURE 10.1
Finding the striking point of a semi-circular curved ramp

Please refer back to Chapter 8 on arches to revise setting out segmental and semi-circular arches.

The most common curves used as a ramp on boundary walls are segmental or semi-circular. They are usually inverted, i.e. opposite to an arch, and in most cases only a quarter of the full circle is used.

If the curve is semi-circular then dividing the span in half will give the centre or striking point (Figure 10.1).

If the curve is segmental then a similar approach again to arches should be taken (Figure 10.2).

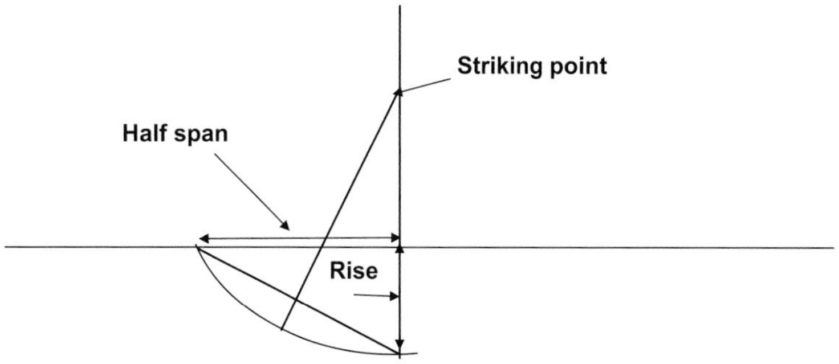

FIGURE 10.2
Finding the striking point of a segmental curved ramp

Sloping ramps

Sloping ramps can be used in the same positions as for circular ramps. Boundary walls, chimney stacks and gable ends are the most common places. The setting out will consist of obtuse angles.

Obtuse angles on elevation

These should be set out full size and angles taken off the drawing with a protractor. See Figure 10.3 for setting-out details.

The following procedure should be followed.

a. The required angle needs to be set out on paper or timber sheet material in full size using a protractor.

b. Bisect the angle using a compass.

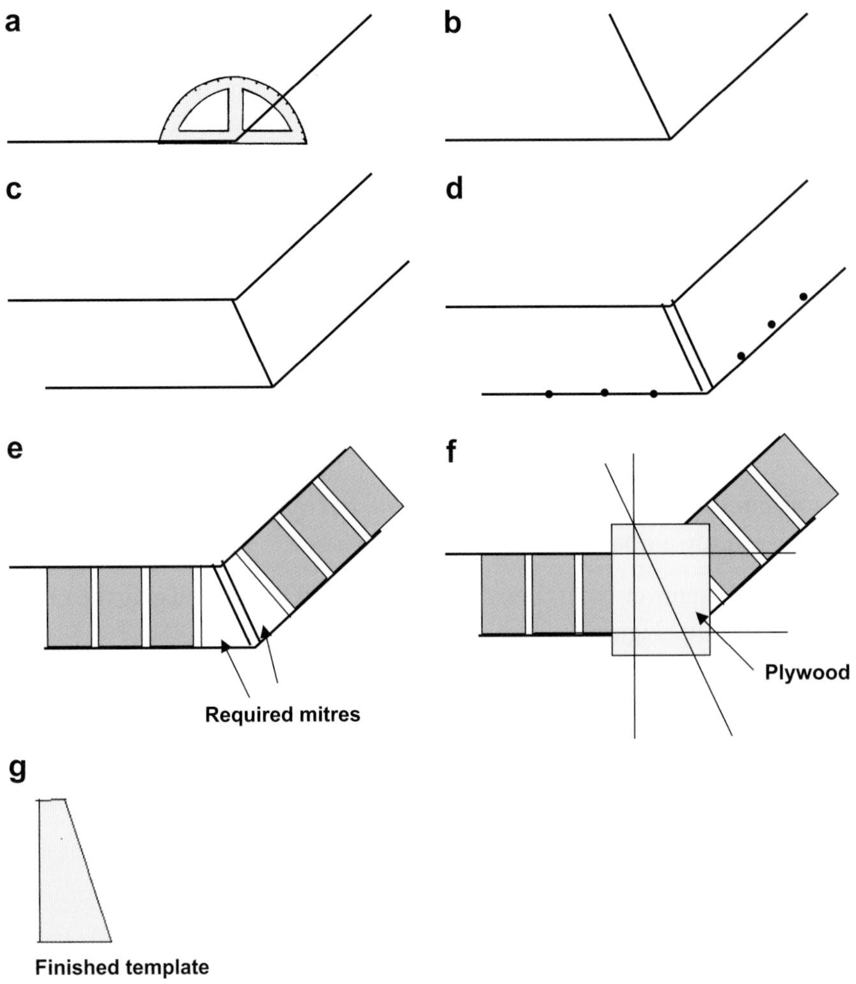

FIGURE 10.3
Setting out the mitres

c. Draw in parallel lines to the depth of the brick on edge.

d. Set out a 10 mm joint on the bisector line and mark off 65 mm on the base line.

e. Square off the base line to give the shape of the mitred brick.

f. Extend lines to allow the shape to be transferred to a piece of hardboard to produce a template.

g. Reverse the template to produce the second mitre.

Curved work on elevation

Curved work on elevation is mainly used in arch construction and has already been dealt with in Chapter 8.

Curved work on elevation will also be required on boundary walls and gable ends.

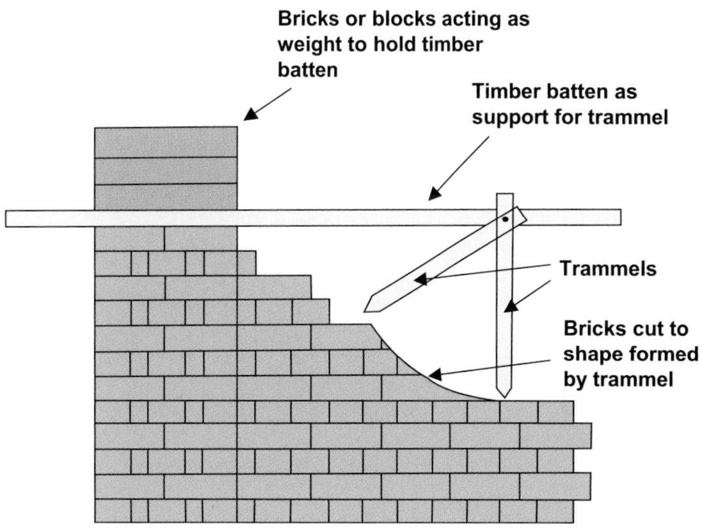

FIGURE 10.4
Constructing a curved ramp

The curves can be either convex or concave. They are set out similarly to curved work on plan (see Chapter 9) and are mainly built with a trammel. This consists of a piece of timber, usually 25 × 50 mm.

Depending on whether the ramp is a concave or convex curve, the trammel needs to be secured either to a timber batten or directly to pre-erected brickwork.

Concave curves

The main brickwork has to be built up to the height of the striking point to support a timber beam (Figure 10.4). The timber beam is placed on the erected wall and bricks are used to provide weight to hold it in place.

A trammel is fixed to the required striking point in such a way as to allow it to swing easily.

The bricks are marked and cut to the shape of the curve formed by the trammel.

The trammel can then be adapted to form the curve of the brick on edge. It is usually shortened by the depth of the brick on edge plus a joint.

The bricks can then be laid using the trammel to check for accuracy (Figure 10.5). If the radius is small the bricks may have to be cut wedge shaped to produce an acceptable appearance. On larger radii, V-shaped joints may be satisfactory.

Curved ramps can be used when reducing brickwork such as in chimney stack and gable end construction.

Figure 10.6 shows a small curved ramp used to reduce a chimney stack. Stainless steel cramps are used to secure the first ramp bricks (Figure 10.7).

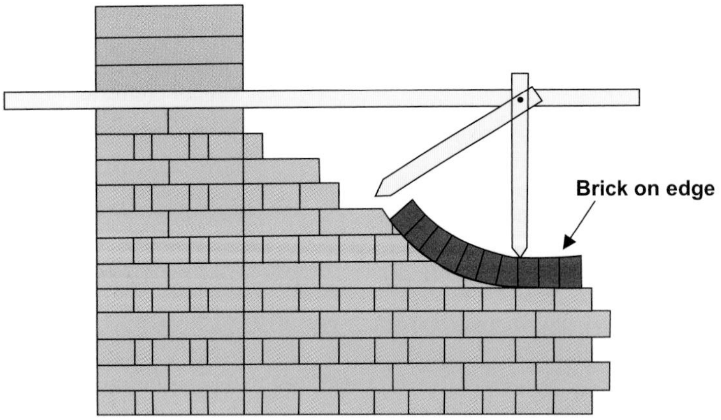

FIGURE 10.5
Constructing a brick-on-edge curved ramp

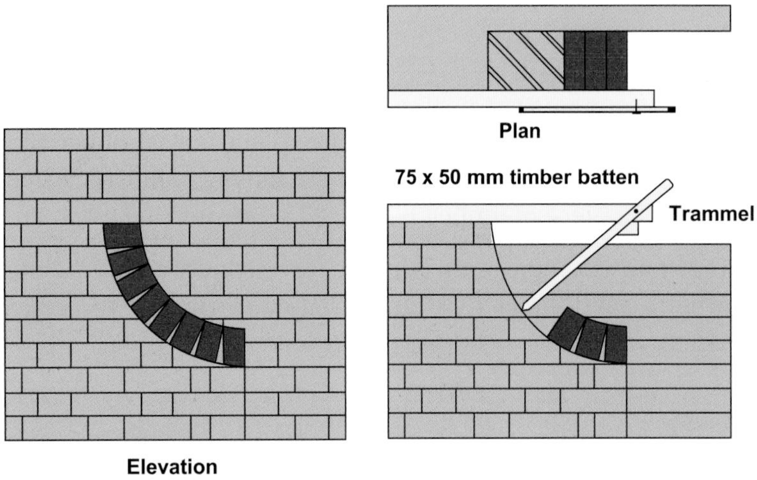

FIGURE 10.6
Using a trammel to construct a concave circular ramp to reduce a chimney breast

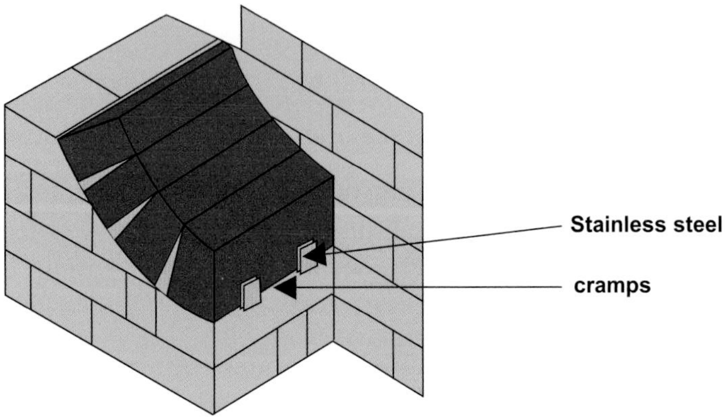

FIGURE 10.7
Stainless steel cramps securing first ramp bricks

Convex curves

The problem with convex curves is that the striking point is in the brickwork below the curve.

The trammel has to be secured to the brickwork mortar joints where possible or by first securing a piece of timber to the brickwork (Figure 10.8).

The trammel is them fixed to the backing timber in such a way as to allow it to swing easily.

When the striking point does not fall in a mortar joint a backing piece of timber has to be fixed, as shown in Figure 10.9.

As with concave curves the trammel can be adjusted, in this case extended, to show the path of the brick-on-edge finish to the curve (Figure 10.10).

Ramped work on elevation

Obtuse angles on elevation

Brickwork laid to form a ramp can be found in gables, boundary walls, chimney stacks and finishing to piers.

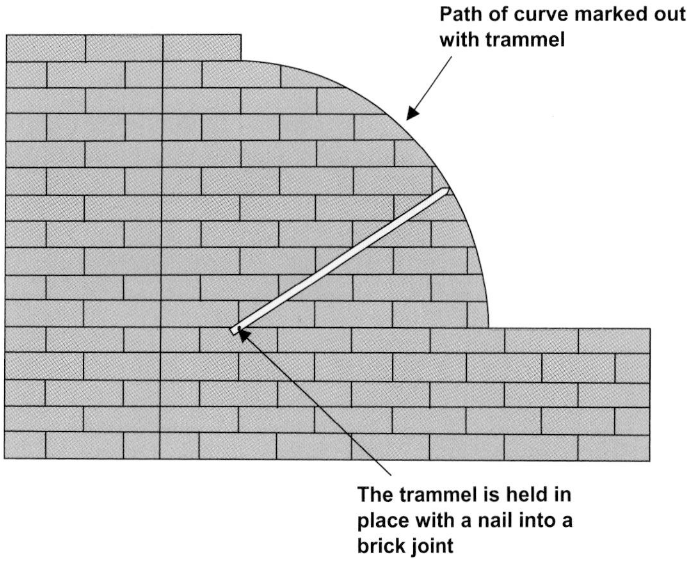

FIGURE 10.8
Using a trammel to mark out a convex curve

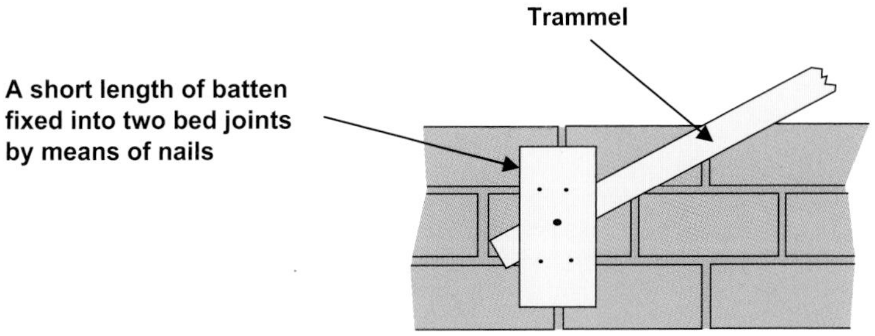

FIGURE 10.9
Trammel fixing

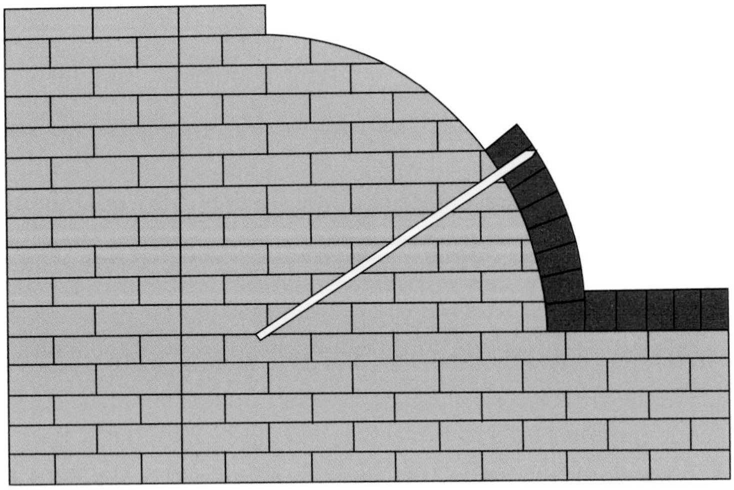

FIGURE 10.10
Trammel adjusted to show path of brick on edge

The main body of brickwork should be constructed first and then the ramp set out using timbers (Figure 10.11).

Lines and pins are used to establish the correct angle to build the ramp. The line is then repositioned to give the correct angle of the brick-on-edge ramp (Figure 10.12).

Note how the bottom brick on edge is allowed to overhang to form a drip, therefore preventing rain from running down the face of the brickwork.

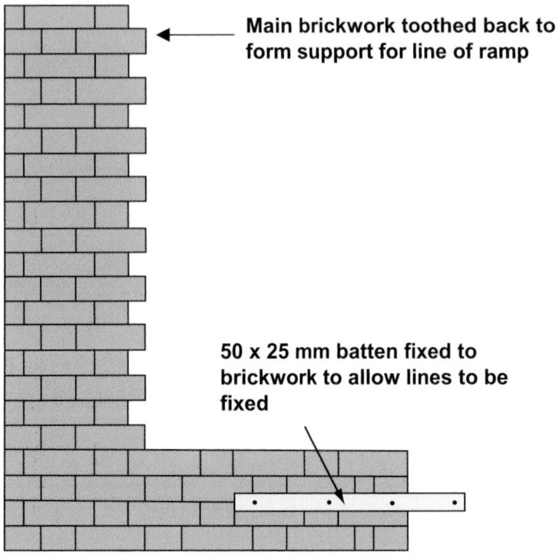

FIGURE 10.11
Preparation for a brick ramp

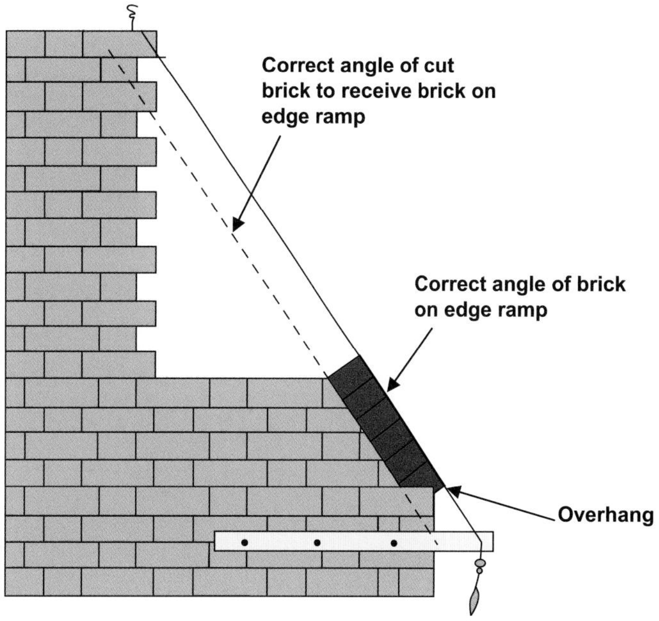

FIGURE 10.12
Maintaining the correct angle with line and pins

Tumbling in

Another method used to finish off the ramp is tumbling in. This practice is used in more advanced work, but an example is shown so that students are aware of the procedure.

Tumbling in is a very decorative method of reducing a wall (Figure 10.13). It allows the face side of the bricks to form the sloping side of the work and provide a good surface to resist the action of rain and frost.

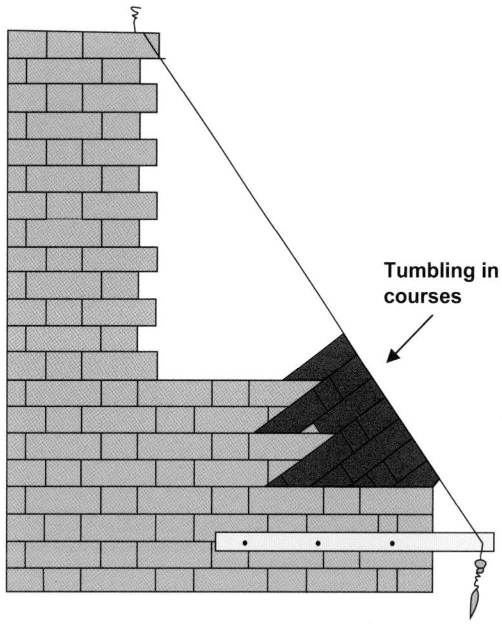

FIGURE 10.13
The use of tumbling in

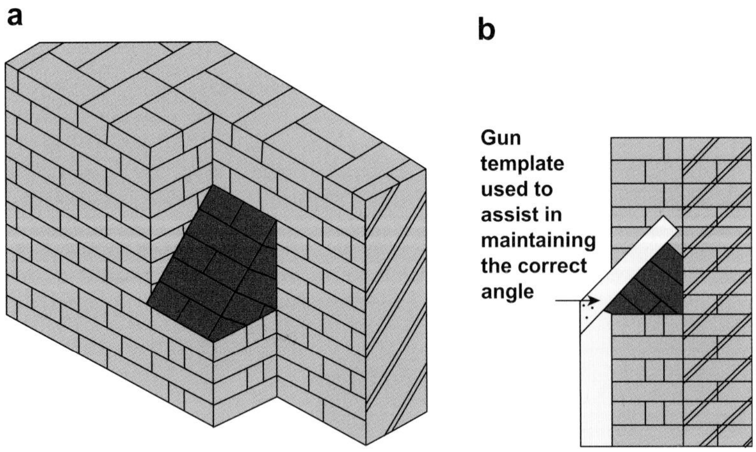

FIGURE 10.14
Tumbling into cap off a pier: (a) isometric view; (b) end elevation

Tumbling in can be used to cap off a wall (Figure 10.14).

A gun template is used to maintain the correct angle of slope. This is made out of two pieces of timber nailed together to form the required slope.

Where brick-on-edge cappings are used on ramps or gable ends, mitred cuts will be required (Figure 10.15).

Boundary walls

These are described as free standing because the brickwork derives no support from floors or roof structures, in contrast to that used in a building.

For this reason local authority building control department approval is usually required if a wall higher than 1 m is to be built next to a public road, or 2 m high elsewhere, to check that the design is stable.

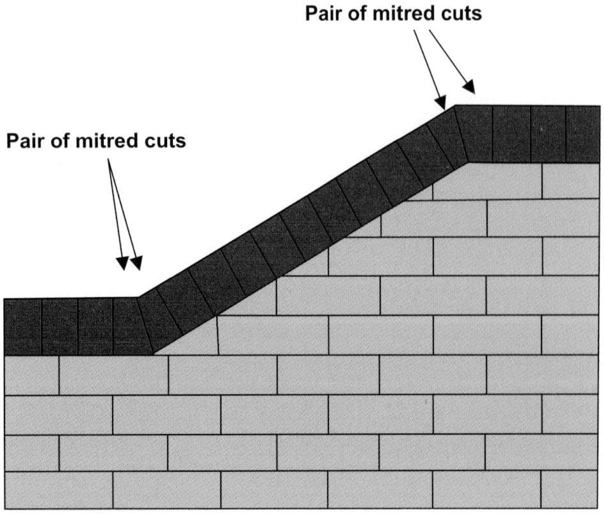

FIGURE 10.15
Mitred cuts to brick-on-edge ramp

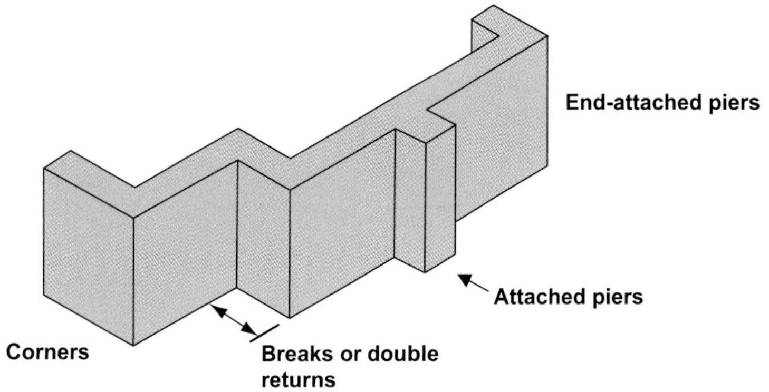

FIGURE 10.16
Ways of improving stability of free-standing walls

Corners, attached piers and 'breaks' in a free-standing wall provide support to resist overturning and failure at the weakest point, where a flexible damp-proof course (DPC) has been used (Figure 10.16).

Attached piers must be properly bonded to the wall (see Figure 10.18).

The DPC for boundary walls is best provided by using two courses of black or red class A engineering bricks, bedded and jointed in cement mortar.

The brickwork above and below is solidly bonded to these DPC bricks (Figure 10.17). (Note also, two more courses of contrast-colour bricks at mid-height, for decorative effect.) The result bonds better than if separated by a flexible DPC, which in a free-standing boundary wall would create a plane of weakness.

Attached piers to boundary walls can be terminated in various ways. All should provide a decorative and weatherproof finish.

The boundary wall shown in Figure 10.18 shows four different finishes to the attached piers. Two show concrete capping and the other two show brickwork, one with tumbling in and the other with plinths.

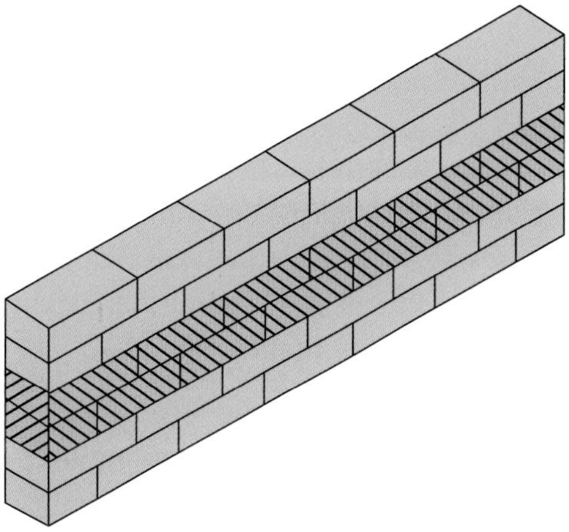

FIGURE 10.17
Engineering bricks used as damp-proof course

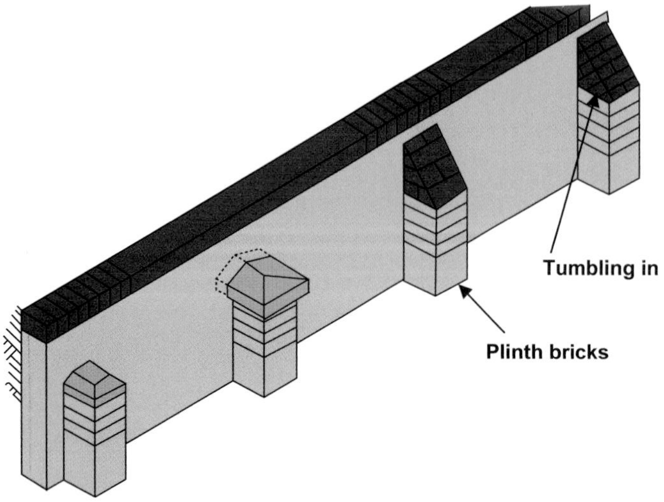

FIGURE 10.18
Attached piers to boundary wall

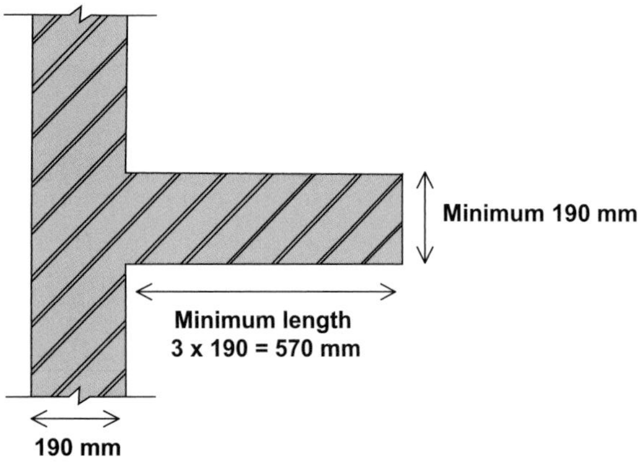

FIGURE 10.19
Building Regulation requirements for buttressing walls

Buttresses

When a wall requires additional lateral support a buttress can be designed and constructed at intervals along its face to provide more stability (Figure 10.19).

Lateral forces can cause a wall to buckle, so buttresses can be attached to the wall at intervals along its length. These buttresses will generally have to resist greater stresses at the base of the wall than at the top of the wall. Therefore, buttresses are wider at the bottom than at the top. The buttress should be properly bonded to the main wall.

The buttress can be built by battering the face or a stepped face, reducing the width with the use of plinth bricks or tumbling in.

A buttress should be built from the base of the main wall to a distance from the top of the main wall equal to three times the thickness of the main wall it is supporting (including the thickness of the main wall itself).

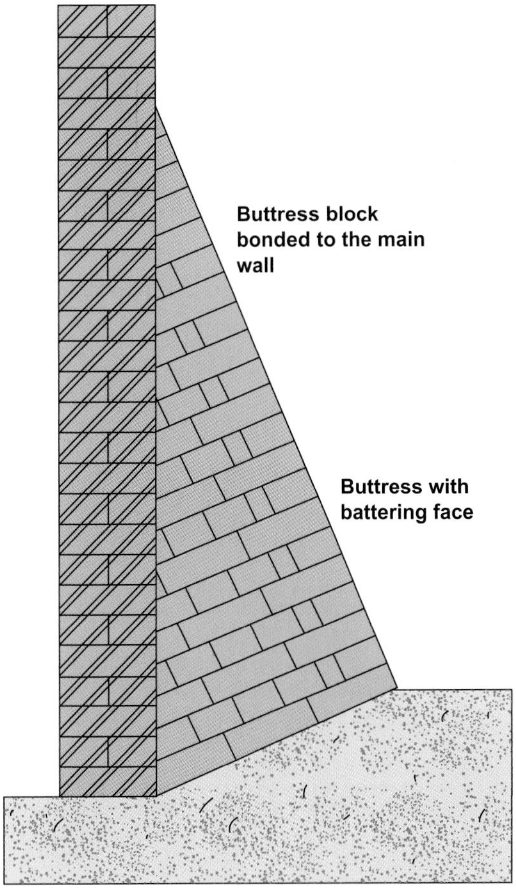

FIGURE 10.20
Section through a buttress with a battering face

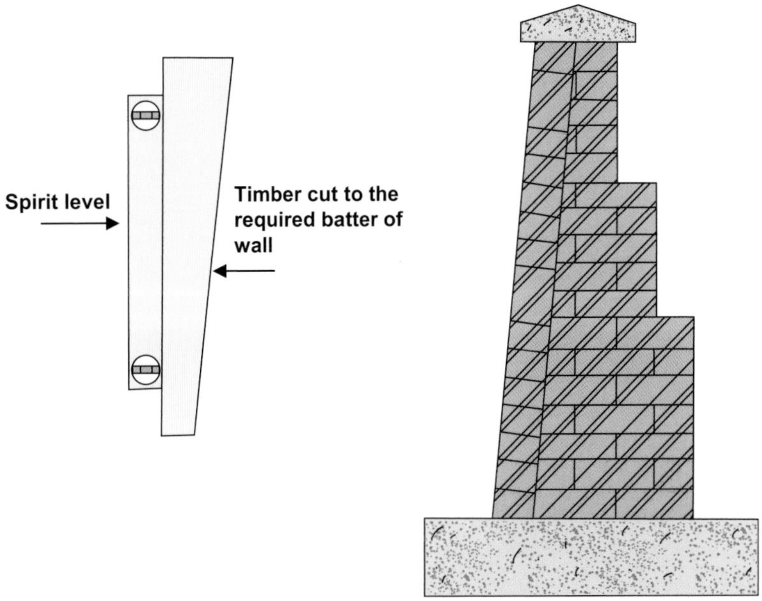

FIGURE 10.21
Section through retaining wall with battered face to front and stepped at rear

The projection of the buttress should be at least three times the thickness of the main wall and at least 190 mm wide.

Battering walls

Retaining walls can be designed with battering faces (Figure 10.20). A special battering plumb rule is required to erect the battering face of the wall. Alternatively, a spirit level with an adjustable vertical tube can be used (Figure 10.21).

Multiple-choice questions

Self-assessment

This section of the book is designed to allow you to check your level of knowledge. The section consists of revision questions for this chapter. The questions are all multiple choice and have four possible answers. The answers are to be found at the end of the book.

The main type of multiple-choice question will be the four-option multiple-choice question. This will consist of a question or statement, known as the stem, followed by a choice of four different answers, called the responses. Only one of these responses is the correct answer; the others are incorrect and are known as distracters.

You should attempt to answer the questions by choosing either (a), (b), (c) or (d).

Example

The person employed by the local authority to ensure that the Building Regulations are observed is called the:

(a) clerk of works
(b) building control officer
(c) council inspector
(d) safety officer

The correct answer is the building control officer, and therefore (b) would be the correct response.

Ramped brickwork

Question 1 Identify the item of equipment shown:

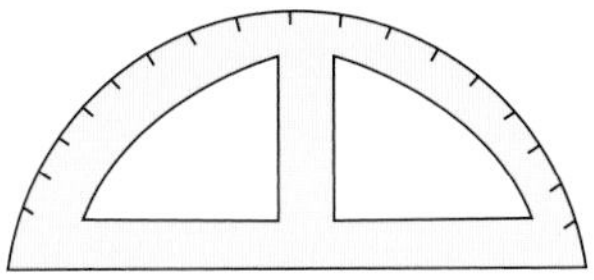

(a) template
(b) trammel
(c) protractor
(d) divider

Question 2 If V-joints are formed on the face side of a brick curved ramp, what type of curve would it be?

(a) a concave curve
(b) a convex curve
(c) an elliptical curve
(d) a segmental curve

Question 3 Name the attached pier that is designed to support a wall at intervals along its length:

(a) buttress

(b) detached pier

(c) ramped wall

(d) battered wall

Question 4 What is the template shown used for?

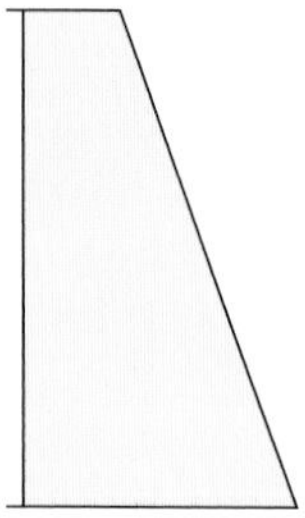

(a) cutting ramped bricks

(b) cutting brick on edge

(c) cutting mitres

(d) cutting arch bricks

Question 5 What is the item of equipment shown used for?

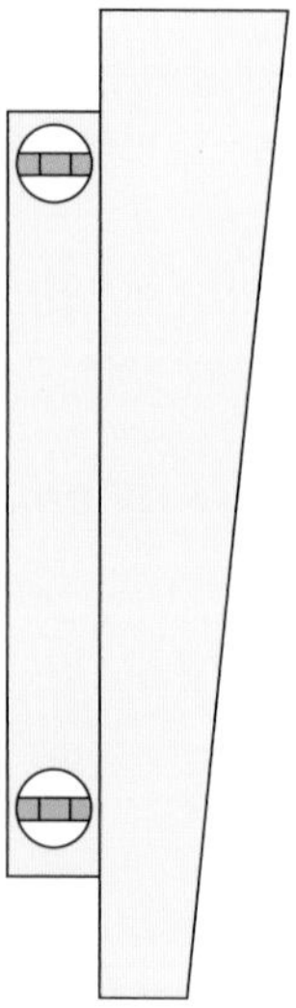

(a) to form angled brickwork

(b) to form battered brickwork

(c) to form straight ramps

(d) to form curved ramps

Question 6 Name the point where the trammel used to form a curved ramp is fixed:

(a) trammel point

(b) radius point

(c) striking point

(d) springing point

Question 7 Identify the two stainless steel items shown:

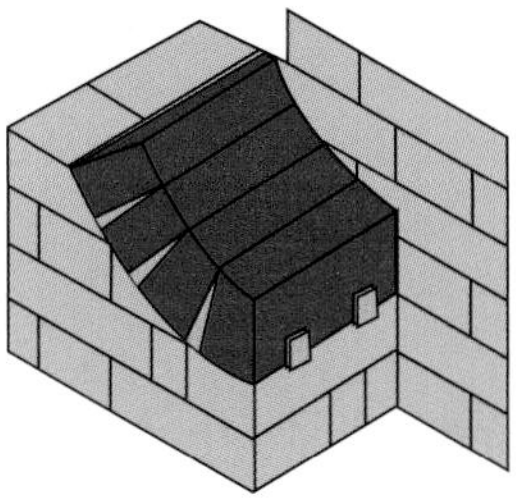

(a) fishtailed cramps

(b) fishtailed wall ties

(c) angled cramps

(d) angled ties

Question 8 Identify the method shown of reducing walls:

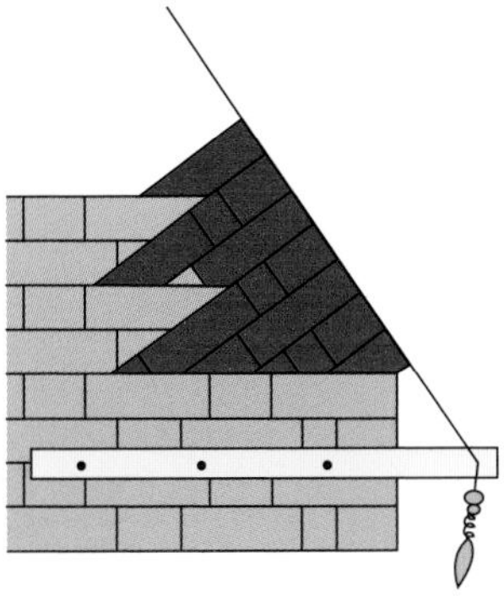

(a) tumbling in

(b) battering

(c) plinths

(d) battering

CHAPTER 11

Splayed Brickwork

This chapter will cover the following NVQ and Diploma units:

- NVQ VR49
- CC 3029

This chapter is about:

- Interpreting building information
- Adopting safe and healthy working practices
- Selecting materials, components and equipment
- Preparing and erecting brickwork and blockwork walls splayed on plan

The following NVQ performance criteria will be covered:

- Interpretation of information
- Safe work practices
- Selection of resources
- Preparing and erecting walls splayed on plan

The following Diploma outcomes will be covered:

- Know how to set out obtuse and acute angled quoins
- Know how to build obtuse and acute angled quoins

Building information

This chapter will deal with the construction of splayed brickwork. Splayed brickwork can be found when obtuse (more than 90°) or acute (less than 90°) quoins need to be built

Information on the type can be gathered from drawings, details and specifications.

Information from drawings

The student should be able to extract information required for setting out from drawings. Drawings have been dealt with in Chapter 3. Please refer back if necessary.

Information from specifications

The specification is a precise description of all the essential information and job requirements that will affect the price of the work but cannot be shown on the drawings.

Safe work practices

As stated before, there are several hidden dangers when working on the building site, especially when working at height. When building splayed angles there will be a great deal of bricks to be cut so special care and attention should be taken to protect the eyes. It is essential to be aware and to ensure that correct personal protective equipment is worn, especially safety boots and goggles.

It is vital that the highest standard of workmanship and the correct interpretation of the current Building Regulations are maintained throughout the construction of all types of splayed walls.

Selection of resources

Various items of equipment will be required when setting out splayed walls, such as tapes, templates and supporting timber.

Along with the normal bricks and mortar there may be a need for special bricks.

Setting out splayed brickwork

Although most buildings are designed with right-angled quoins, occasionally you will be required to set out walls with angles either greater than or less than 90 degrees (90°).

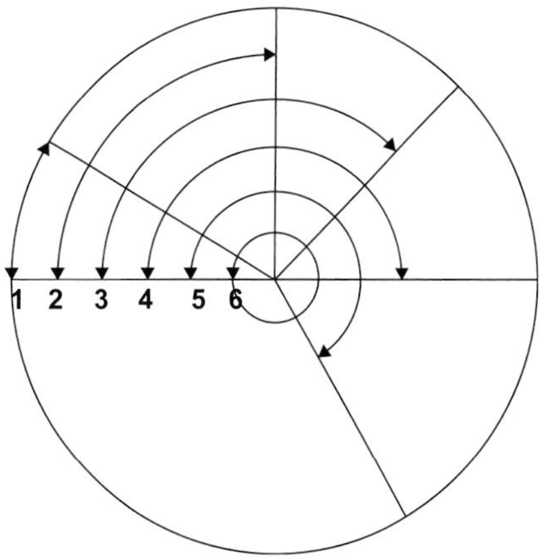

FIGURE 11.1
Angles by bisection: 1, acute angle; 2, right angle; 3, obtuse angle; 4, straight angle; 5, reflex angle; 6, revolution

Angles less than 90° are known as acute angles, and those greater than 90° as obtuse angles.

Geometry

Angles can be produced by bisection.

A complete revolution consists of 360°. The division of the whole revolution into a series of separate angles is shown in Figure 11.1.

Most of the angles required on a building site can be set out geometrically by the bisection of others.

The following details show the various angles being set out in drawing office practice.

It is important to set out a right angle first (Figure 11.2).

The radius of a circle is 1/6 of the circumference. If an arc is drawn on the curve (Figure 11.3) this will produce a 60° angle.

Once a 60° angle has been produced this can be bisected to produce a 30° angle (Figure 11.4).

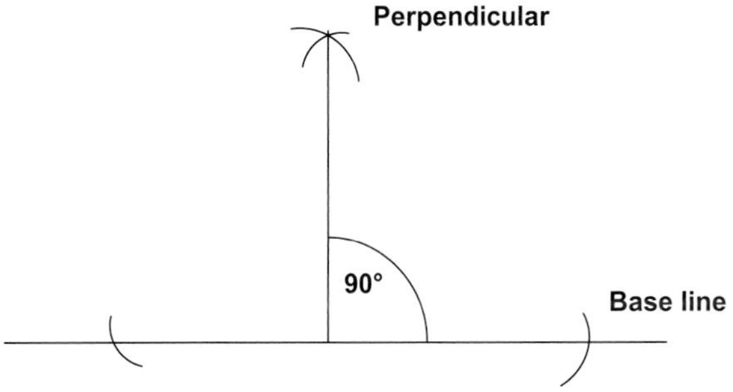

FIGURE 11.2
Drawing a right angle (90°)

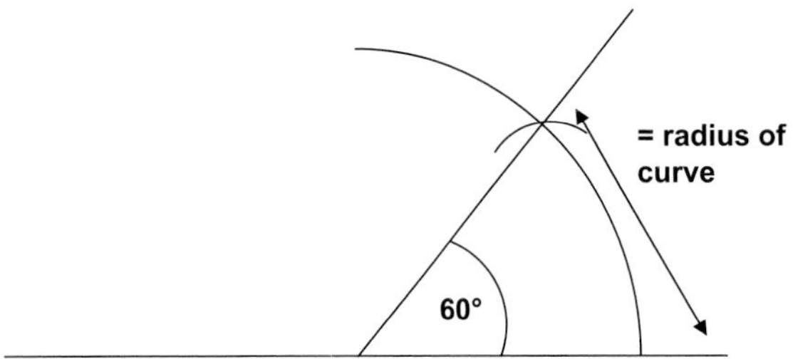

FIGURE 11.3
Drawing a 60° angle

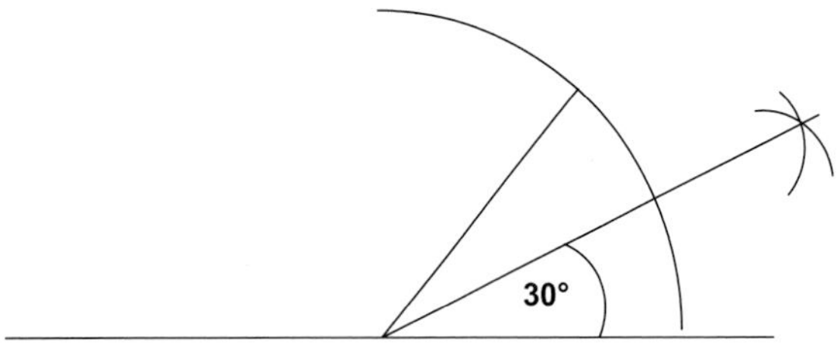

FIGURE 11.4
Drawing a 30° angle

A 45° angle can be drawn by bisecting a 90° angle (Figure 11.5).

RIGHT ANGLES

When setting out the above angles on site again it is essential to start with a right angle.

There are three methods of setting out a right angle which the trainee should know at this stage of training:

- builder's square
- 3:4:5 method
- using a tape.

Use the method with which you are most comfortable.

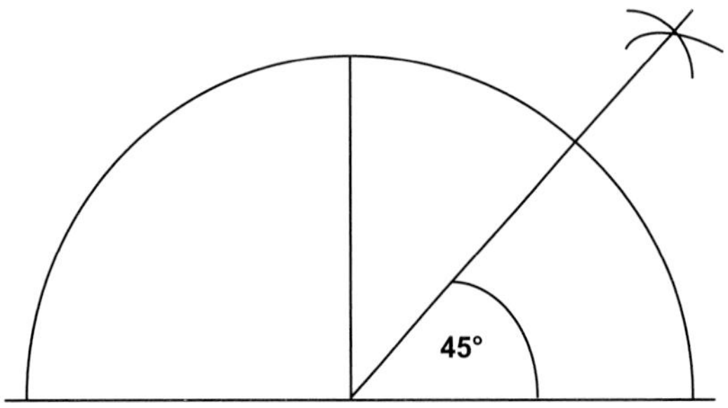

FIGURE 11.5
Drawing a 45° angle

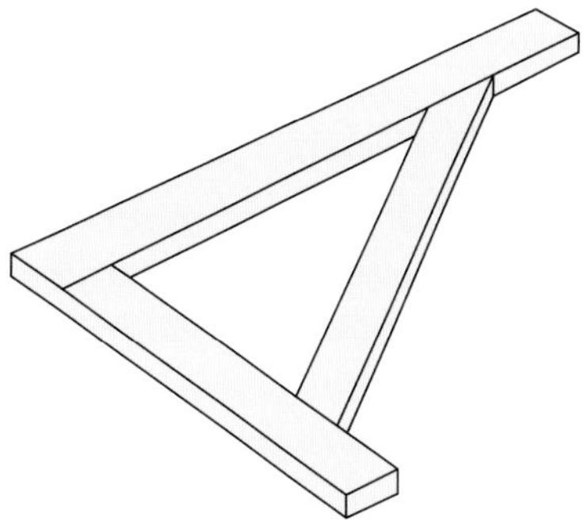

FIGURE 11.6
Builder's square

Builder's square

Most buildings have square corners; that is, corners set at 90° or a right angle. The builder's square (Figure 11.6) is the most commonly used method of setting out a right-angled corner.

The builder's square can be constructed of 75 × 25 mm timbers, half-jointed at the 90° angle with a diagonal brace, tenoned or dovetailed into side lengths.

The builder's square is laid to the previously fixed front line and the end wall line is placed to the square to produce a right angle (Figure 11.7).

The 3:4:5 method

A right angle can also be set out using the 3:4:5 method (Figure 11.8).

A peg should be fixed exactly 3 m from the corner peg on the fixed frontage line.

A tape is then hooked to the nail on the corner peg and another tape hooked to the peg on the front line.

Stretch both tapes towards the side wall line. A third peg should be fixed at the point where 4 m on one tape and 5 m on the other cross.

This will establish the end line at 90° to the front line.

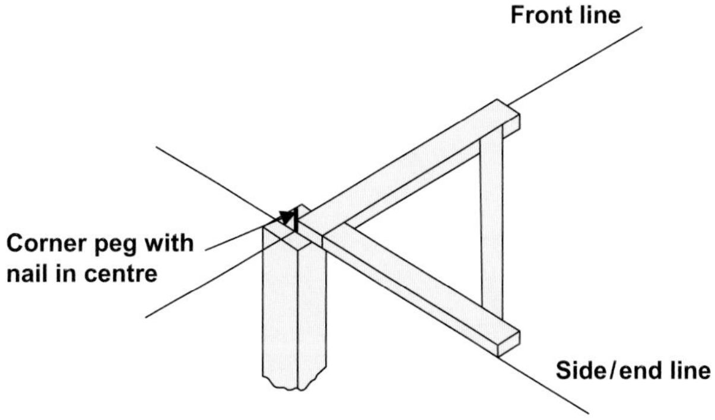

FIGURE 11.7
Setting out a right angle with a builder's square

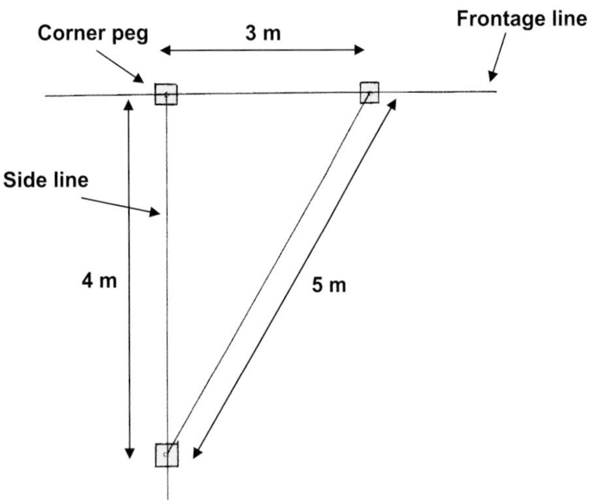

FIGURE 11.8
Setting out a right angle with the 3:4:5 method

Unless a builder's square with sides at least 2 m long is available, the 3:4:5 method is the most accurate method of setting out right angles on the building site.

Using a tape

A peg should be driven into the ground where the right angle is required. Using the tape, measure out an equal distance either side of the main peg A, to give pegs B and C (Figure 11.9).

Use two tapes as shown in Figure 11.9 and set with a measurement equal to or slightly greater than the distance between peg A and peg B.

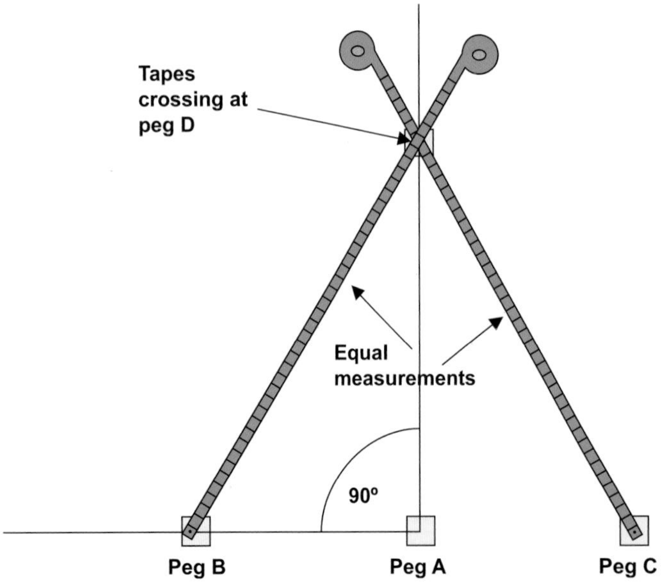

FIGURE 11.9
Setting out a right angle with a tape

Place the hooks of the tapes onto the nails on pegs A and B. The point where they come together marks the position of peg D.

Once a right angle has been completed it can usually be adapted to form most other angles.

Exactly as on the drawing board, the right angle can be bisected to form a 45° angle. If the 45° angle was again bisected it would give a 22.5° angle, and so on.

An angle of 60° can be set out in a similar way to setting out a right angle, as in Figure 11.9.

If the distance between pegs B and C were used to set the tape to mark out peg D, the resulting angle would be 60°.

To construct an angle greater than 90°, first produce a right angle, then add the extra angle to it. For example, a 135° angle would be produced from a 90° angle and a 45° angle (Figure 11.10).

Bonding splayed brickwork

Acute angles

Note the method of planning quoin brick to obtain quarter bond along a wall. Owing to its small size, a closer is omitted from this corner; it would be difficult to cut and awkward when laying (Figure 11.11).

When using standard size bricks to bond an acute quoin, the header face is greater than half a brick and must be reduced or it will affect the quarter bonding pattern.

An alternative method of arrangement of bonding at the acute corner to avoid the sharp corner is to produce a 'birdsmouth'. With this arrangement the bricks are set back from the acute corner as shown in Figure 11.12.

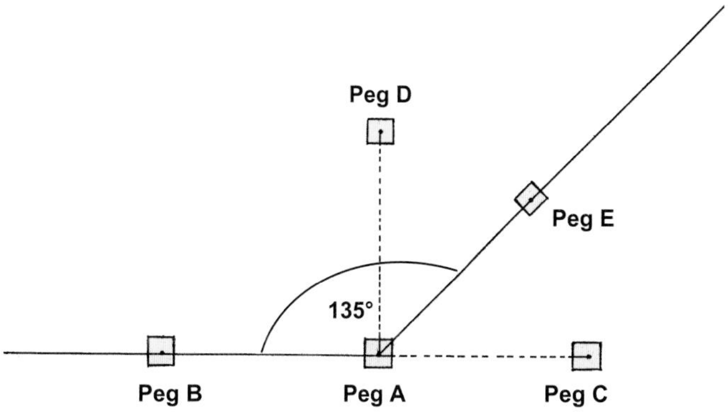

FIGURE 11.10
Setting out an obtuse angle

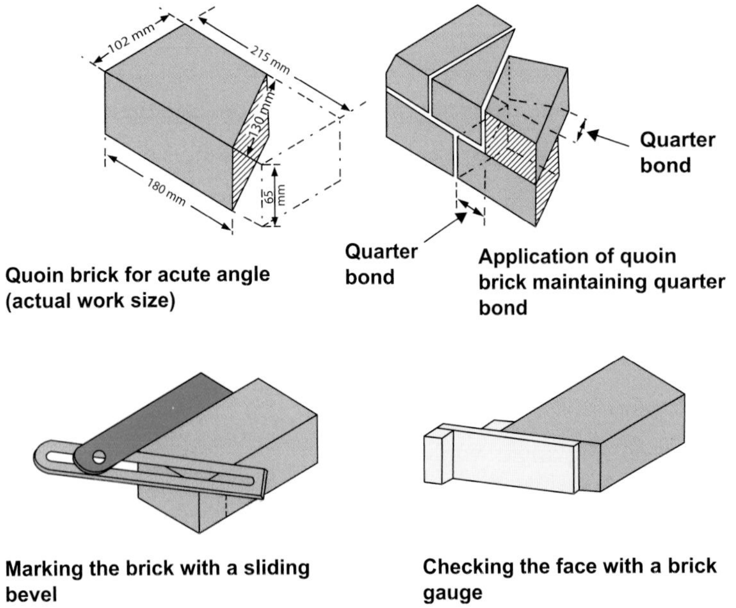

FIGURE 11.11
Acute quoin bricks

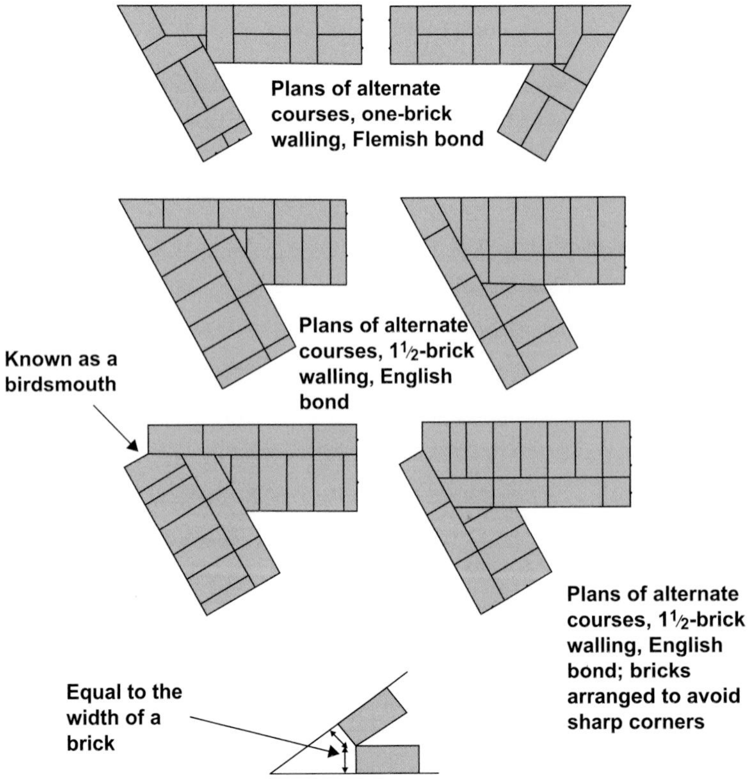

FIGURE 11.12
Bonding arrangements of acute angles

Obtuse angles or squint corners

Note the planning of the quoin brick. Both acute and obtuse angle bricks are planned to conform to standard brick sizes and are most generally used. Others bricks are normally oversized and purpose-made to the architect's specification (Figure 11.13).

Note the alternate courses of brickwork at internal angles, 'A', in the bottom element of Figure 11.14. Careful craftsmanship is needed here to avoid the appearance of a straight joint, by emphasizing the tie bricks to the left and right when jointing up.

Internal and external dogleg bricks are available for building splayed walls that are faced both sides, as in boundary walls (Figure 11.15).

An obtuse corner can be built using dogleg bricks and squint bricks to form the angle (Figure 11.16). Obtuse angles can also be constructed without using special bricks on the external face of the quoin (Figure 11.17).

Building splayed brickwork

Building splayed brickwork requires more skill than building basic straight walling.

It is essential to keep the work plumb and level and to the same gauge as the main wall.

As mentioned before in the bonding of acute and obtuse quoins, it is important to maintain bond at the corners.

Obtuse quoins are most commonly used in building bay windows. Figure 11.18 shows the setting out for a bay window. The template rests on the ground to allow the bay to be set out at foundation level.

Building acute quoins

It is essential that the acute quoin is built plumb (Figure 11.19).

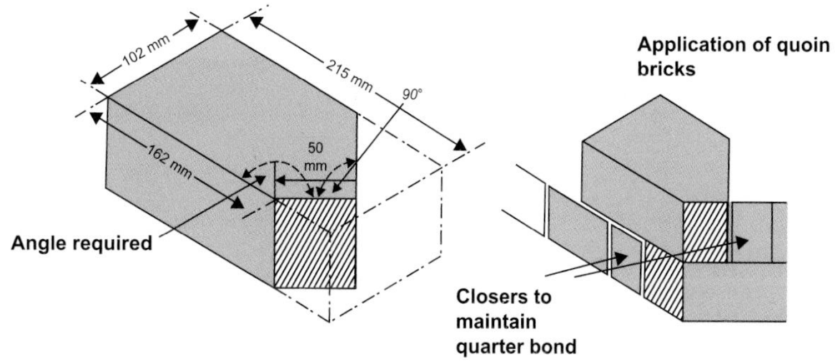

FIGURE 11.13
Quoin brick for squint angle (actual work size)

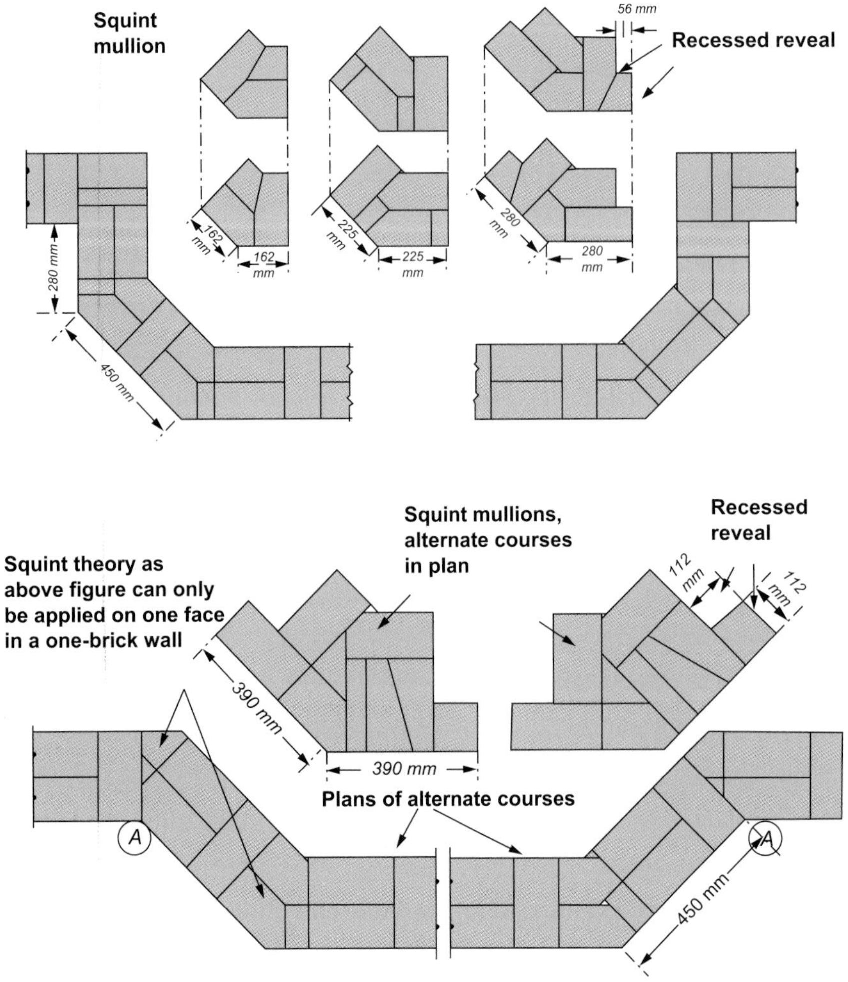

FIGURE 11.14
Obtuse or squint angles

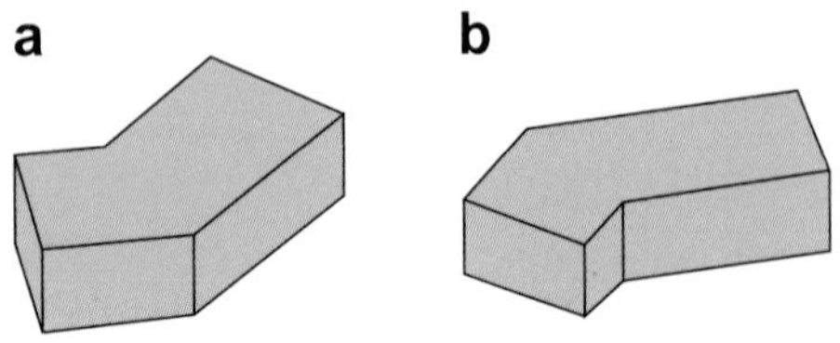

FIGURE 11.15
Dogleg bricks: (a) external dogleg; (b) internal dogleg

When plumbing up the quoin take special care when tapping back the specially cut brick as tapping one face can also move the opposite face owing to the shape of the special brick. A typical acute quoin is shown in Figure 11.20.

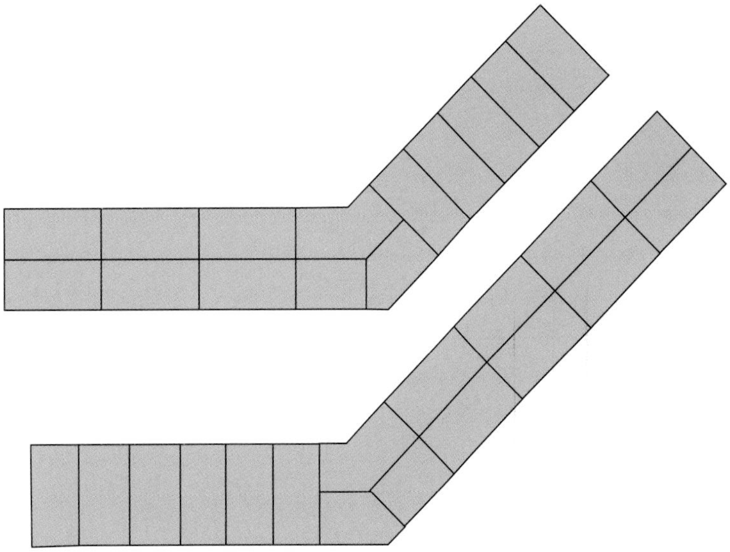

FIGURE 11.16
Obtuse corner using dogleg bricks and squint bricks

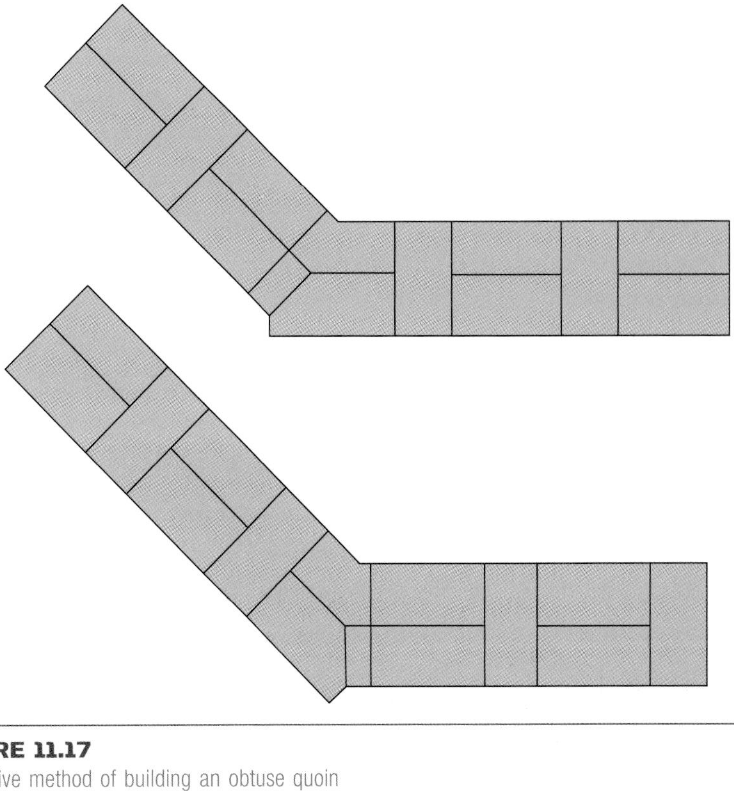

FIGURE 11.17
Alternative method of building an obtuse quoin

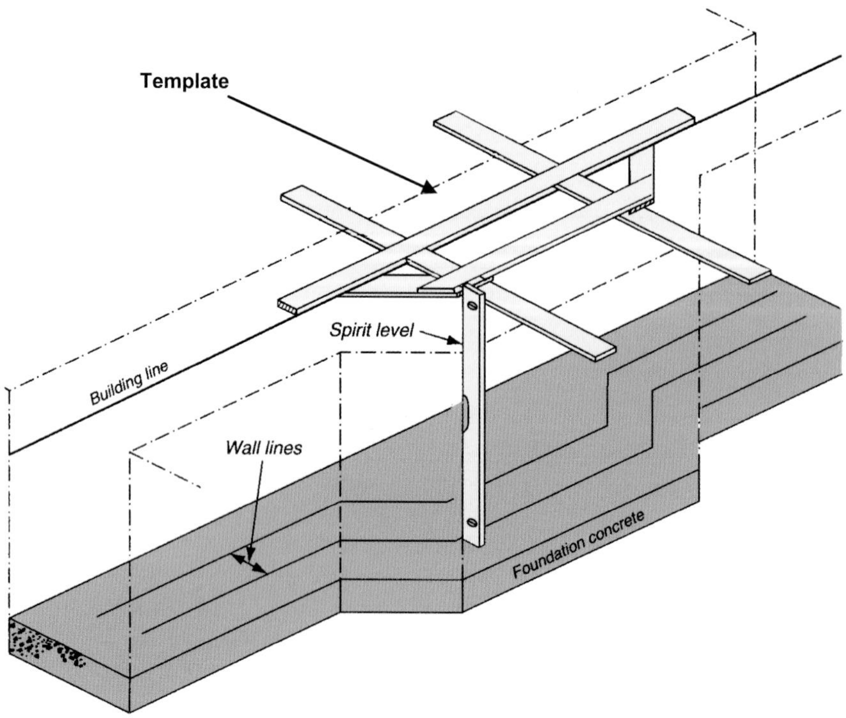

FIGURE 11.18
Preparation for building an obtuse angled bay window

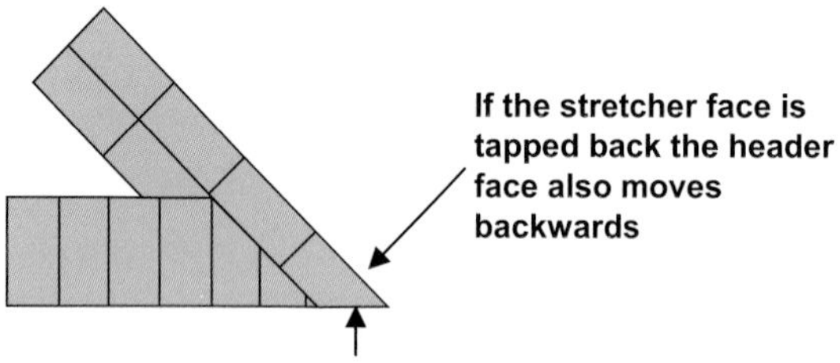

FIGURE 11.19
Plumbing an acute quoin

Building obtuse quoins

Unlike acute quoins, obtuse quoins are built using special squint bricks.

Again, the obtuse quoin must be built plumb.

The same problem as with acute quoins will occur when plumbing up the quoin: take care when tapping back the special quoin brick as tapping one face can also move the opposite face owing to the shape of the special brick.

Squint bricks are available as left or right handed. Both types will be required (Figure 11.21).

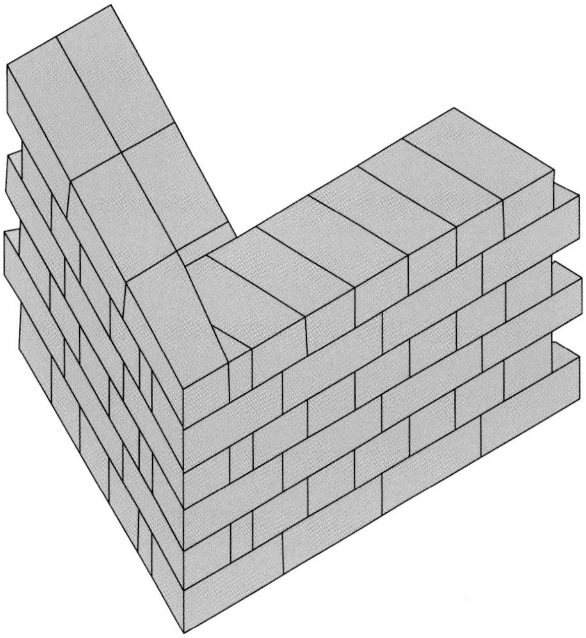

FIGURE 11.20
Typical acute quoin

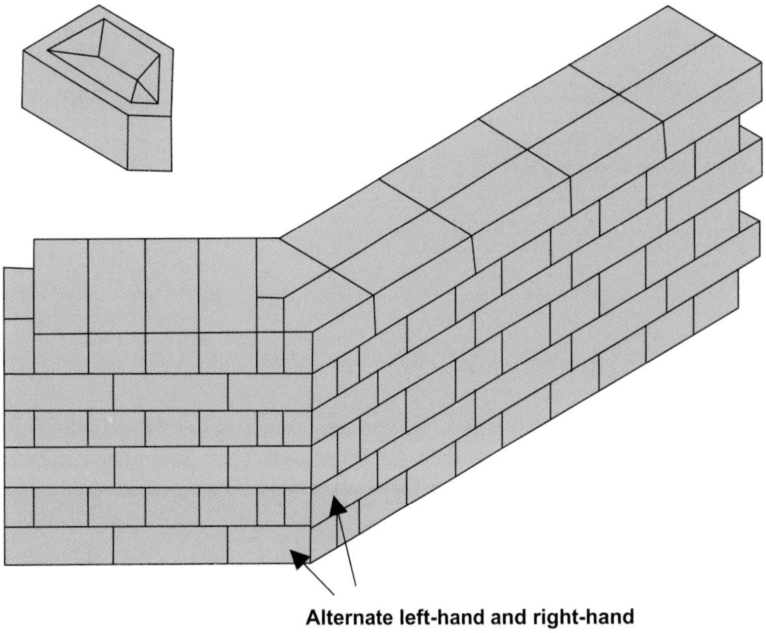

FIGURE 11.21
Typical obtuse quoin

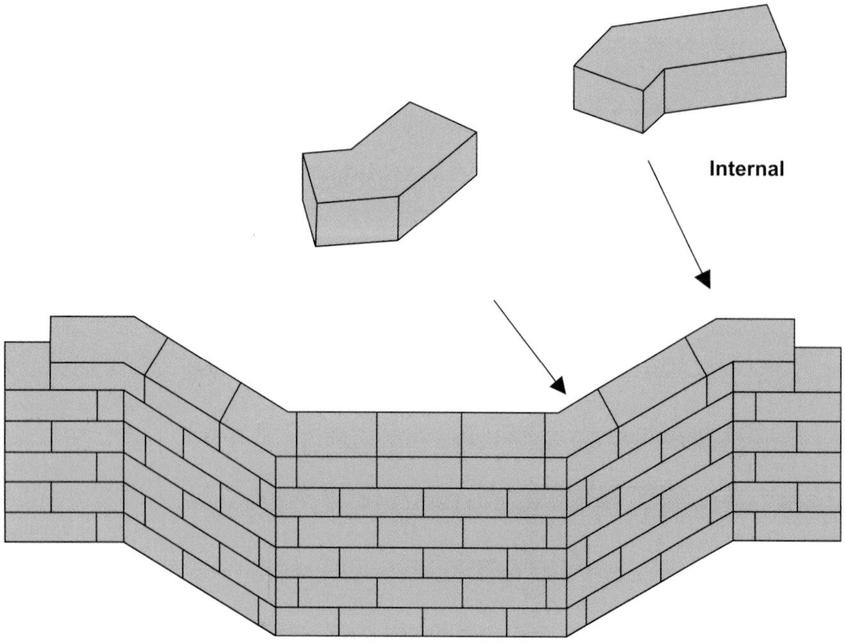

FIGURE 11.22
Constructing a bay window using dogleg bricks

Building bay windows with dogleg bricks

Dogleg bricks are available as left or right handed as well as internal or external. Again, it is essential to maintain plumb at all corners.

The bay shown in Figure 11.22 contains eight external plumbing points.

Multiple-choice questions

Self-assessment

This section of the book is designed to allow you to check your level of knowledge. The section consists of revision questions for this chapter. The questions are all multiple choice and have four possible answers. The answers are to be found at the end of the book.

The main type of multiple-choice question will be the four-option multiple-choice question. This will consist of a question or statement, known as the stem, followed by a choice of four different answers, called the responses. Only one of these responses is the correct answer; the others are incorrect and are known as distracters.

You should attempt to answer the questions by choosing either (a), (b), (c) or (d).

Example

The person employed by the local authority to ensure that the Building Regulations are observed is called the:

(a) clerk of works
(b) building control officer
(c) council inspector
(d) safety officer

The correct answer is the building control officer, and therefore (b) would be the correct response.

Splayed brickwork

Question 1 Identify the following special used for building splayed quoins:

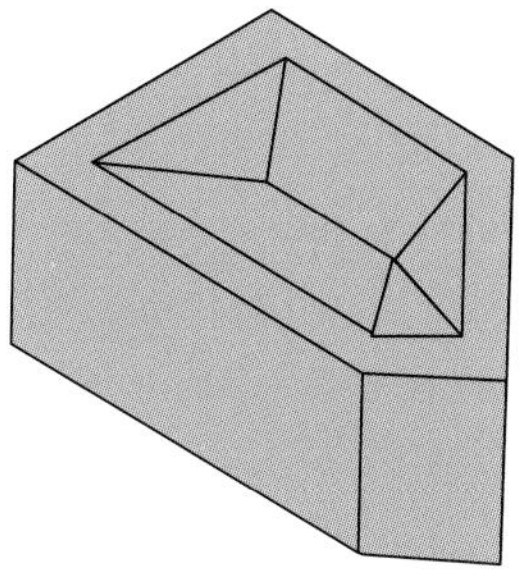

(a) dogleg brick
(b) squint brick
(c) cant brick
(d) acute brick

Question 2 Identify the following special used for building splayed quoins:

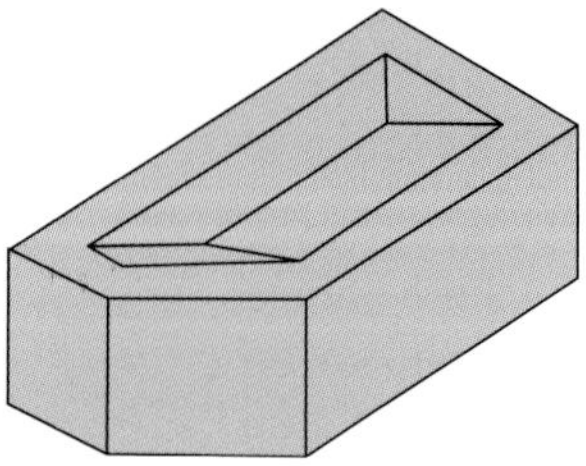

(a) dogleg brick
(b) squint brick
(c) cant brick
(d) obtuse brick

Question 3 Quoins that have an angle greater than 90° are known as:
(a) obtuse angle
(b) straight angle
(c) reflex angle
(d) acute angle

Question 4 Quoins that have an angle less than 90° are known as:
(a) obtuse angle
(b) straight angle
(c) reflex angle
(d) acute angle

Question 5 Identify the following special used for building splayed quoins:

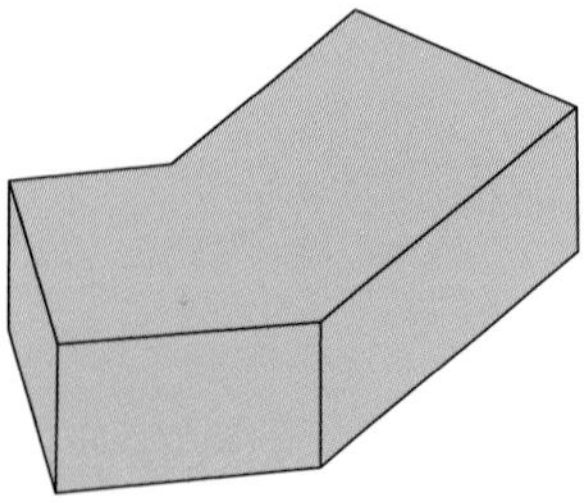

(a) squint brick
(b) dogleg brick
(c) cant brick
(d) obtuse brick

Question 6 Which of the following activities would require squint bricks?

(a) building right-angled quoins

(b) building an segmental arch

(c) building an angled bay window

(d) building an acute quoin

Question 7 An angled bay window can be set out using:

(a) a template

(b) lines and pins

(c) boning rods

(d) a trammel

Question 8 Identify the angle shown:

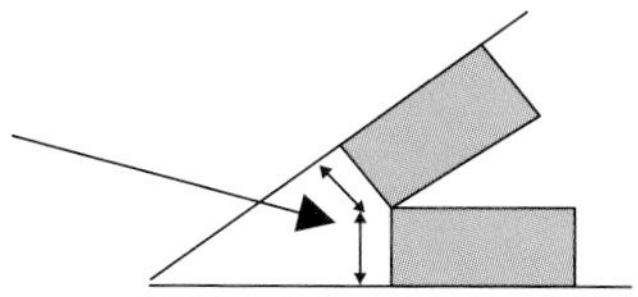

(a) acute angle

(b) obtuse angle

(c) bull's eye

(d) birdsmouth

CHAPTER 12

Reinforcement

This chapter will cover the following NVQ and Diploma units:

- NVQ VR49
- CC 3029

This chapter is about:

- Interpreting building information
- Adopting safe and healthy working practices
- Selecting materials, components and equipment
- Preparing and erecting reinforced brickwork

The following NVQ performance criteria will be covered:

- Interpretation of information
- Safe work practices
- Selection of resources
- Preparing and erecting reinforced walls

The following Diploma outcomes will be covered:

- Know how to set out reinforced walling
- Know how to build reinforced walling

Building information

This chapter will deal with the construction of reinforced brickwork. Reinforced brickwork can be found in constructions designed for extra strength.

Information on the type can be gathered from drawings, details and specifications.

Information from drawings

The student should be able to extract information required for setting out from drawings. Drawings have been dealt with in Chapter 3. Please refer back if necessary.

Information from specifications

The specification is a precise description of all the essential information and job requirements that will affect the price of the work but cannot be shown on the drawings.

Safe work practices

As stated before, there are several hidden dangers when working on the building site, especially when working at height. Reinforced walling will involve the fixing of steel and laying concrete. It is therefore essential to be aware and to ensure that correct personal protective equipment is worn, especially safety boots and goggles.

It is vital that the highest standard of workmanship and the correct interpretation of the current Building Regulations are maintained throughout the construction of all types of reinforced walling.

Selection of resources

Various items of equipment will be required when setting out reinforced walls, such as bricks, reinforcement and concrete.

Along with the normal bricks and mortar there may be a need for special quality bricks.

Bricks

The bricks recommended for reinforced walls are engineering quality bricks (Figure 12.1). These are exceptionally hard, dense bricks which have a low porosity and therefore absorb very little water.

Engineering bricks are intended for walls that are heavily loaded or very exposed to a risk of frost damage.

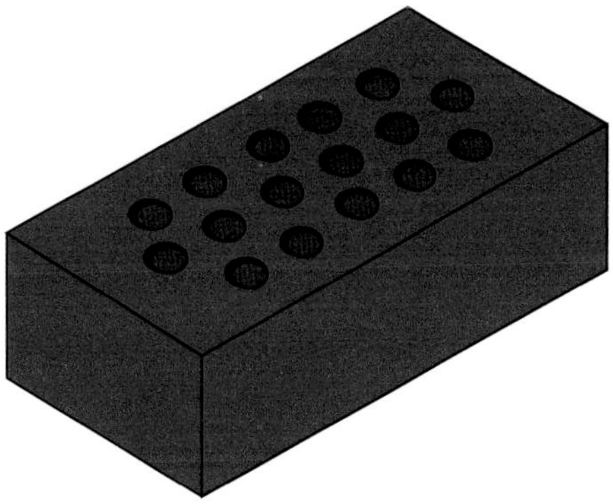

FIGURE 12.1
Engineering brick

Reinforcement

There are various types of reinforcement available depending on the design of the wall (Figure 12.2).

Hoop iron was used in many old reinforced walls. The metal reinforcing strips were bedded along the courses of the wall spaced at the centre of a brick and built into the horizontal joints of the brickwork.

The hoop iron was joined together with a welt lap on straight lengths and a hooked joint at corners. The hoop iron was well galvanized or tarred and sanded to prevent corrosion.

A modern method involves building in expanded metals. These are available in coils in various widths. The metal is pressed out in diamond-shaped

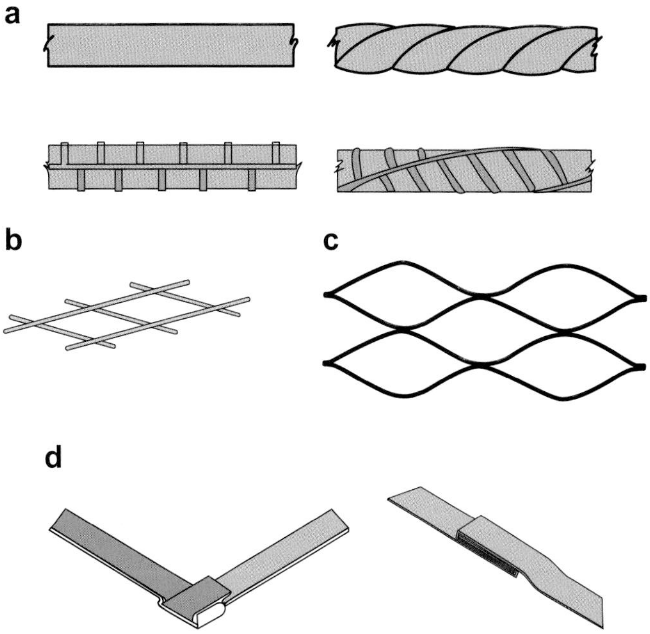

FIGURE 12.2
Types of reinforcement: (a) steel bars; (b) mesh; (c) expanded metal; (d) hoop iron

meshes of various gauges. The diamond meshes, when horizontally laid on the mortar, become embedded and give an excellent bond. This bond it transforms the mortar into compression, making the wall stronger.

Welded mesh is also available for horizontal reinforcement. It usually consists of two 3 mm rods with connecting rods welded at set intervals along its length. This provides similar strength to expanded metals.

Steel rods are also available, mainly for vertical reinforcement. These are built into purpose-made voids in the wall and concreted in. This produces a very strong design.

Reinforced walling

Buildings designed with normal brickwork are strong enough to withstand the loads placed upon them.

Brickwork will carry all loads that are placed directly downwards onto the walls. These loads tend to crush the wall and place the wall in compression (Figure 12.3).

Brickwork is very strong in compression but weak in tension. If loads are applied to the wall from the side, such as wind loads or pressure from the ground, these loads put the wall in tension.

Where tension is possible the wall has to be designed to resist the tension or it is liable to crack and fall over (Figure 12.4). The placing of the reinforcement will depend on the amount of pressure on the wall and the likely location of tension in the wall.

Bonding

As well as designing a wall with steel reinforcement, special bonds can be used that allow reinforcement to be added to give it increased strength.

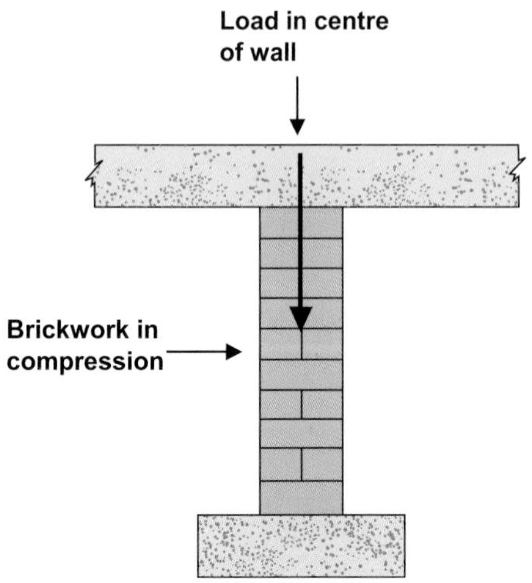

FIGURE 12.3
Wall in compression

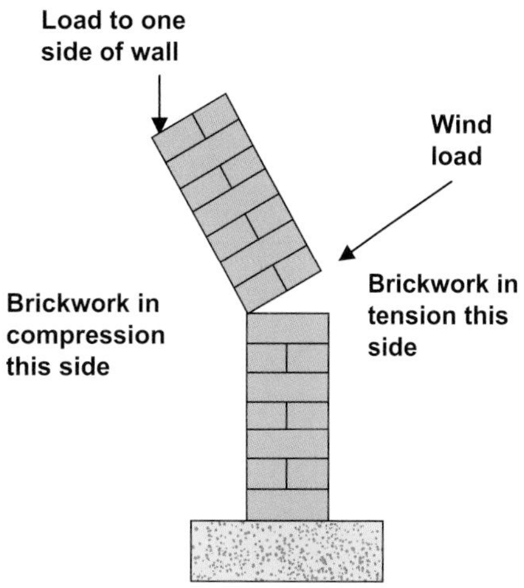

FIGURE 12.4
Wall in tension

Quetta bond

Quetta bond is used on one-and-a-half-brick walls to give extra strength. Quetta bond shows Flemish bond on the face, but it is not sectional like ordinary Flemish bond.

Vertical steel bars are cast into the concrete foundation and the wall is built around the steel bars.

The bonding arrangement leaves quarter-brick pockets where vertical steel can be placed and then the void filled in with concrete. This is shown in detail in Figure 12.5.

Rat trap bond

Rat trap bond is a brick-on-edge bond. It is used mainly to save on materials. The finished appearance is again not unlike Flemish bond on both sides.

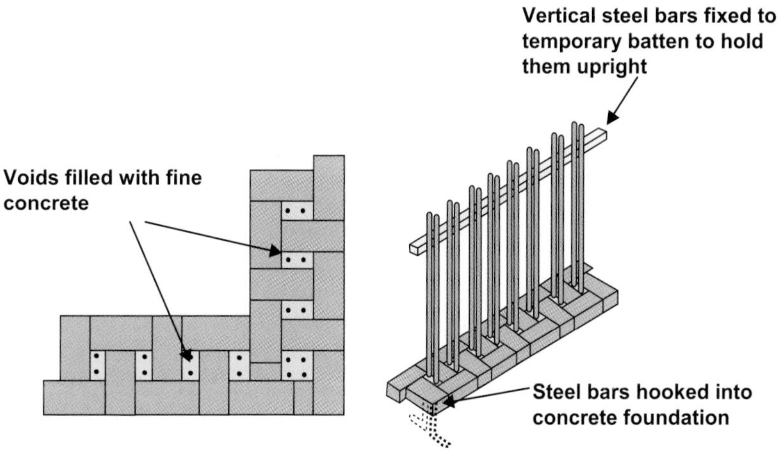

FIGURE 12.5
Quetta bond

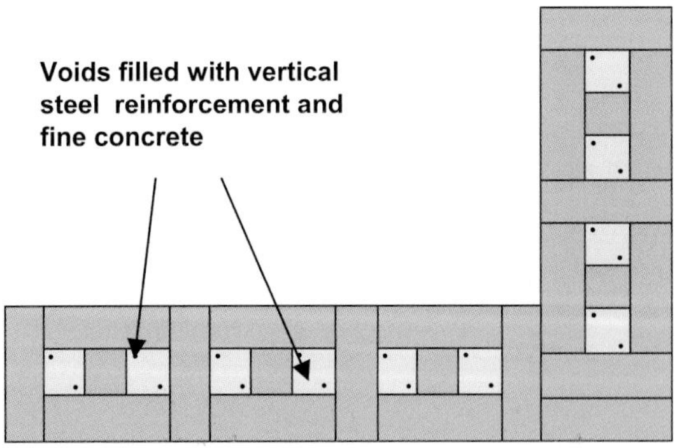

FIGURE 12.6
Rat trap bond

The bonding arrangement, like quetta bond, leaves voids that can be filled with vertical steel reinforcement and fine concrete to provide extra strength for the wall (Figure 12.6).

Hollow blocks are also ideal when extra strength is required. The blocks are laid half-bond so that the hollow part is maintained upwards through the wall. This hollow part can be filled in with steel and concrete (Figure 12.7).

Bed joint reinforced brickwork

Welded mesh

Steel welded mesh reinforcement is laid in the horizontal bed joint to add strength to the brick wall and prevent movement. Depending on the design the reinforcement can be placed on every third or sixth bed joint.

The reinforcement should be placed on a bed of mortar with a minimum of 15 mm from the face of the brickwork (Figure 12.8).

It can also be used to tie junctions together.

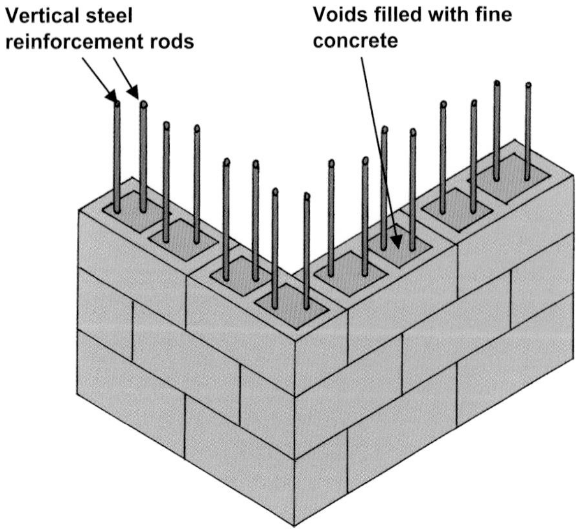

FIGURE 12.7
Reinforced hollow concrete block walling

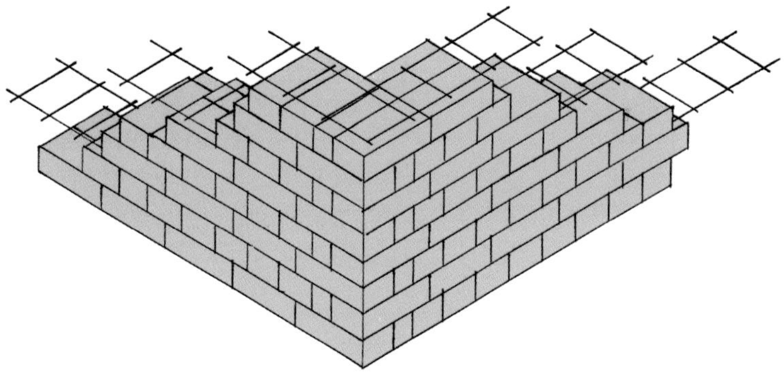

FIGURE 12.8
Welded mesh bed joint reinforcement

Hoop iron

Placing of reinforcement in the wall is important so that the greatest strength can be produced.

When hoop iron is being placed it should be in the centre of each half-brick (Figure 12.9).

Vertical reinforcement

Vertical reinforcement can be used in other areas as well as for plain walling. Gate piers are one such situation, when the pillar is required to resist the pressure applied by the swinging of the gate.

As mentioned in previous chapters on detached and attached piers, gates and ornamental ironwork are often built into piers. It is essential that the boundary wall is thickened at any point where gates are being fixed so that it is strong enough to withstand the extra pressures applied.

Occasionally the pier supporting a gate will be a detached pier, but more often it is attached to the boundary wall. Because of the stresses placed on the piers by the gates the piers have to be extra strong.

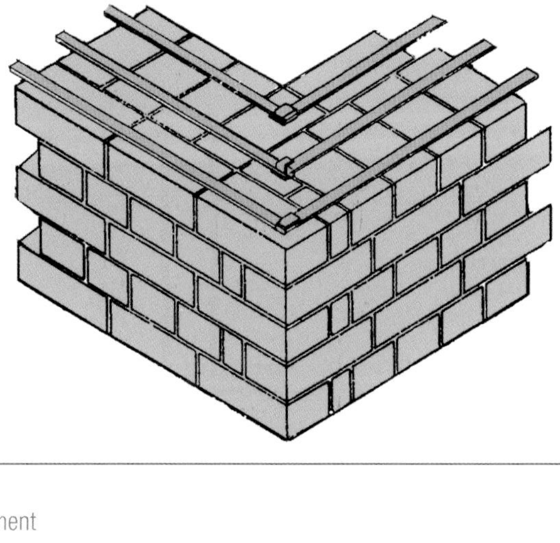

FIGURE 12.9
Hoop iron reinforcement

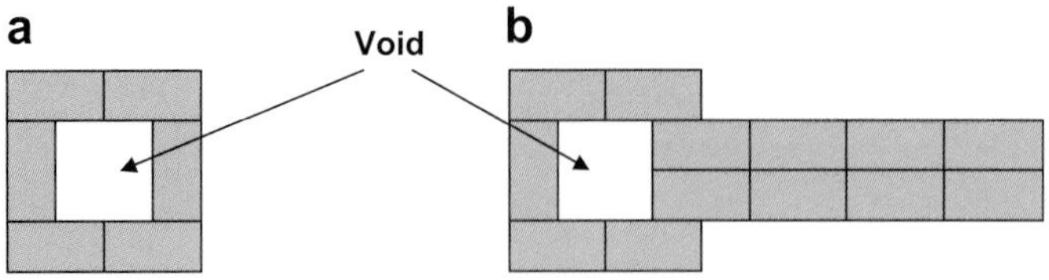

FIGURE 12.10
Hollow gate piers: (a) detached pier; (b) attached pier

One method of achieving this is to construct the piers hollow and infill with concrete (Figure 12.10). The concrete can also be reinforced and even attached to the foundation for extra resistance to side movement. Alternatively, the piers may be built of solid brickwork (Figure 12.11).

Hollow piers

Detached and attached piers can be strengthened by the introduction of reinforced concrete into the void (Figure 12.12).

The brick pier is constructed as a hollow pier, allowing reinforced concrete to be placed into the pier to provide additional strength.

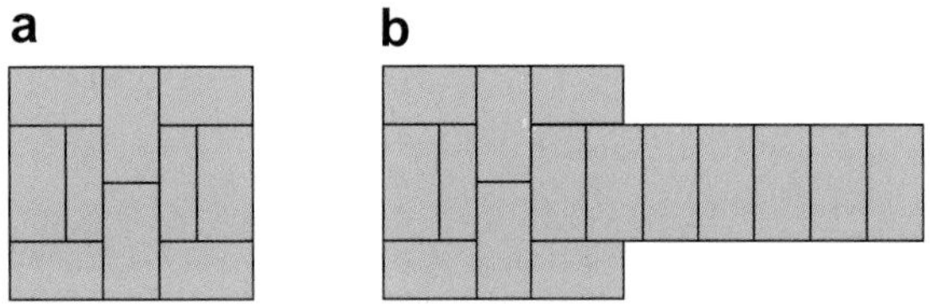

FIGURE 12.11
Solid gate piers: (a) detached pier; (b) attached pier

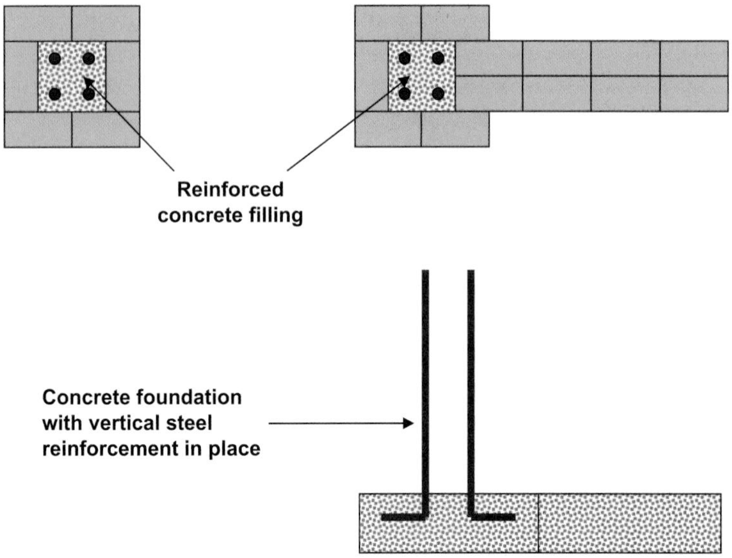

FIGURE 12.12
Reinforcing hollow gate piers

It is important to allow the brick pier to set hard before pouring the concrete, or the pier may be damaged. The concrete will require compacting into the pier to achieve maximum strength.

Steel reinforcement should be placed into the concrete foundation when it is being laid. The pier will then have to built around it.

It is essential to ensure that the void in the pier is kept free from mortar droppings, which would affect the finished strength of the pier.

Fixing for gates

When piers are being built to receive gates it is essential to plan beforehand. Gates can be fixed after the brickwork has hardened or built in as the work proceeds (Figure 12.13).

Fixings for gates can be fitted after the piers have been built by either:

- building in sand courses
- drilling and fixing hinges.

To fix the brackets after the brickwork has set, the sand courses are removed and all the sand is brushed out to leave a clean recess. The recess should be dampened to ensure adhesion of the mortar.

The gates could be erected in position to identify the correct position of the brackets. These can then be fitted and the brickwork replaced as solidly as possible to ensure no movement when the gate swings.

FIXING AS THE WORK PROCEEDS

When gates are being fixed as the work proceeds they can be erected in position and supported in their final position by timbers.

The brackets can then be built in as the work proceeds and could also be allowed to pass into the void so they would be further attached to the concrete infill.

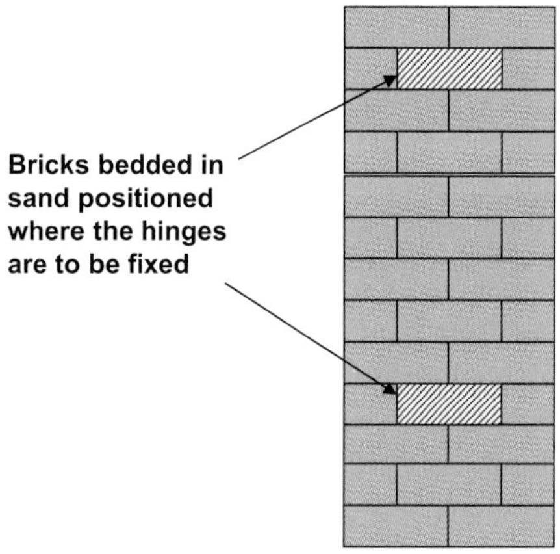

FIGURE 12.13
Building in sand courses

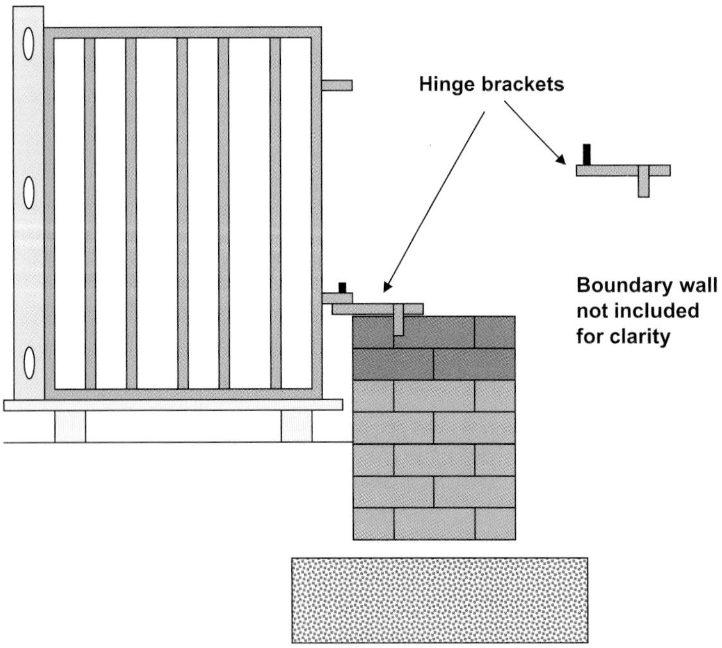

FIGURE 12.14
Fixing the gates

The piers must be constructed absolutely plumb otherwise the gates would not swing correctly. Once the gate has been placed in its final position on timbers it can be plumbed and supported with raking struts (Figure 12.14).

Normally, two brackets are built into the pier.

The brickwork can be built up to the first bracket; this may include the damp-proof course. The bracket is then built in, allowing it to extend into the void, and could be fixed to the vertical reinforcement. The brickwork is then built up to the next bracket, which is fitted in exactly the same way.

Double gates are fixed in a similar way (Figure 12.15). Care should again be taken to set the gates in their required position on battens and support in an upright position with raking struts.

Alternatively, sand courses could be built in and the brackets fixed afterwards.

Brick retaining walls

Any walling intended to hold back soil must be properly designed by a structural engineer, so as to resist overturning from sideways pressure. Even a 215 mm thick wall that is only six courses high and intended to hold back a soil bank may begin to topple after a couple of years (Figure 12.16).

The line AB in Figure 12.16 represents a typical natural angle of repose for soil, say 45 degrees. Motorway embankments are frequently seen left at this natural slope with the soil completely stable with grass and bushes growing on them. The angle of repose, of course, depends on the type and condition of the soil.

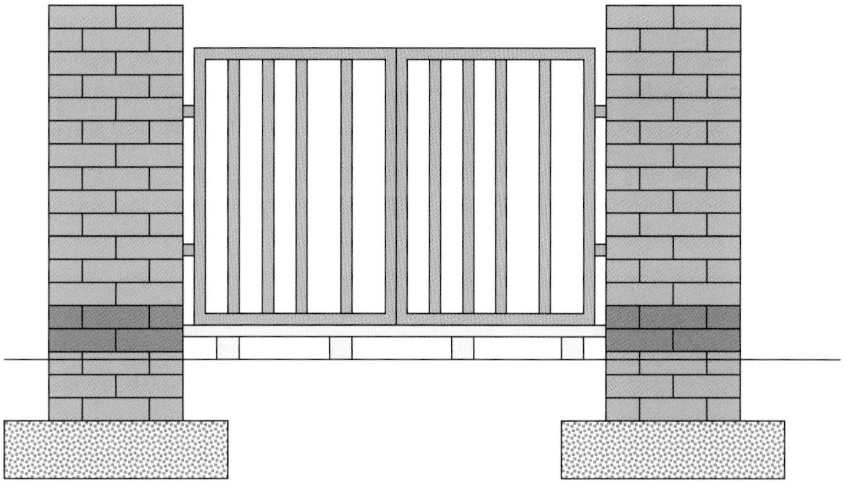

FIGURE 12.15
Fixing double gates

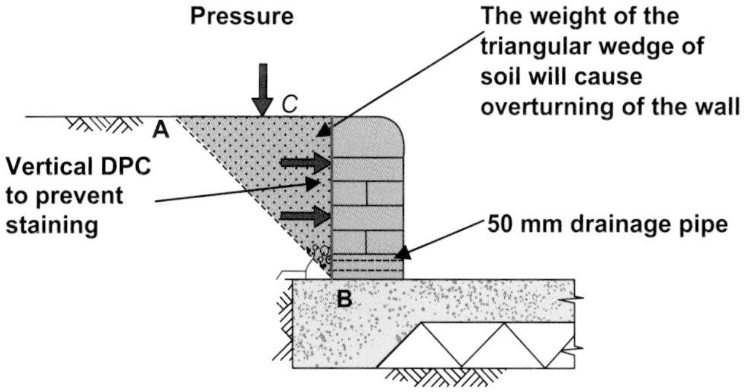

FIGURE 12.16
Cross-section through low retaining wall, showing typical sideways pressure from the soil behind

It is the 'soil wedge' above line AB in Figure 12.16 that presses against the vertical back of a retaining wall, along its full length, which is tending to push over the brickwork. Retaining walls must be able permanently to resist the pressure of this triangular mass of soil and remain plumb.

Any extra load at point C in Figure 12.16, e.g. stacks of materials or vehicles, is called a surcharge, and increases the overturning pressures on the retaining wall.

The retaining wall can be reinforced with vertical steel rods which are anchored into the concrete foundation (Figure 12.17).

Retaining wall construction

Class A or B engineering quality bricks, or other frost-proof bricks, should be specified, using a group (i) or group (ii) mortar, to ensure long-term durability.

The soil side of a brick retaining wall should be covered with a damp-proof membrane (DPM), to prevent through-penetration of ground water, which

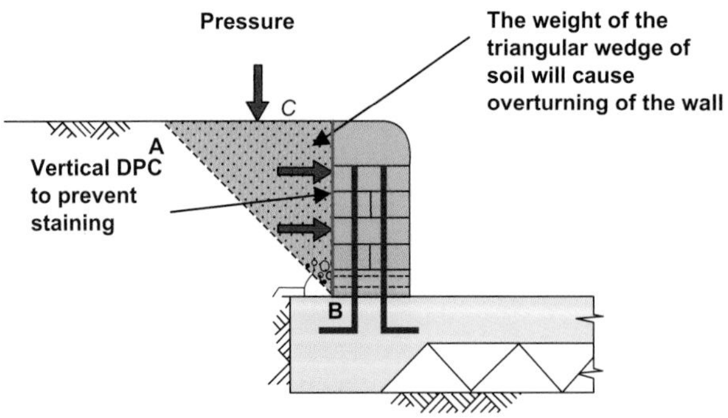

FIGURE 12.17
Reinforcement added to retaining wall for extra strength to resist overturning of the wall

can cause unsightly efflorescence or other salt staining of the outer surface. This DPM can be formed from 900 mm wide rolls of paper-backed self-adhesive bitumen, pressed firmly on to the brickwork on the soil side of the wall.

Alternatively, three separate coats of bituminous emulsion paint may be applied to the back surface of the brickwork to form a vertical DPM.

Multiple-choice questions

Self-assessment

This section of the book is designed to allow you to check your level of knowledge. The section consists of revision questions for this chapter. The questions are all multiple choice and have four possible answers. The answers are to be found at the end of the book.

The main type of multiple-choice question will be the four-option multiple-choice question. This will consist of a question or statement, known as the stem, followed by a choice of four different answers, called the responses. Only one of these responses is the correct answer; the others are incorrect and are known as distracters.

You should attempt to answer the questions by choosing either (a), (b), (c) or (d).

Example

The person employed by the local authority to ensure that the Building Regulations are observed is called the:

(a) clerk of works

(b) building control officer

(c) council inspector

(d) safety officer

The correct answer is the building control officer, and therefore (b) would be the correct response.

Reinforcement

Question 1 Identify the type of reinforcement shown:

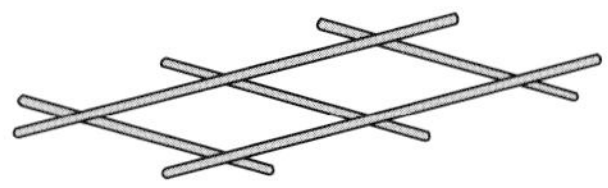

(a) welded mesh

(b) expanded metal

(c) hoop iron

(d) steel fabric

Question 2 Why is steel reinforcement laid in the bed joint mortar?

(a) to save on mortar

(b) to strengthen to mortar

(c) to be able to use cheaper bricks

(d) to strengthen the brickwork

Question 3 What is used to fill in the pockets of quetta bond?

(a) vertical steel rods and insulation

(b) fine concrete

(c) vertical steel rods

(d) vertical steel rods and fine concrete

Question 4 What is the maximum diameter of horizontal bed reinforcement?

(a) 10 mm

(b) 8 mm

(c) 6 mm

(d) 4 mm

Question 5 Identify the type of reinforcement shown:

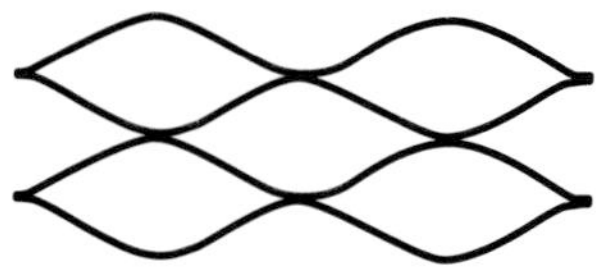

(a) welded mesh

(b) expanded metal

(c) hoop iron

(d) steel fabric

Question 6 What type of brick is best suited for retaining walls?

(a) engineering brick

(b) facing brick

(c) common brick

(d) glazed brick

Question 7 Brickwork is very strong in compression but weak in what?

(a) stress

(b) waterproofing

(c) tension

(d) wind resistance

Question 8 Identify the type of reinforcing bond shown:

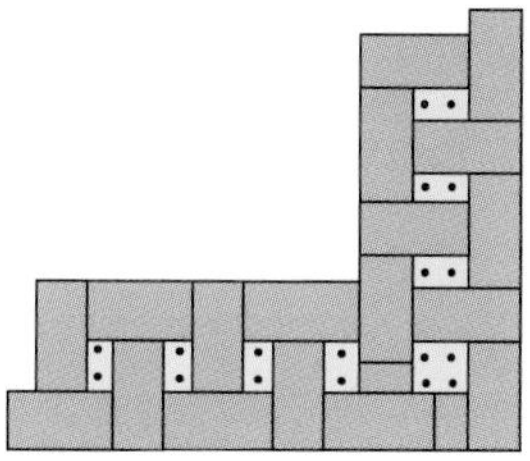

(a) rat trap bond

(b) quetta bond

(c) Flemish garden wall bond

(d) English garden wall bond

CHAPTER 13

Decorative Brickwork

This chapter will cover the following NVQ and Diploma units:

- NVQ VR49
- CC 3029

This chapter is about:

- Interpreting building information
- Adopting safe and healthy working practices
- Selecting materials, components and equipment
- Preparing and erecting decorative features

The following NVQ performance criteria will be covered:

- Interpretation of information
- Safe work practices
- Selection of resources
- Preparing and erecting decorative brickwork

The following Diploma outcomes will be covered:

- Know how to set out brickwork incorporating features
- Know how to build brickwork incorporating features

Building information

This chapter will deal with the construction of decorative brickwork. Decorative brickwork can be found in numerous positions around a building, from floor to roof.

Information on the type can be gathered from drawings, details and specifications.

Information from drawings

The student should be able to extract information required for setting out from drawings. Drawings have been dealt with in Chapter 3. Please refer back if necessary.

Information from specifications

The specification is a precise description of all the essential information and job requirements that will affect the price of the work but cannot be shown on the drawings.

Safe work practices

Decorative brickwork can take place at any point on a building, especially at height. As stated before, there are several hidden dangers when working on the building site, especially when working at height. It is therefore essential to be aware and ensure that correct personal protective equipment is worn, especially safety boots and hard hats.

It is vital that the highest standard of workmanship and the correct interpretation of the current Building Regulations are maintained throughout the construction of all types of decorative brickwork.

Selection of resources

Various special bricks will be required when building decorative brickwork, such as plinths, bullnoses and cants.

Special bricks

The bulk of production by brick manufacturers is of plain, basic metric bricks, with a work size of 215 × 102.5 × 65 mm. These are produced as solid, frogged or perforated, from clay, sand lime or concrete, as described in Chapter 4 of Level 2.

Manufacturers do, however, make a wide range of other brick shapes, and some examples are shown in Figure 13.1.

The brick standard describes these variants as 'bricks of special shapes and sizes'. The majority are based on the dimensions of the basic metric brick

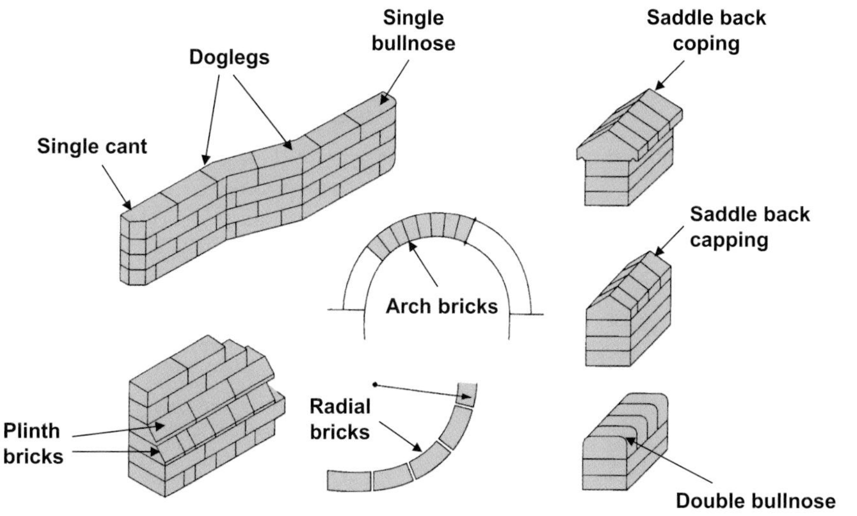

FIGURE 13.1
Uses of special bricks

(Figure 13.2), so they will fit in with normal bonding arrangements and vertical gauge. These are given a type number and prefix number to make ordering a simple matter.

Bricklayers know the most commonly used ones as 'standard specials', intended for the decorative and functional purposes required of brickwork (Figure 13.3). They usually refer to all other shapes that might be specified by an architect as 'special specials'. These require detailed drawings to be sent to the brick manufacturer, so that a mould can be made.

AVAILABILITY OF SPECIAL SHAPED BRICKS

Although a brick of special shape might be referred to as a 'standard special', it will not necessarily be held in stock by the brick manufacturer or builders' merchant. Since the hundreds of different colours and surface textures of facings each have their own respective range of special shapes, stocking them all would be a very expensive business.

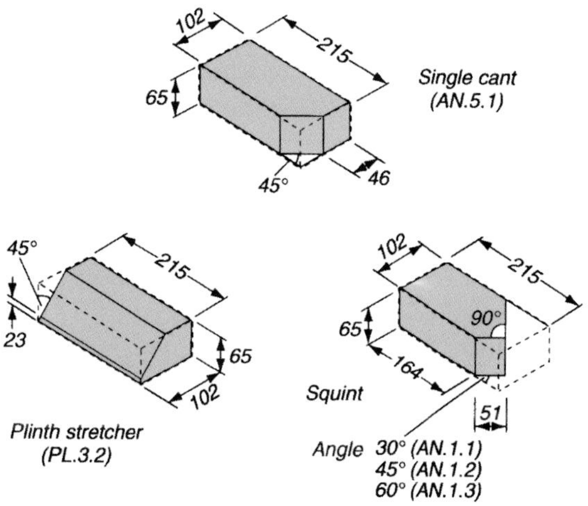

FIGURE 13.2
Dimensions of typical special bricks

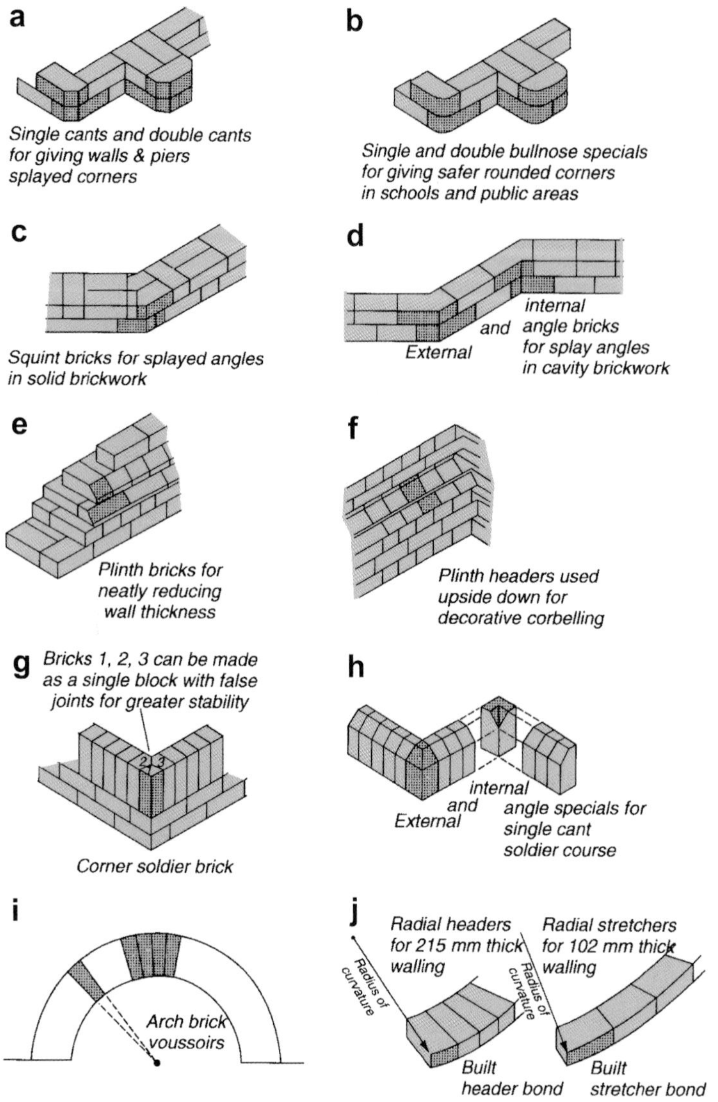

FIGURE 13.3
Application of special bricks

Special bricks are divided into 10 groups for convenience of classification. Each group has recognizable prefix letters to make the system user-friendly. Table 13.1 shows a selection from the total range available.

Stop bricks

Certain special bricks are shaped so that, for example, a bullnose effect can be changed neatly back to a square corner. Examples of different kinds of stop bricks are shown in use in Figures 13.4 and 13.5.

Bonding with obtuse angle specials

SQUINT BRICKS (AN.1)

Traditionally, squint bricks are produced specifically for bonding obtuse angle quoins in solid brick walling (Figure 13.6). The dimensions of each

Table 13.1 Standard classification of specials

Prefix	Stands for	Examples
BD	Bonding bricks	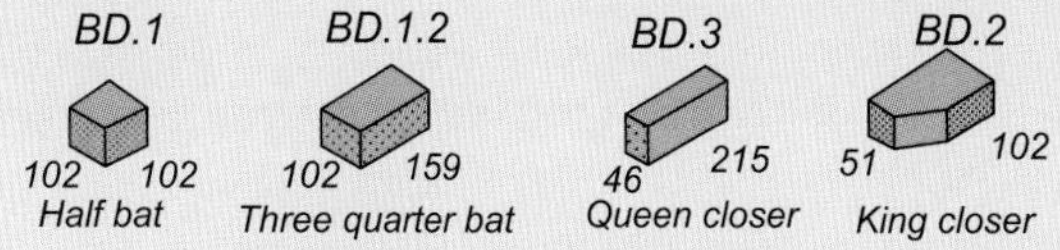 BD.1 102 102 Half bat; BD.1.2 102 159 Three quarter bat; BD.3 46 215 Queen closer; BD.2 51 102 King closer
CP	Copings and cappings	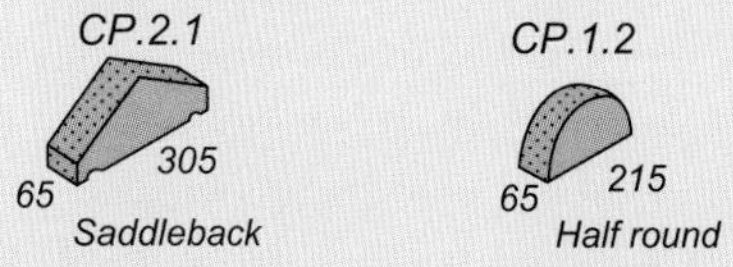 CP.2.1 65 305 Saddleback; CP.1.2 65 215 Half round
BN	Bullnose bricks	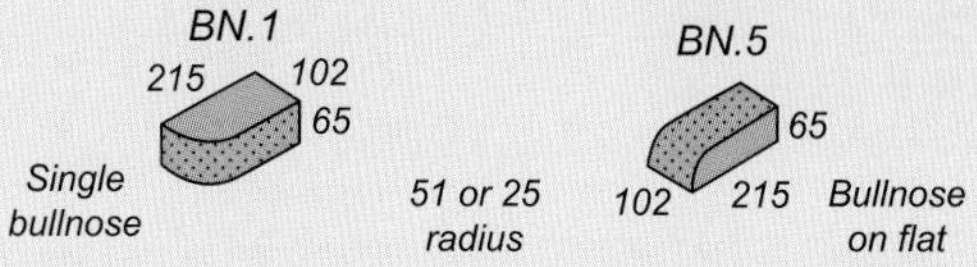 BN.1 215 102 65 Single bullnose; 51 or 25 radius; BN.5 65 102 215 Bullnose on flat
AN	Angle bricks	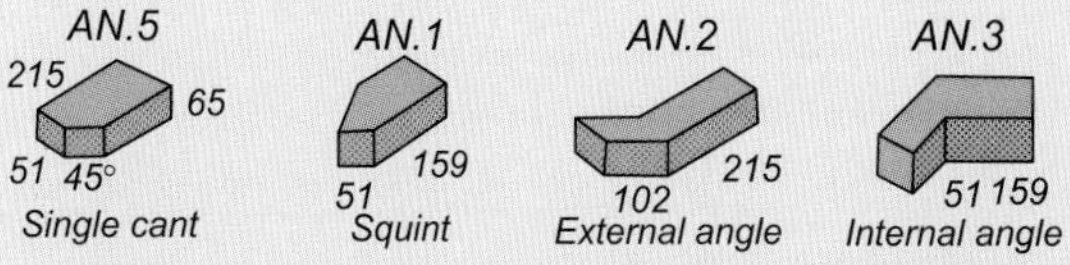 AN.5 215 65 51 45° Single cant; AN.1 159 51 Squint; AN.2 102 215 External angle; AN.3 51 159 Internal angle
PL	Plinth bricks	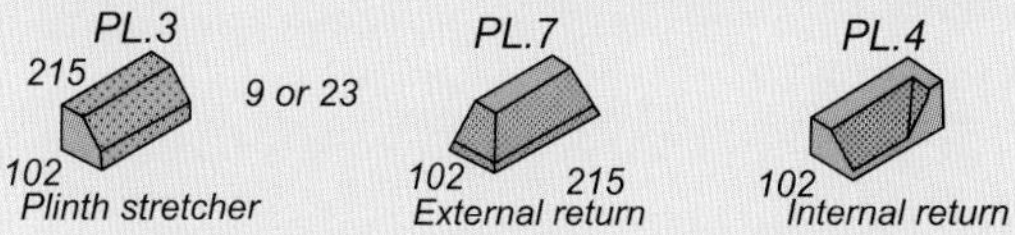 PL.3 215 102 Plinth stretcher; 9 or 23; PL.7 102 215 External return; PL.4 102 Internal return
AR	Arch bricks	 AR.2 75 102; AR.1 75 102
RD	Radial bricks	 RD.1 102 Radial header; RD.2 215 Radial stretcher
SL	Slip bricks	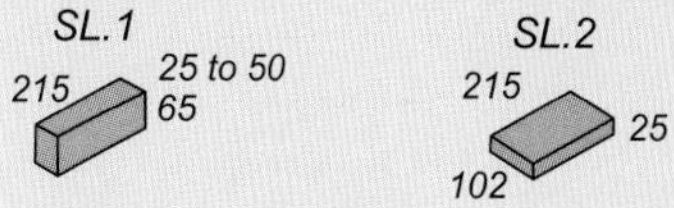 SL.1 215 25 to 50 65; SL.2 215 25 102
SD	Soldier bricks	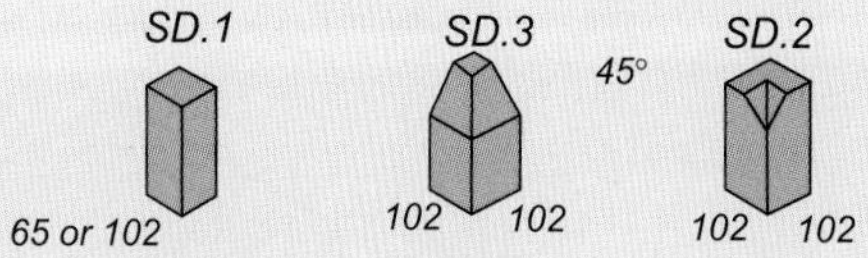 SD.1 65 or 102; SD.3 102 102; 45°; SD.2 102 102
NS	Non-standard bricks	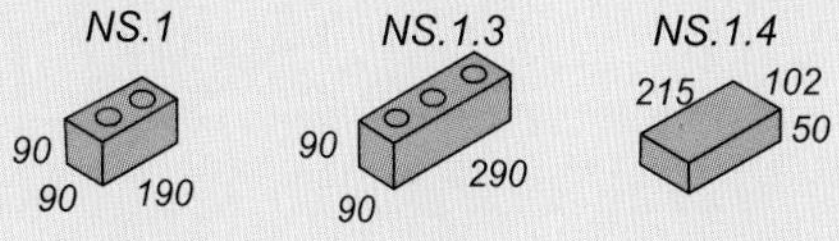 NS.1 90 90 190; NS.1.3 90 290 90; NS.1.4 215 102 50

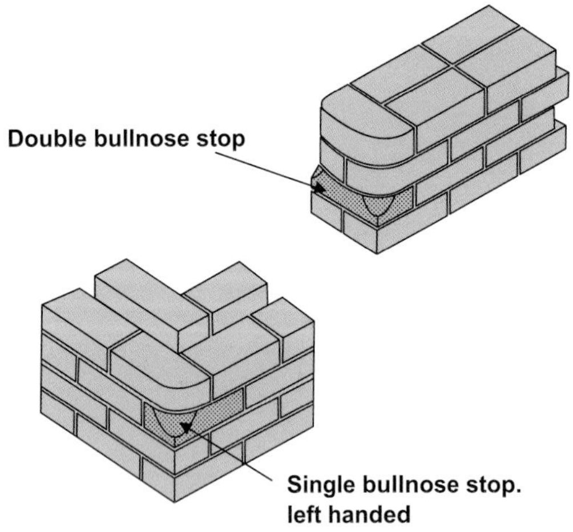

FIGURE 13.4
Uses of special shape bullnose stop bricks

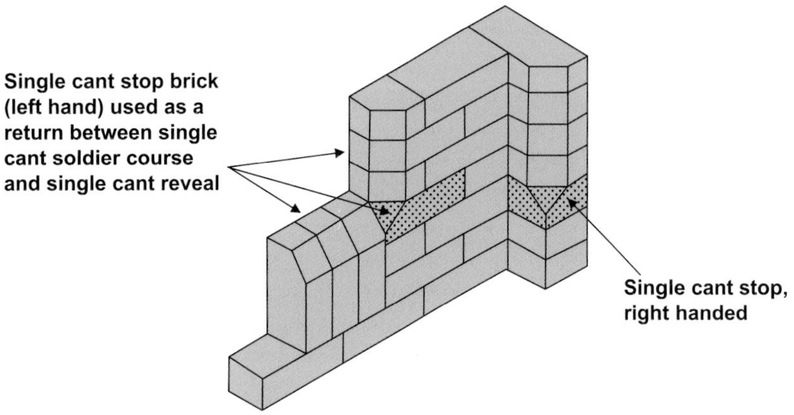

FIGURE 13.5
Uses of special shape cant stop bricks

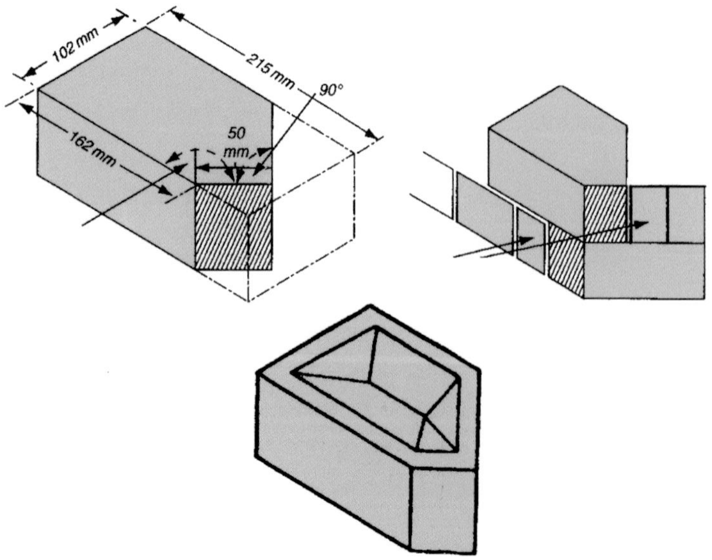

FIGURE 13.6
Squint bricks

face of the squint brick shown allow quarter-lap to be maintained on face. However, in some modern buildings the bricklayer may be called upon to use squint bricks in 102 mm thick walling, although this is not good practice because the half-lap of stretcher bonding is not being maintained around the obtuse angle quoin.

EXTERNAL ANGLE (AN.2) AND INTERNAL ANGLE (AN.3) SPECIALS

These special shaped bricks, illustrated in Figure 13.3(d), are intended for bonding obtuse angle quoins in stretcher bond (cavity) walling. The face dimensions of each allow half-lap to be continued around splayed angles.

Internal angle specials (AN.3), sometimes referred to as doglegs, are faced on the sides opposite to external angle special bricks (AN.2).

Arch bricks

A trained bricklayer can set out an axed arch and cut the required number of voussoirs from basic size bricks using a hammer, bolster and comb hammer.

There are four standard shape arch bricks, however, that manufacturers will mould to a tapering shape, suitable for set spans for semi-circular arches of 910, 1360, 1810 and 2710 mm (Figure 13.7).

Arches of any other span or shape will require special shape arch bricks, to obtain parallel mortar joints between voussoirs, and would need to be specially ordered.

Brick-on-edge quoin blocks, angles and stopped ends

To avoid cutting mitres at the point where a brick-on-edge capping turns a corner or obtuse angle, a range of fired clay blocks is made for the purpose. Figure 13.8 shows that these specials are made to suit bullnose and cant as well as plain brick-on-edge finish. All are very secure when firmly bedded, providing solid support for intermediate bricks.

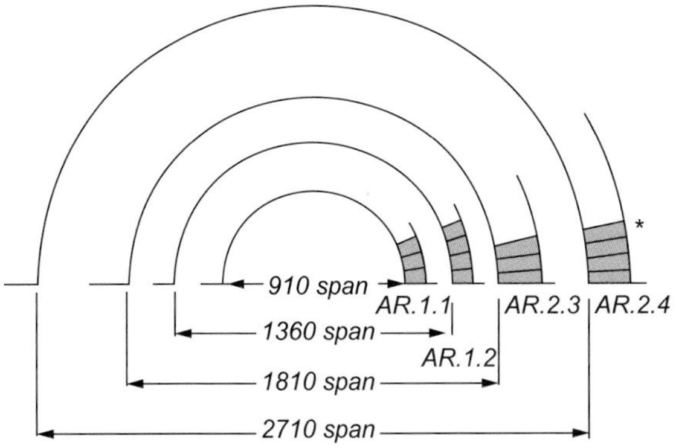

FIGURE 13.7
Arch bricks

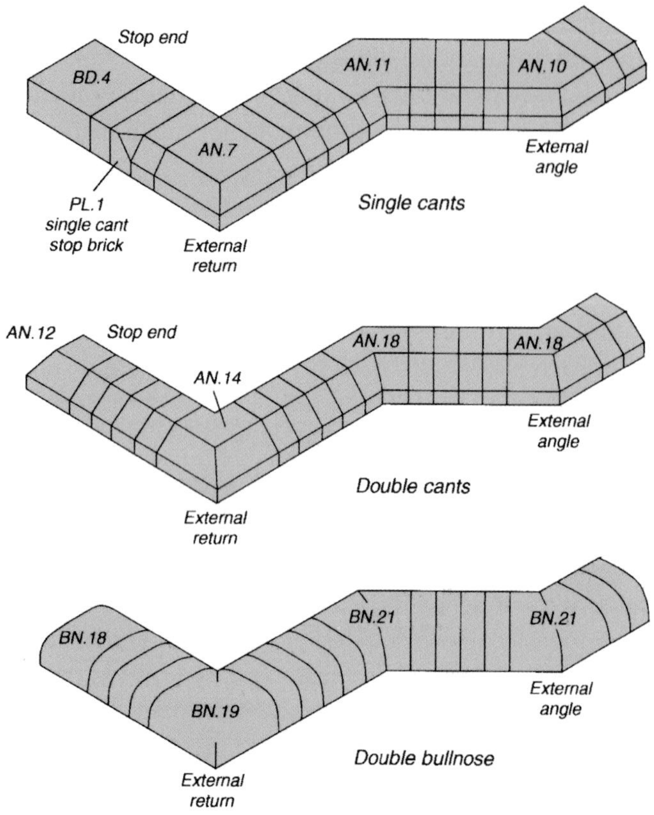

FIGURE 13.8
Brick cappings

'Handing' of special shaped bricks

If the single cant special bricks, shown in Figure 13.3(a), have a smooth face and are solid wirecuts, then they can be used as the 'quoin bricks' in every course simply by being turned over in alternate courses. They do not have a top and a bottom as such.

Depending on the method of manufacture, some specials will have a single frog, and should always be laid frog-up. Other facing bricks may be perforated wirecuts with a dragwire surface texture, and these must not be turned over either. This is because the surface texture is directional, and the bricks must not be bedded so that rain is retained at the surface.

Therefore, special shape bricks, like those shown in Figure 13.3(a)–(d) inclusive, have a definite top and bottom, owing to the presence of a single frog or a dragwire surface texture. Left-handed and right-handed versions must be ordered, as indicated in Figure 13.9.

Similarly, can you imagine trying to use the single bullnose and single cant stop bricks illustrated in Figures 13.4 and 13.5, respectively, if the change back to a square angle was required on the course below that shown in the illustrations? For this reason, left- and right-handed stop bricks must also be ordered to suit requirements (Figure 13.10).

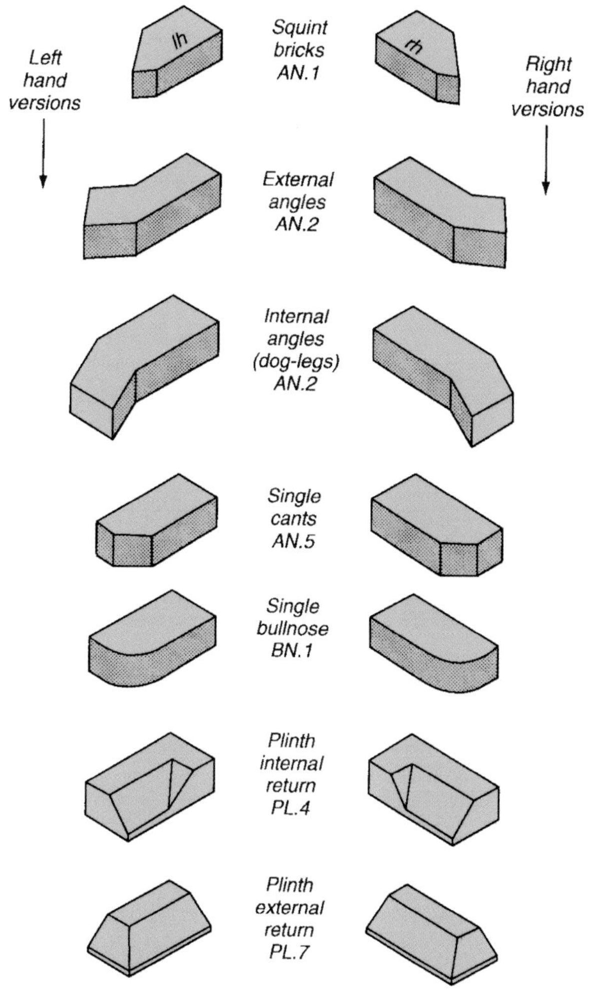

FIGURE 13.9
Handing of specials

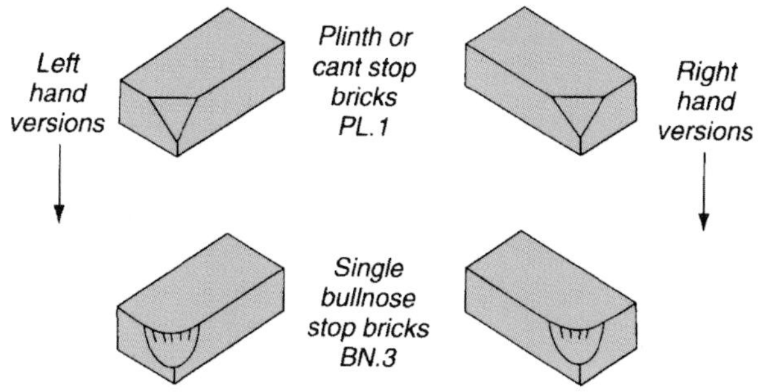

FIGURE 13.10
Handing of special shape stop bricks

Radial specials

In a similar way to arch bricks, the brick standard details some commonly used radial bricks with type numbers that can be quoted when ordering (Figure 13.11). If a different radius of curvature is required to construct curved brickwork, then detailed drawings, from which special moulds can be made, must be sent to the brick manufacturer.

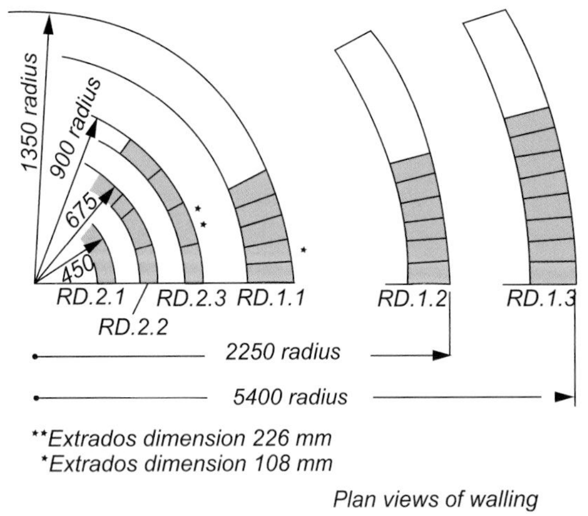

FIGURE 13.11
Radial bricks

Decorative features

A bricklayer will have to create many decorative features during work activities, which will require many of the special bricks available. These decorative features may include:

- string courses
- soldier courses
- dog toothing
- oversailing courses
- corbels
- dentil courses.

String courses

These are horizontal courses built into the face of a wall to form a decorative feature. A soldier course is a form of string course, when it continues around a building (Figures 13.12 and 13.13).

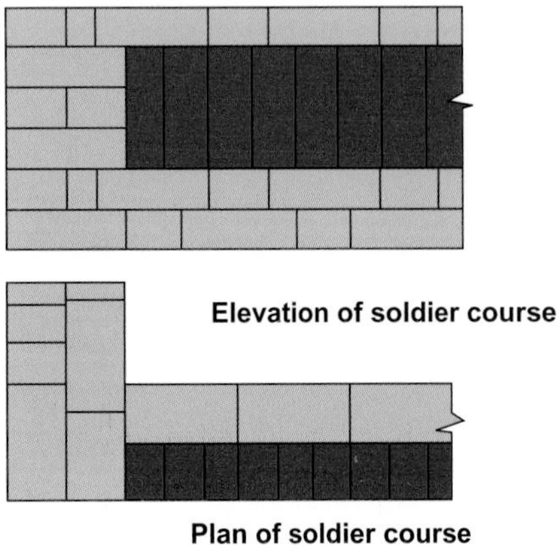

FIGURE 13.12
Constructing a soldier string course

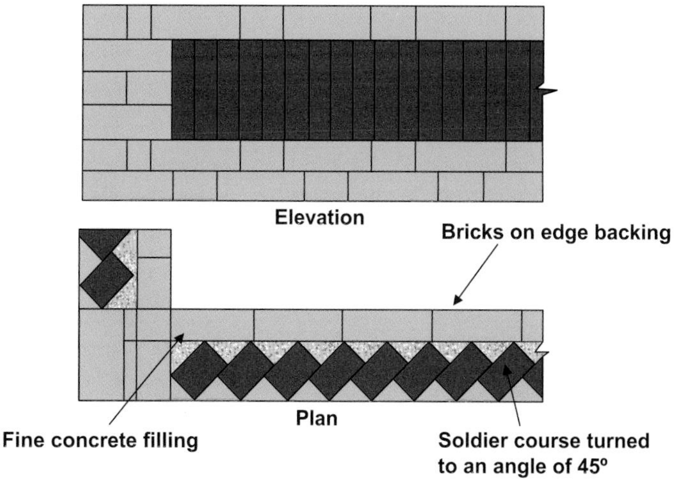

FIGURE 13.13
Soldier bricks turned to an angle of 45°

Take great care to lay the soldier course plumb otherwise a very poor effect will be produced and the bricks will appear to be leaning. A small boat level should be used for checking the work (Figure 13.14).

Dog toothing

Dog toothing is when bricks are laid 45 degrees to the face line, and can be either vertical (Figure 13.13) or flat (Figure 13.15).

The correct angle must be maintained throughout and alignment must be kept. 'A' shows the treatment in the centre of a wall, the work having been carried out from both ends.

To achieve accuracy a template can be made at 45 degrees with a stop batten lined up with the face.

Dog toothing can also be laid either projecting or recessed. If it is designed to be projecting from the face of the main wall then it is known as oversailing work.

Oversailing courses

Oversailing courses or corbels is the term given to a brick or several bricks which project from the face of a wall.

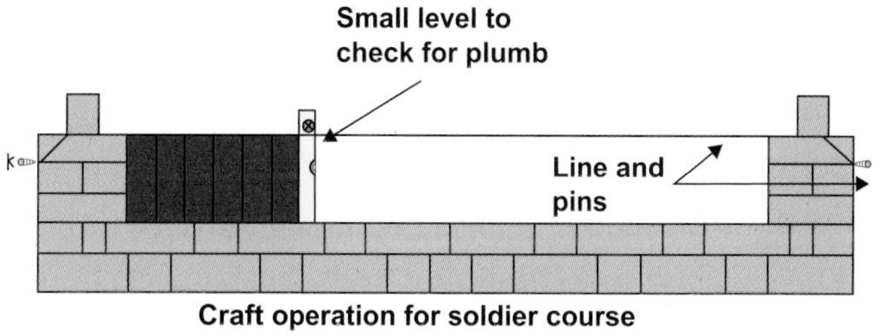

FIGURE 13.14
Checking soldier course bricks

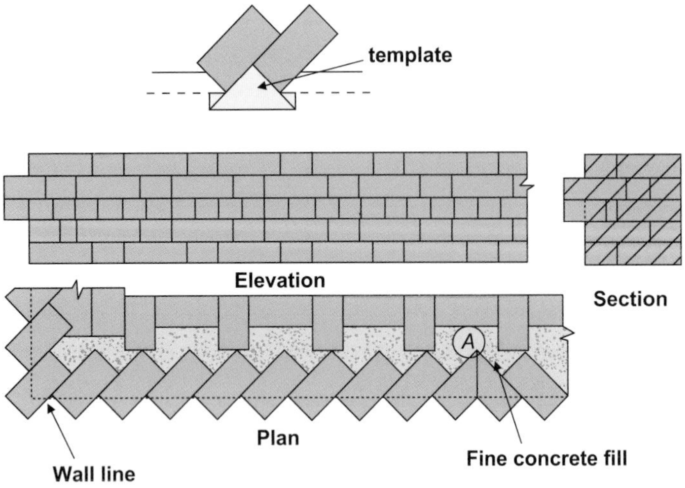

FIGURE 13.15
Creation of a dog toothing oversailing course

Oversailing or corbelling work may be used to thicken a wall or form a decorative feature. When this type of work is undertaken it is important that the underside arris of each course is level because it is this line that is seen by the eye.

It is good practice to use headers for corbelling out and restrict the projection to a quarter of a brick.

The Building Regulations state that the extent to which any part of a wall overhangs the wall below shall not be such as to impair the stability of the wall or any part of it. To prevent collapse, under no circumstances should a wall overhang more than the thickness of the wall below the corbel.

Brick corbels

The building of oversailing courses forming corbels – in 56 mm oversailers for laying and in 28 mm for bond arrangement – is not an easy operation to perform. In the case of the 56 mm corbels, headers should be used wherever possible; if the use of stretchers cannot be avoided, they should be the last bricks laid on that course and should be well bedded, a cross-joint being placed throughout the length of the brick (Figures 13.16 and 13.17).

Each corbel course must be laid to line, but fixed to the bottom arris, as this forms the sight line.

In the case of 28 mm corbels, the greatest possible lap must be maintained; if special attention is paid to this point, and common sense applied, no difficulty should be experienced in carrying out this work.

Both internal and external straight joints are sometimes unavoidable, but they should be reduced to a minimum and the specified bond should be adhered to, if possible. The final shape of the brickwork must be considered; this will continue to a greater height than the corbel and must be practical and not involve unnecessary cutting.

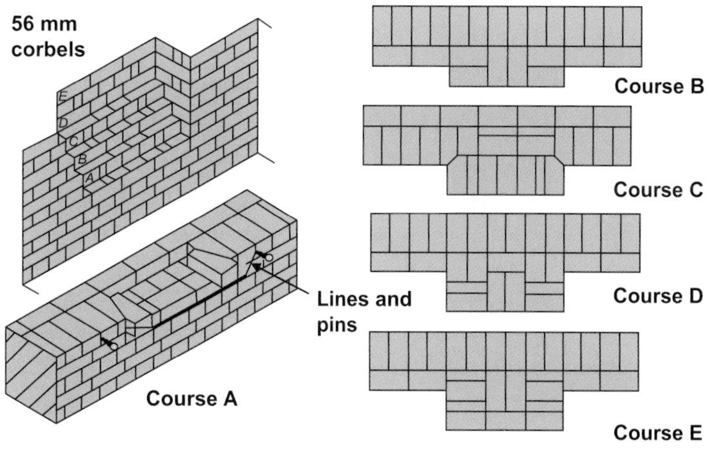

FIGURE 13.16
Setting corbel courses to line

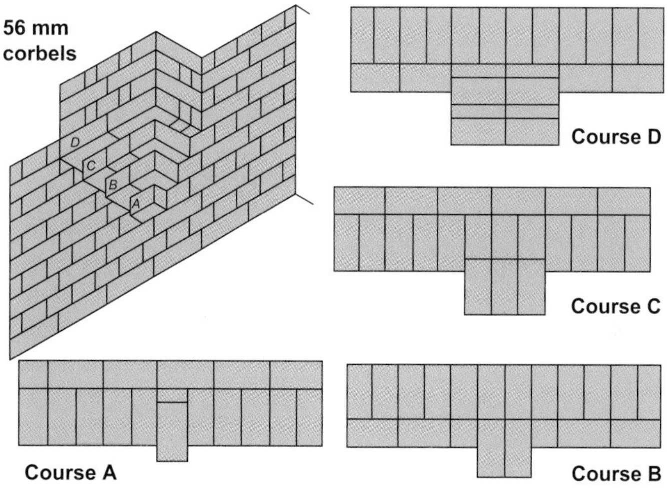

FIGURE 13.17
56 mm (1¼ brick) corbelling

When a corbel is being built it must be kept well tailed down to prevent overturning; that is, the back bricks must be laid first to bond the previous corbel before attempting to lay the next oversailing course (Figure 13.18).

Dentil course

String or band courses are a decorative feature and are quite separate from normal bonding arrangements. Straight joints between main brickwork and the string course must be avoided wherever possible.

Dentil courses provide a decorative feature at the upper surface of a wall, usually at eaves level. They are formed by projecting bricks, usually 28 mm, in various patterns (Figure 13.19). It is preferable to project headers.

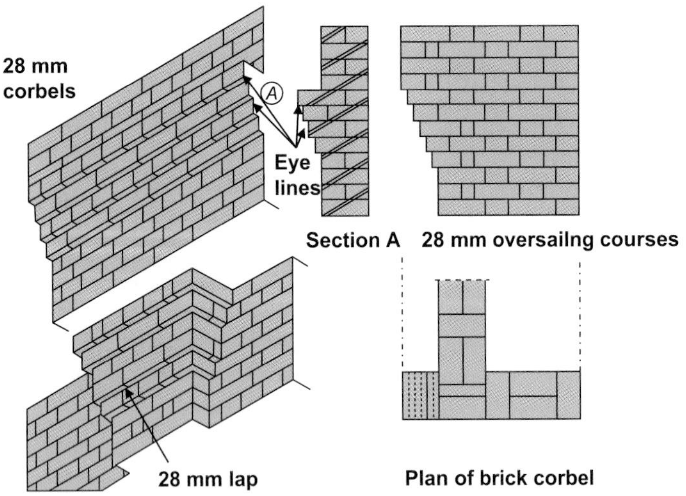

FIGURE 13.18
28 mm (1/8 brick) corbelling

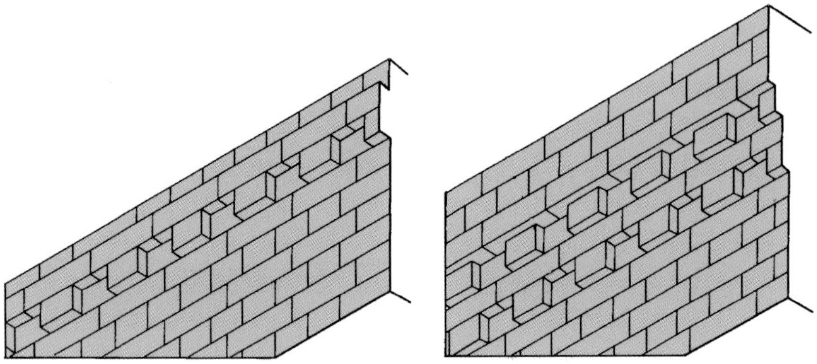

FIGURE 13.19
Single and double dentil courses

The eye or sight lines must be maintained.

Many other designs are possible for string courses by the introduction of basket weave and herringbone bonds, tiles or bullnose and cant special bricks set as soldiers. All these are laid in the manner already described. Eye-lines must be maintained and bonding knowledge applied.

Decorative panels

Decorative bonds are used in panel work for floors and paths, or vertically in walls. When used in walls they can be flush, projecting or recessed.

Flush panels

Figure 13.20 shows the necessary operations.

1. Preparation – The main brickwork is erected and provision is made for the insertion of the panel. The backing is built as the main work

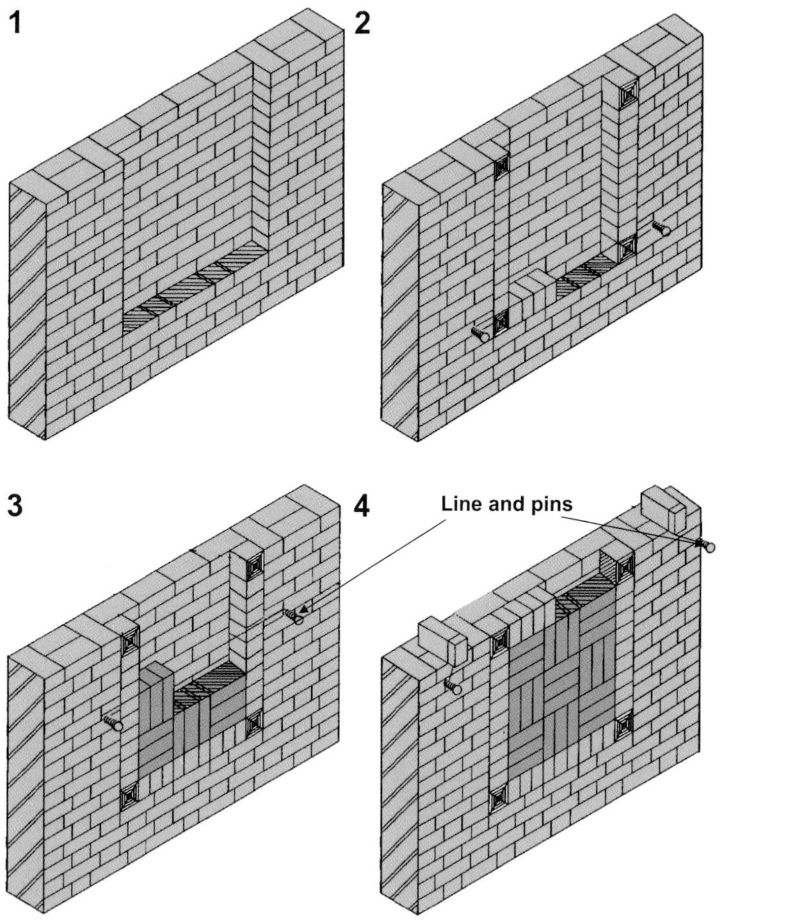

FIGURE 13.20
Building a flush panel: (1) preparation; (2) building the surround; (3) inserting the panel; (4) finished basket weave panel

proceeds, as this gives adequate bond and assists when the panel is erected.

2. Erecting the surround – Tile corners are shown. The plumbing of the side must be watched and line and pins used for the base. Surrounds are not a necessity in this type of panel and have been added in this case as a decorative feature.
3. Insertion of panel – Whatever the type of bond adopted, it is always wise to make free use of line and pins, although a straight edge may be used for final adjustment.
4. Finish – Make sure that the top of the surround is level with the main brickwork. If it is low it gives an unsightly joint and if it is high, unnecessary cutting is involved.

Projecting panels

Operations 1 and 3 are similar to those for flush panels. In erecting the surround, mitred brick corners have been adopted (Figure 13.21).

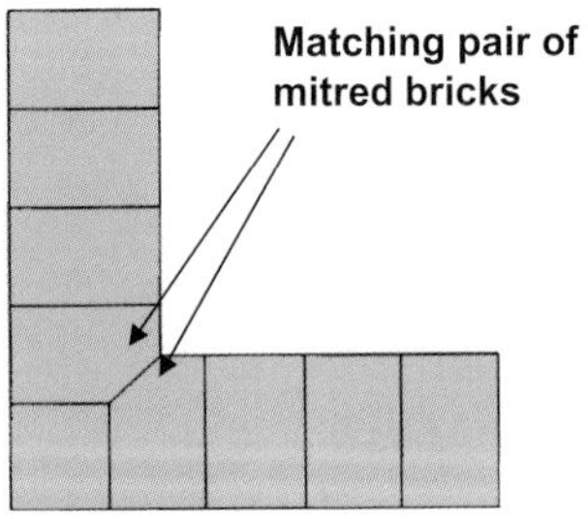

FIGURE 13.21
Mitred corner to panel surround

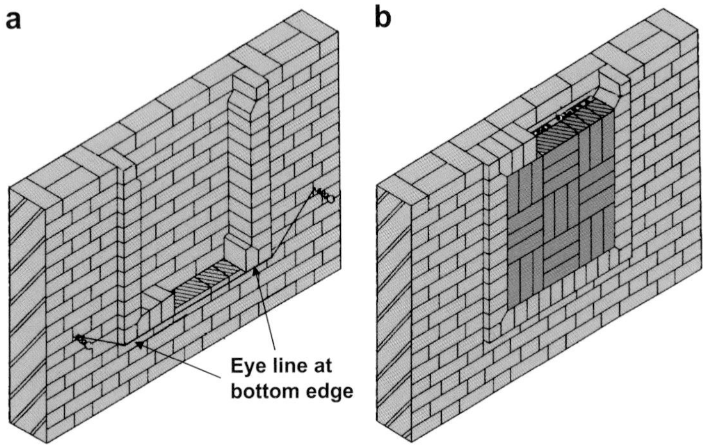

FIGURE 13.22
Building a projecting panel: (a) erecting the surround; (b) finished basket weave panel

Note that the line and pins have been used on the bottom edge of the surround. This is an eyeline. With regard to the finish, a 25 mm projection of the panel leaves a wide joint between the backing and the panel. A filling of fine concrete is the best practical method of dealing with this, as it saves excessive use of mortar and obviates the hollow gaps caused by attempting to insert too large a piece of brick (Figure 13.22).

Recessed panels

Notice the backing in this case. A one-brick wall is illustrated and the backing is brick on edge, which can be block bonded in every 225 mm. If the wall is thicker than one brick, a brick-on-edge backing is unnecessary (Figure 13.23).

Types of decorative panel

Decorative bonds used in panel work include:

- stack bond
- basket weave
- herringbone.

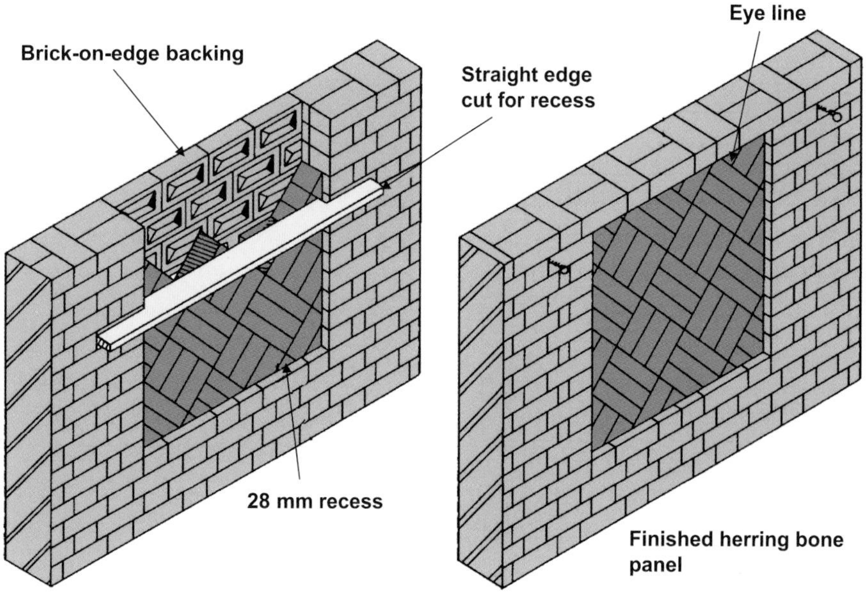

FIGURE 13.23
Building a recessed panel

STACK BOND

This is the simplest form of decorative panel. As the name suggests, it consists simply of stacking one brick on top of another without bonding the bricks.

It is essential to make sure that the vertical and horizontal joints match the surrounding brickwork. To help with all panel work the bricks should be selected for size and shape. Figure 13.24 shows a typical stack bond panel.

BASKET WEAVE

These panels can be built vertically or diagonally.

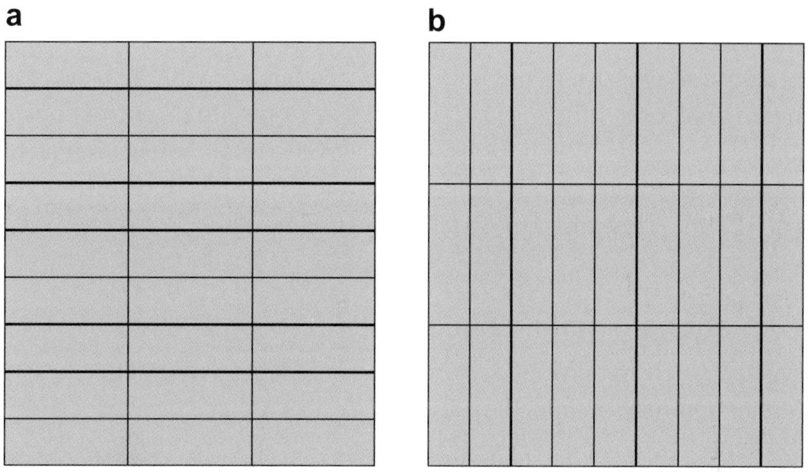

FIGURE 13.24
Stack bonded panels: (a) stack bond with bricks laid flat; (b) stack bond with bricks laid on end

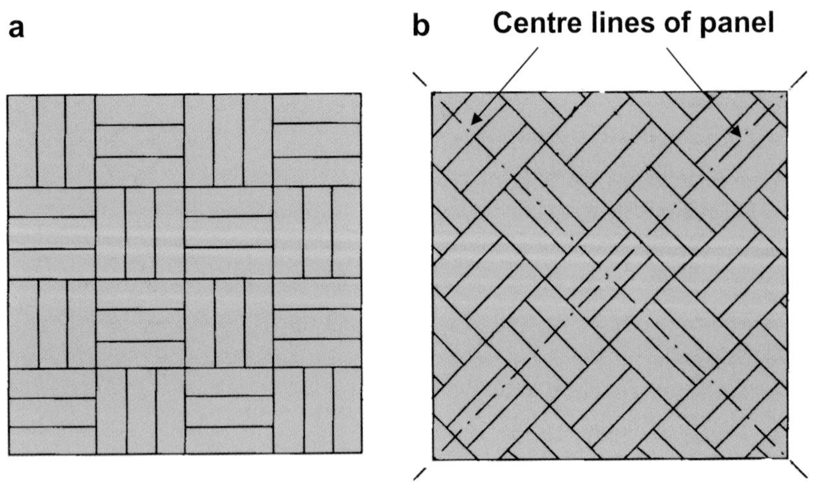

FIGURE 13.25
Basket weave panels: (a) vertical basket weave; (b) diagonal basket weave

When building a basket weave panel it is very important to ensure that all the vertical and horizontal bricks are individually plumbed and levelled, and the horizontal bricks are exactly to course with the main wall (Figure 13.25). Again, it is essential to select the bricks for size.

HERRINGBONE

This consists of a series of patterns of bricks which are laid at 90 degrees to each other, but at 45 degrees to the horizontal plain. Such patterns may be laid vertically or horizontally (Figure 13.26).

There are several variations of herringbone, two of which are shown in Figure 13.27.

Inserting the panel

In diagonal basket weave or herringbone bonds the position of the bottom cuts is important. If the angle is lost it will be necessary to shorten the length of a brick. This, in turn will lead to a search for bricks of excessive length and eventually it may prove impossible to proceed further with the work, owing to the original loss of the angle. If a brick fails to fit into its correct position without cutting when the bottom cuts have been made, the work has been carried out incorrectly and the fault must be found before further work proceeds. When finishing, the eyeline must be noted.

Setting out herringbone panels

Whenever panels in herringbone are being built it is good practice to set out the panel first on a suitable flat surface, such as a piece of plywood.

Deduct 20 mm from the length and 10 mm from the height to allow for the first bed joint. Do not deduct 10 mm for the joint at the top of the panel as the top of the herringbone panel will be level with the top bricks of the wall.

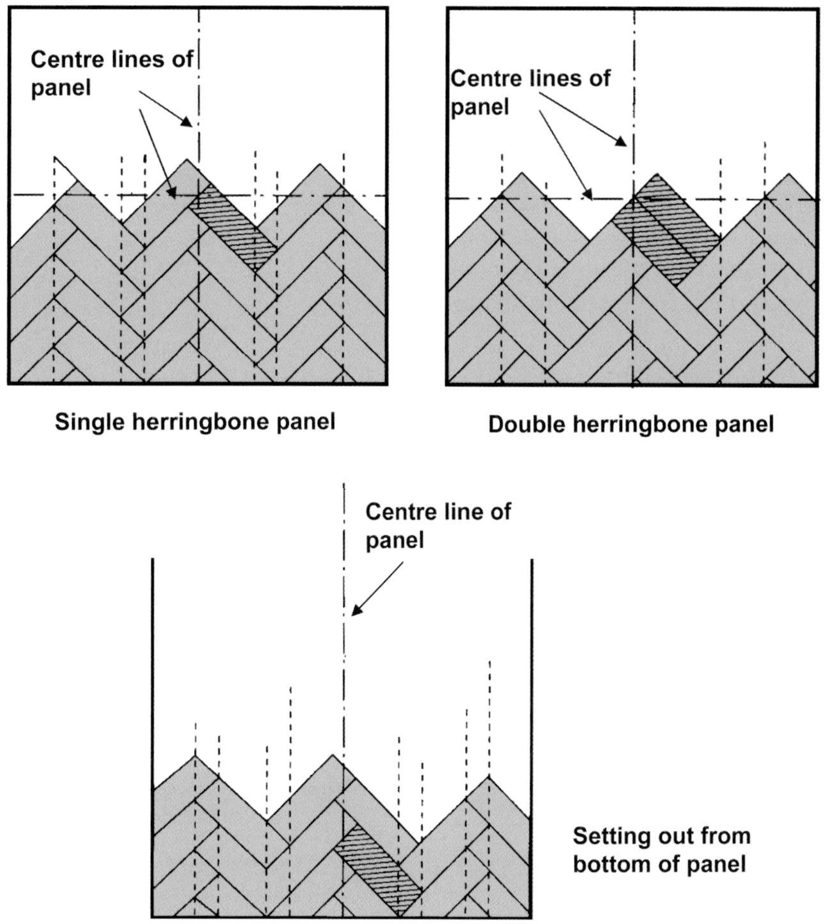

FIGURE 13.26
Herringbone panels

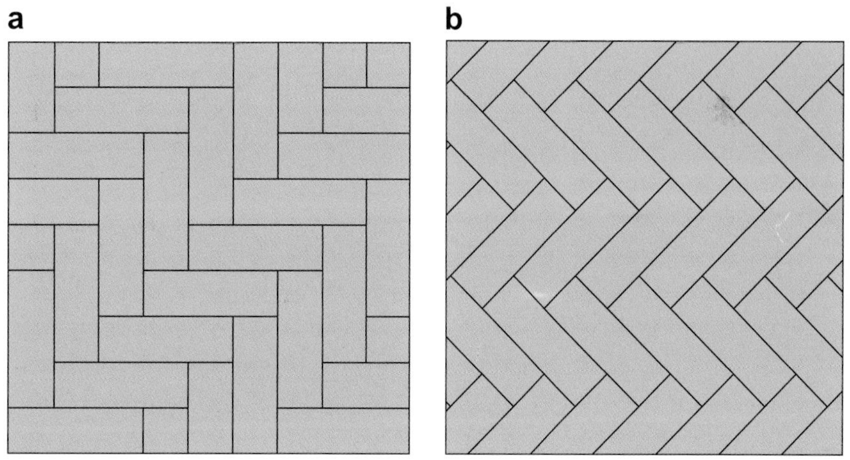

FIGURE 13.27
Variations of herringbone panels: (a) a variation which requires only square cutting; (b) horizontal herringbone

On the board mark horizontal and vertical lines and lines at 45 degrees to them through the centre point (Figure 13.28). This gives the position of the first bricks to be placed in the setting out process.

Starting with two bricks, place all the bricks dry and accurately on the board (Figure 13.29). Any inaccuracy at this stage will be reflected in the built panel.

Next, transfer the outline of the board to the bottom and two sides of the dry bricks and cut the bricks to shape.

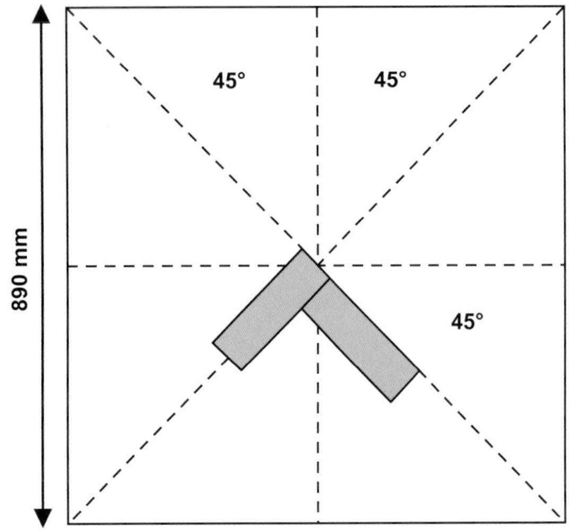

FIGURE 13.28
Setting out the first bricks in a herringbone panel

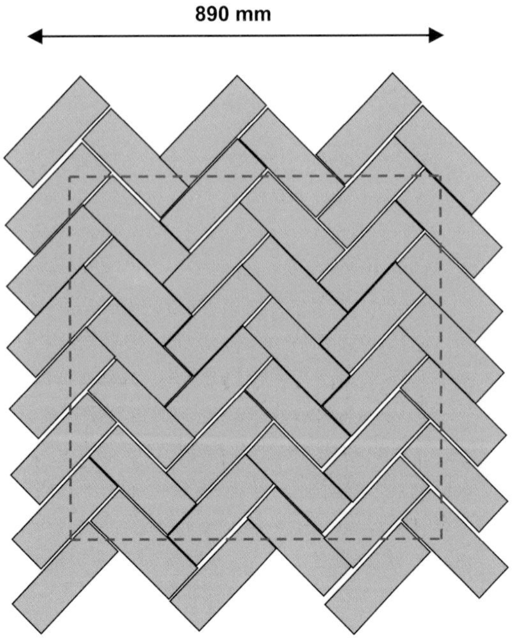

FIGURE 13.29
Dry bonding the herringbone panel

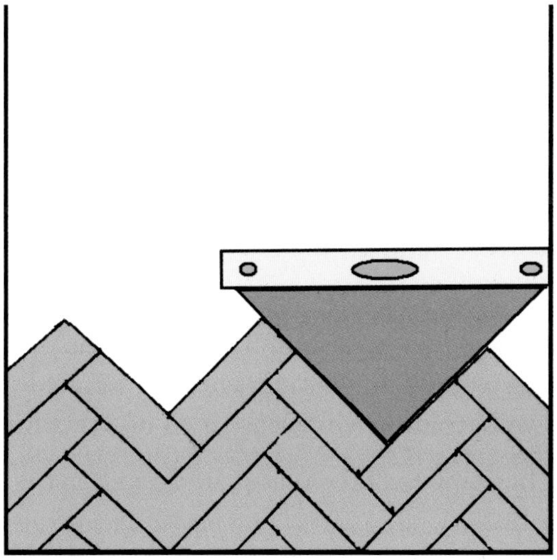

FIGURE 13.30
Checking the accuracy of a single vertical herringbone panel

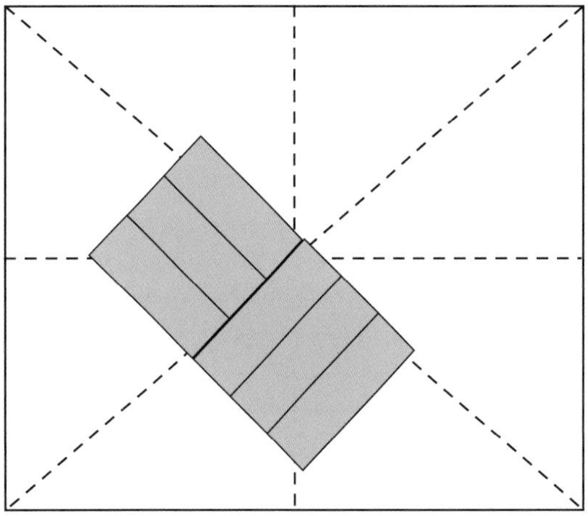

FIGURE 13.31
Setting out the first bricks in a double herringbone panel

The bricks are then carefully marked out and cut to their required shapes before building into the panel.

The building in of the panel must be controlled at an angle of 45 degrees. Traditionally, a boat level and a 45 degree set square were used, but boat levels now have adjustable vials (Figure 13.30).

Figure 13.31 shows the setting out of the centre bricks for a double herringbone panel

Plinths

Plinth bricks are specially produced to reduce or extend brickwork. They are available in headers and stretchers as well as internal and external returns.

They have a 45 degree sloping surface, and are used to achieve this set-on of face brickwork for a few courses above ground level, by 56 mm per course.

These plinth headers and stretchers are based on the standard 215 × 102 × 65 mm size to fit in with the bonding and gauge of the main walling.

Planning the bonding

Plinths are always a source of anxiety for the bricklayer, as difficulties arise in maintaining correct bond on the face while avoiding unnecessary cuts and straight joints.

The bonding arrangement for any plinth brickwork must be planned well in advance of starting work. A little time spent with squared paper, drawing it out beforehand, will avoid the embarrassment of straight joints later on.

Draw in the face bonding required on squared paper in the numerical order shown in Figure 13.32.

Start above the plinth courses on both faces of a quoin.

Plumb bonding perpends down through the plinth courses. Adjust the face bonding from (3) towards (4) in Figure 13.32, to avoid straight joints on both faces of the quoin, using broken bond and bevelled closers.

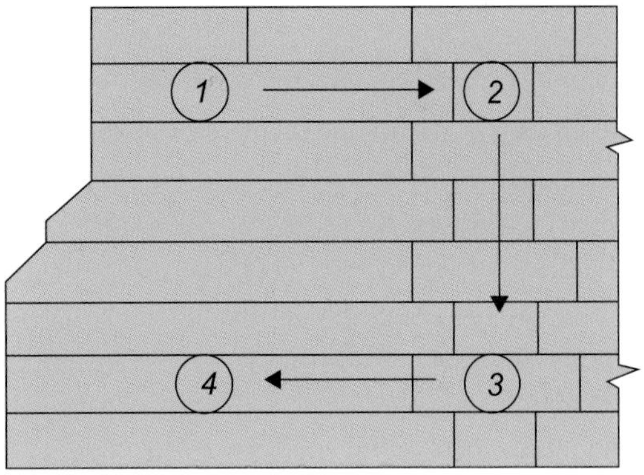

FIGURE 13.32
Planning the bonding procedure for plinth brickwork

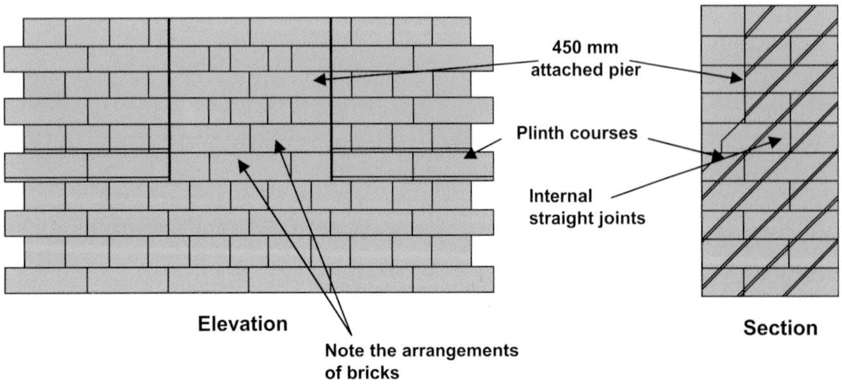

FIGURE 13.33
Bonding of an attached pier above and below a plinth course

Figure 13.32, if carefully followed, should enable an apprentice to set out most plinth problems.

Figure 13.33 shows the setting out of an attached pier above plinth courses. The arrangement of bricks is sound and avoids unnecessary cutting.

Plinth bricks may also be used upside down to construct the brick corbel courses to give a more decorative effect (Figure 13.34).

The bonding arrangements at quoins are shown in Figure 13.35 for a double plinth course and at Figure 13.36 for a single plinth course

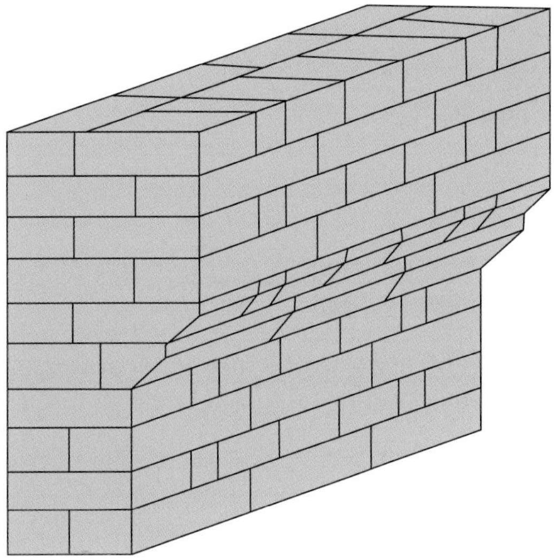

FIGURE 13.34
Plinths being used inverted to widen a wall

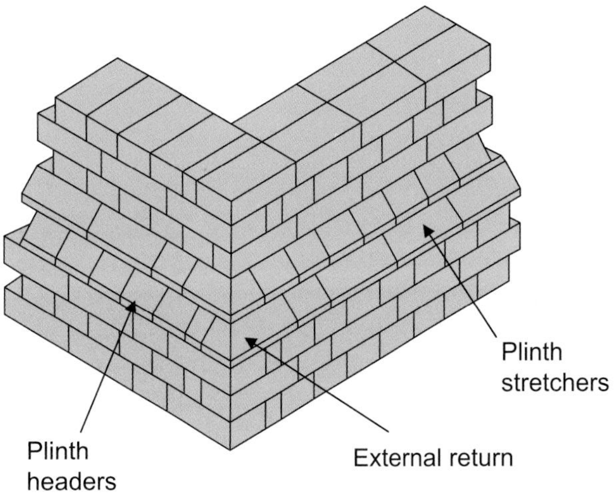

FIGURE 13.35
Double plinth course in English bond

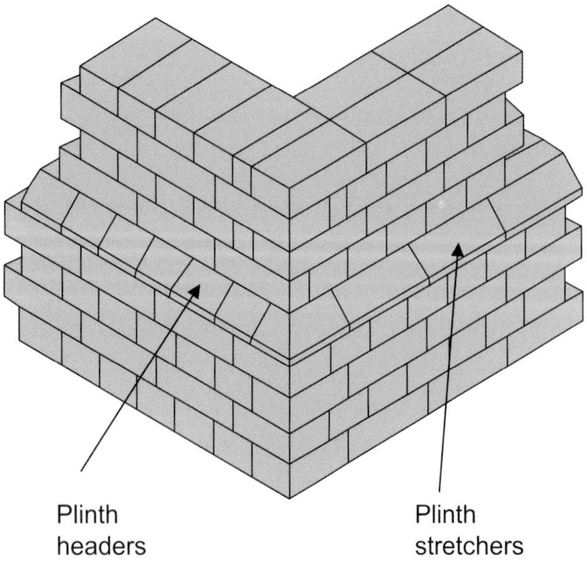

FIGURE 13.36
Single plinth course in English bond

Contrasting bricks

Decorative effects can be achieved by mixing a contrasting brick with the main colour brickwork to produce patterns.

Decorative brick quoin

Figure 13.37 shows other ways of making the angles of a building look more obvious; this is sometimes described as rusticated blocked corners and dressed corners.

Patterned brickwork can be achieved by inserting contrasting bricks in various patterns. Figure 13.38 shows a diamond pattern known as 'diapers' being used.

There are also bonds which are very decorative in themselves. One such bond is Dutch bond (Figure 13.39). This is similar to English bond in

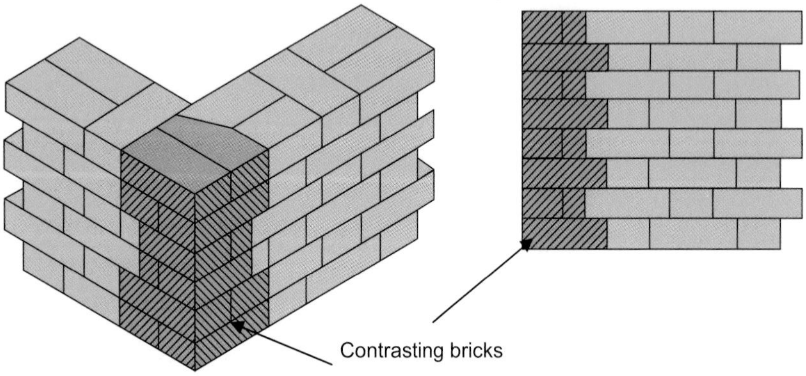

FIGURE 13.37
Decorative brick quoins

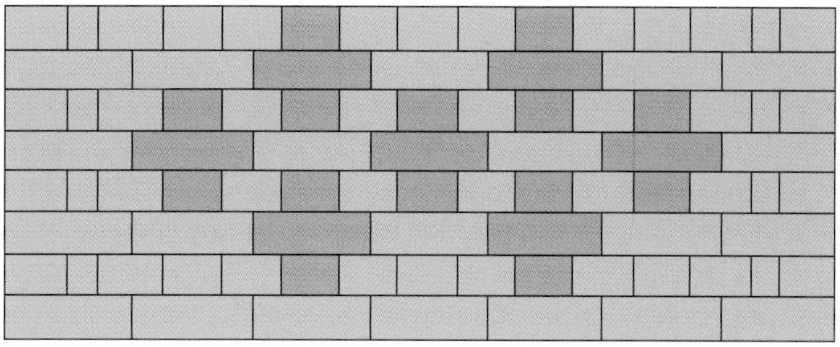

FIGURE 13.38
Diapers

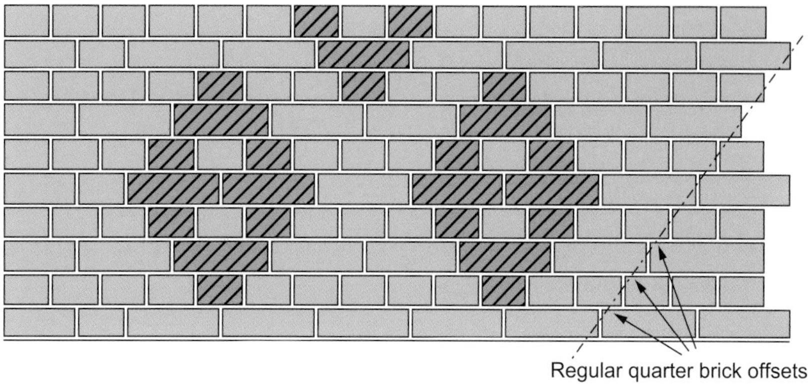

FIGURE 13.39
Dutch bond

appearance, the difference being the external corner. A three-quarter bat takes the place of the header and closer, and a half-bat or header is introduced into every other stretcher course next to the three-quarter.

Brick on edge

Brick on edge is very often required to finish off a boundary wall.

If normal quality bricks were used they would not stand up to the weather. Special quality bricks are available for terminating the tops of walls (Figure 13.40).

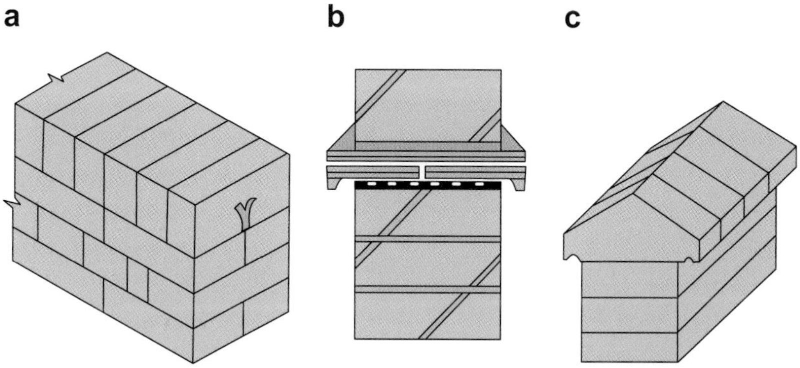

FIGURE 13.40
Types of brick capping: (a) brick on edge; (b) brick on edge with tile creasing; (c) brick coping

Multiple-choice questions

Self-assessment

This section of the book is designed to allow you to check your level of knowledge. The section consists of revision questions for this chapter. The questions are all multiple choice and have four possible answers. The answers are to be found at the end of the book.

The main type of multiple-choice question will be the four-option multiple-choice question. This will consist of a question or statement, known as the stem, followed by a choice of four different answers, called the responses. Only one of these responses is the correct answer; the others are incorrect and are known as distracters.

You should attempt to answer the questions by choosing either (a), (b), (c) or (d).

Example

The person employed by the local authority to ensure that the Building Regulations are observed is called the:

(a) clerk of works
(b) building control officer
(c) council inspector
(d) safety officer

The correct answer is the building control officer, and therefore (b) would be the correct response.

Decorative brickwork

Question 1 A pattern of contrasting bricks in the shape of a diamond is known as:

(a) diamond bond
(b) diaper bond
(c) chevron bond
(d) Dutch bond

Question 2 Identify the pattern of brickwork shown:

(a) Dutch bond
(b) stack bond
(c) herringbone
(d) basket weave

Question 3 What are plinth bricks used for?

(a) to produce a natural curve to the wall

(b) to save on facing bricks

(c) to make the wall more decorative

(d) to reduce the width of the wall

Question 4 Radial bricks are used in the construction of:

(a) curved walls

(b) decorative walls

(c) sleeper walls

(d) fender walls

Question 5 Identify the special brick shown:

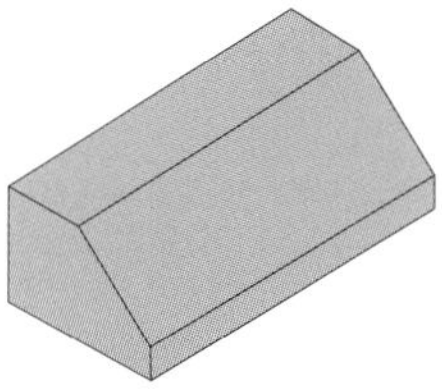

(a) plinth header

(b) bullnose header

(c) plinth stretcher

(d) bullnose stretcher

Question 6 What is a continuous course of bricks around a building known as?

(a) string course

(b) dentil course

(c) decorative course

(d) oversailing course

Question 7 When bonding the plinth course where should the bonding start?

(a) from the quoin header

(b) above the plinth course

(c) below the plinth course

(d) on the plinth course

Question 8 Identify the special brick being used:

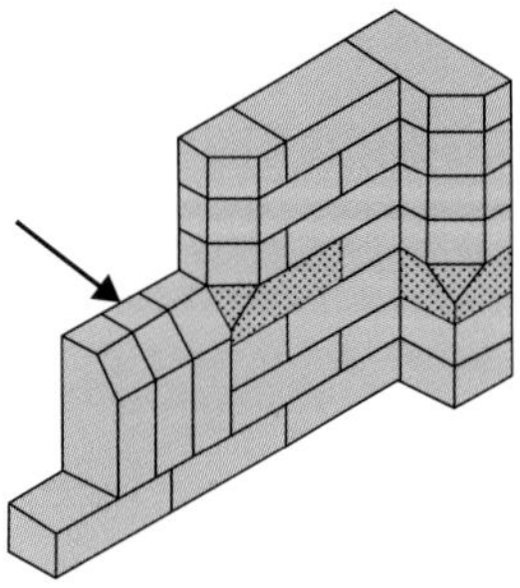

(a) single squint

(b) double squint

(c) single cant

(d) double cant

CHAPTER 14

Repairing and Maintaining Masonry

This chapter will cover the following NVQ and Diploma units.

- NVQ VR50
- CC 3028

This chapter is about:

- Interpreting building information
- Adopting safe and healthy working practices
- Selecting materials, components and equipment
- Repairing and maintaining existing brick and block structures

The following NVQ performance criteria will be covered:

- Interpretation of information
- Safe work practices
- Selection of resources
- Repairing and maintaining structures

The following Diploma outcomes will be covered:

- Know how to select required resources
- Know how to carry out repairs and maintain existing brickwork

Building information

This chapter will deal with the repair and maintenance of masonry structures. It is important with this type of work to identify the source of the problem in order to rectify the problem before attempting to repair or maintain the structure.

Information on repair and maintenance of masonry structures is difficult to obtain as most buildings will be of considerable age and contain work and materials that may have become outdated. Old books and reference material may be useful, as the original drawings may not be available. Depending on the age and state of the property information may also be required from conservation groups.

Regulations

It is essential to research the building before any work commences as there are certain rules and regulations to conform to.

Buildings that are listed or lie within a conservation area are protected by law. This does not mean that you can never alter or demolish one, but carrying out relevant work without the appropriate consent is a criminal offence

The alteration of a listed building and historic structures within its grounds requires listed building consent.

Alterations to the exterior of any building may require planning permission and an application may be needed for some works to houses in conservation areas.

Safe work practices

Repairing and maintaining brickwork can take place at any point on a building, especially at height. As stated before, there are several hidden dangers when working on the building site, especially when working at height. It is therefore essential to be aware and to ensure that correct personal protective equipment is worn, especially safety boots and hard hats.

It is vital that the highest standard of workmanship and the correct interpretation of the current Building Regulations are maintained throughout the repair and maintenance of masonry structures.

Selection of resources

The resources required to repair and maintain masonry structures include various tools and equipment plus bricks, blocks, natural stones, mortars, damp-proof barriers, lintel fixings and ties (Figure 14.1).

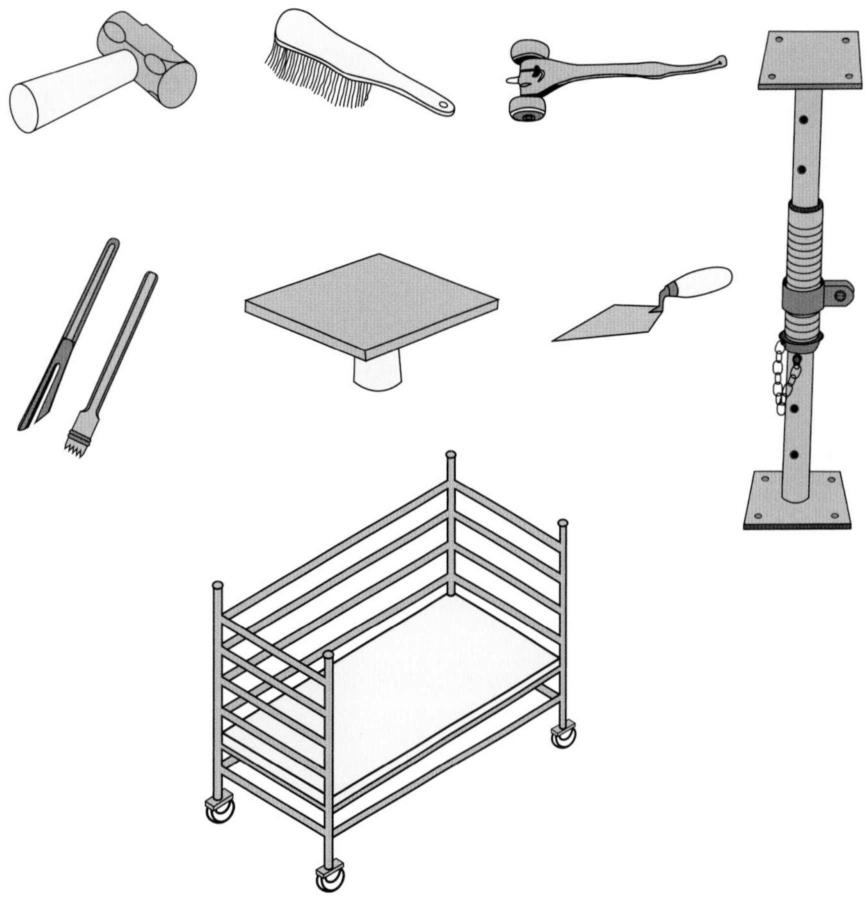

FIGURE 14.1
Selection of resources

Various special bricks will be required to match the existing brickwork and these may be of the old imperial size. It may be necessary to visit a reclamation centre.

Definitions

All buildings deteriorate to some extent as they get older. The rate at which a building will deteriorate depends on several factors, but this chapter deals with their repair and maintenance.

There are several definitions of maintenance but generally it means preventing the building from deteriorating and maintaining its usefulness

Maintenance can be either planned or unplanned.

Planned or routine maintenance keeps to a programme of work required to retain the building to its required standard. Simple routine maintenance includes the painting of the doors and windows.

Unplanned maintenance is work that has been ignored until it has become unsafe and is the most expensive type of maintenance. Replacing doors and windows can be classed as unplanned maintenance.

Although most buildings are designed to last for a minimum of 50 years several require repairs and maintenance before this period ends.

There are several environmental terms which require defining:

- Conservation – the protection of existing buildings to enable them to be usable. It involves keeping their appearance as they were intended.
- Preservation – maintaining buildings exactly as they were and protecting them from decay.
- Refurbishment – bringing an existing old building back to its original state.
- Restoration – carrying out repairs to return a building to a good state of repair. Restoration and replacement are similar
- Replacement – removing old and worn out materials and components from a building and fitting new materials and components.
- Renovation – doing up a building to bring it back to its original state.

Defects

Most buildings are designed to a specification by the client. If a building fails to meet its design specification it can prove to be very costly.

Ten per cent of the work done by the construction industry is involved with repairs and maintenance.

There can be various defects in buildings that are either designed or off plan. These defects can be caused by:

- poor material choice
- poor design
- poor workmanship
- external forces

Poor material choice

A building could have the wrong materials and components specified.

If materials and components are used in the wrong positions it could affect their performance.

An example is the wrong selection of facing bricks for external walls and capping. This can result in the bricks spalling owing to extreme weather conditions on all faces of the external wall. When normal quality bricks are used below damp-proof course (DPC) level these can also have their faces blown off (Figure 14.2).

Remember

When designing a building use standard sizes for all components where possible.

Standard windows and doors are much easier and cheaper to replace than components that are not a standard size.

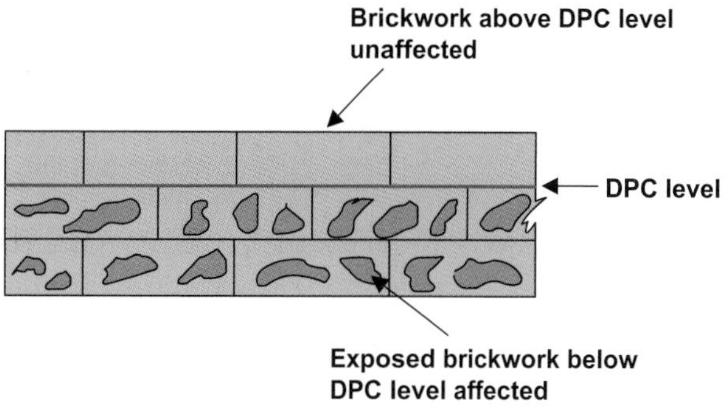

FIGURE 14.2
Spalling of individual bricks

Poor design

CRACKS IN THE STRUCTURE

If the building has not been designed correctly it can become overloaded.

Tell-tale clues of poor design are cracks appearing close to openings for doors and windows, in ceilings and corners of rooms. Sloping upper floors and gaps appearing between floor boards and skirtings are signs of shrinkage or settlement.

Some older properties have wooden lintels and these will deteriorate with age.

If a building is being used for activities other than it was designed for it can cause structural defects if extra loading is applied. This can cause lintels to bend or sag. External brickwork may start to show cracks (Figure 14.3). The building may need to be supported while new and stronger lintels are fitted.

Poor workmanship

Poor supervision can lead to poor workmanship, and both can cause defects in the construction.

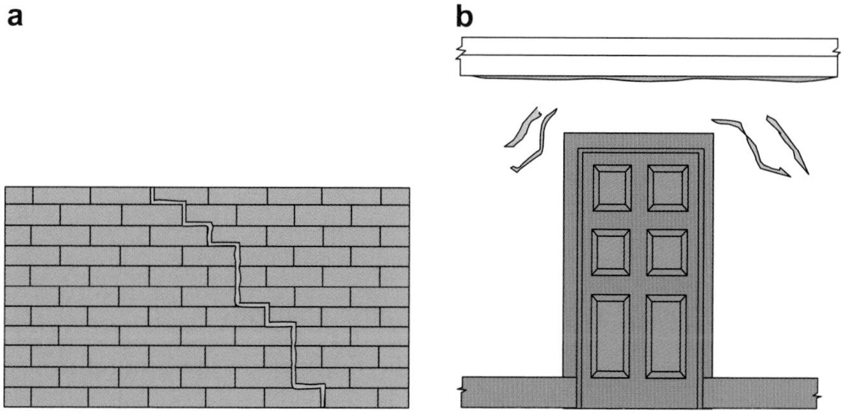

FIGURE 14.3
External and internal cracking: (a) external cracking of brickwork; (b) internal cracks around lintels

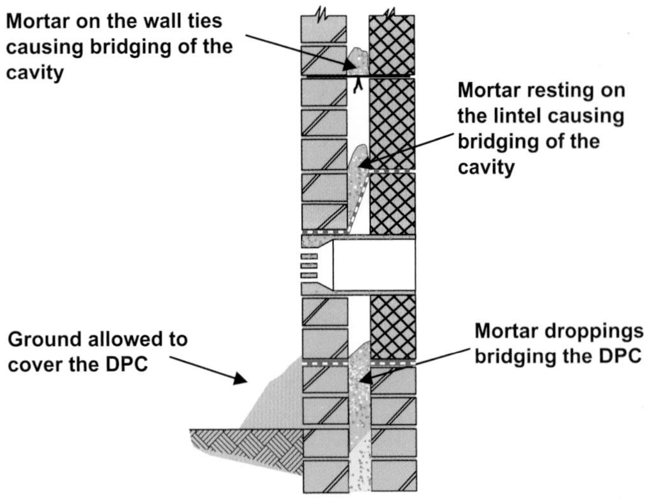

FIGURE 14.4
Poor workmanship causing defective cavities

Poor workmanship can result in dampness in cavities if the work is not completed correctly. Figure 14.4 shows how poor workmanship, allowing mortar to fall down the cavity and causing bridging of the two walls, can cause dampness in the building.

When property owners allow garden soil to cover the DPC this can also cause dampness in buildings.

External forces

The weather is the main problem, with extreme rain penetrating buildings and causing dampness. Penetrating damp is caused by water passing through the external brickwork owing to bridges in the cavity (Figure 14.4), or in older buildings through the solid wall. Percolating damp is when water gets into the building owing to faulty flashings around chimney stacks, etc. (Figure 14.5). This can lead to other problems such as condensation, mould growth and fungal attack.

Other problems are caused by rising damp.

Many older properties did not have a horizontal DPC built into the structure 150 mm above ground level. Some of the older methods have become fragile and allow moisture to penetrate the building by rising damp (Figure 14.6).

In these extreme cases it is necessary to chop out and insert two courses of blue bricks (see the following section). A modern approach is to inject the affected areas with liquid silicone to provide a DPC barrier.

Movement around a building can also cause problems.

Defective services or trees can cause the moisture in the ground to vary, especially in clay soils. These soils can be very wet or extremely hard and cause the building to move up and down (Figure 14.7). This swelling can be caused due to changes in the moisture content of the soil. Broken drains

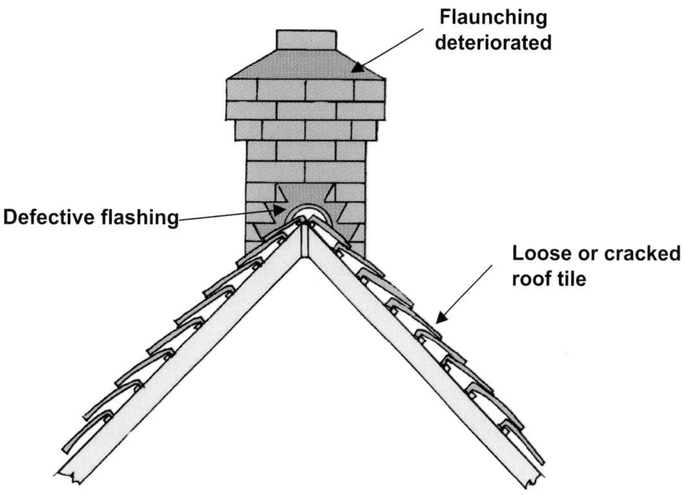

FIGURE 14.5
Rain penetration

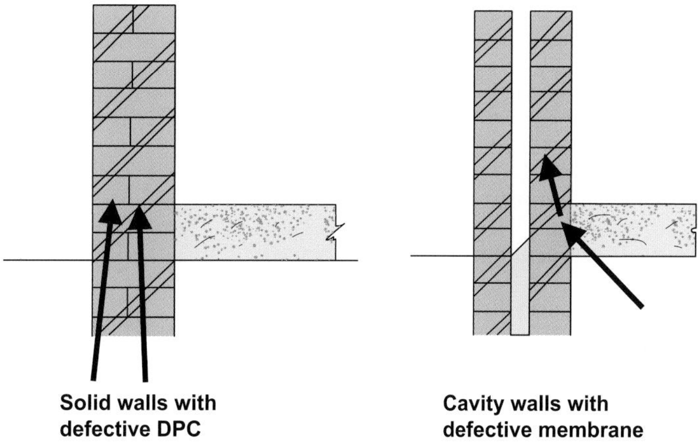

FIGURE 14.6
Rising damp: (a) solid walls with defective damp-proof course; (b) cavity walls with defective membrane

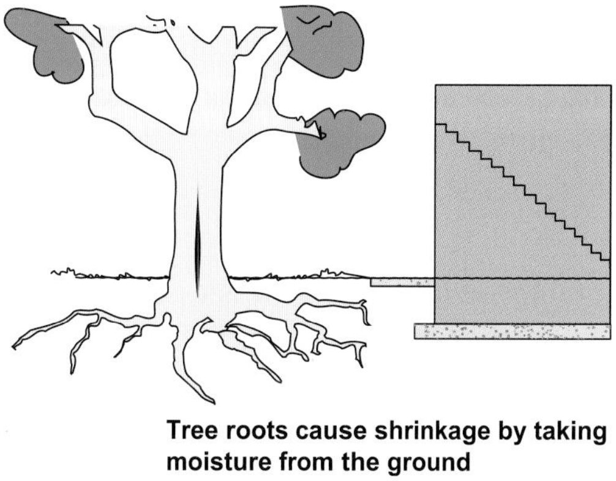

FIGURE 14.7
Ground movement

can cause an increase in moisture content and then hot weather can dry the subsoil out. Surrounding trees can extract moisture from the ground, causing the ground to sink.

Most of the above problems can be dealt with by removing the problem trees and replacing any defective drains.

If the area is liable to subsidence and the building has sunk, causing cracks, it may be necessary to underpin the building and take the foundation down to a level where there is less movement.

Types of repairs and maintenance

Having discussed the causes of the problems, each will be discussed in turn.

Many types of repairs and maintenance will be required. The following list contains just a few covered in this chapter:

- shoring up a building before work is carried out
- underpinning
- replacing/inserting a DPC
- forming an opening and fitting a lintel
- fitting replacement doors and windows
- maintaining and extending masonry walls
- chopping out and replacing defective bricks
- repointing

Shoring up a building before work is carried out

Whenever working on a property where the work involves removing part of the structure it is essential to prevent any movement of the structure.

This can be achieved in many ways, such as bracing, propping and shoring.

BRACING

Temporary bracing can be applied to all windows and doors to prevent any more movement (Figure 14.8).

PROPPING

If walls are to be removed in buildings it is essential to support the upper floors to relieve the loading from these areas.

If the ground floor is hollow timber construction then the props need to go through to the oversite concrete.

Timber or steel adjustable props can be used, with suitably sized head and sole plates (Figure 14.9).

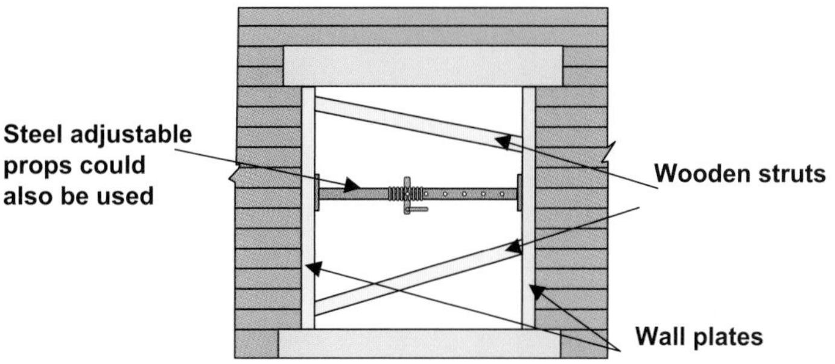

FIGURE 14.8
Bracing to a window

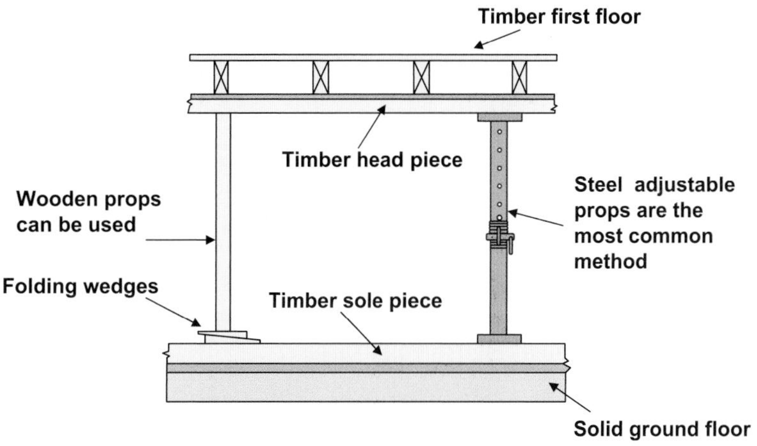

FIGURE 14.9
Support for first floor by propping

SHORING

To prevent the possible collapse of a building while work is being carried out it is essential to provide temporary support to the structure to prevent any movement and possible damage to people or property.

There are three types of shoring systems available, each one providing a particular solution:

- dead shores
- raking shores
- flying shores.

They can be used together to provide a combined solution to a supporting problem.

Dead shores

These are used mainly to support vertical loading from the structure, which includes both live and dead loads.

Their main usage is to support the loading from above while work is being carried out below. Such work could include the removal of a wall or part of a wall to provide a door opening.

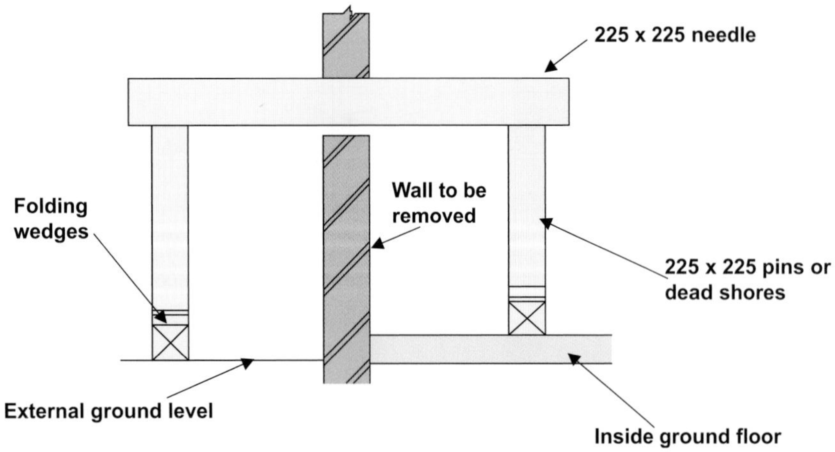

FIGURE 14.10
Needles and pins

They could also be used in repair work when a problem has been discovered with a lintel and it has to be replaced.

Dead shoring consists of needles and pins. The vertical members are the pins, and the horizontal members which are threaded through existing walls are known as the needles (Figure 14.10).

Raking shores

These also provide temporary support to a structure, especially if the building has started to bulge.

They are designed to support both vertical and horizontal loadings from the structure.

The angle of rake will depend on the availability of space around the building, but a minimum rake of 45 degrees is required (Figure 14.11).

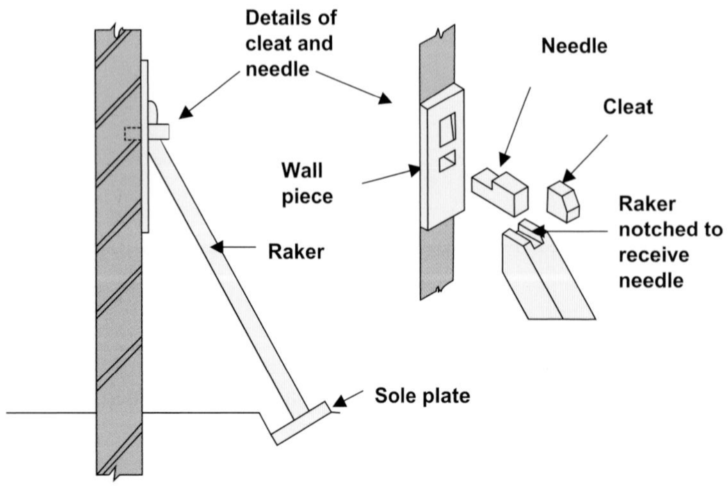

FIGURE 14.11
Raking shoring

Flying shores

Occasionally flying shores may be required to support buildings while demolition takes place between two buildings or there is too little space on the ground for raking shoring.

This method of supporting buildings is totally off the ground. It allows traffic and pedestrians to pass underneath (Figure 14.12).

Because of the practical implications flying shores are not convenient over 10 m.

Underpinning

Whenever there is movement in the ground for whatever reason there will be damage to the structure of a building, and it may even cause structural failure of the foundation. Movement can be caused by subsidence, landslip or changes in the moisture content of the subsoil.

Damaged foundations can be strengthened by underpinning.

It is essential to find out the cause of the movement first. If it has been caused by trees close to the structure then they must be removed.

Underpinning is when a new concrete foundation is constructed underneath the existing one, to a depth where movement is restricted. This new foundation is designed to carry the load of the existing structure down to a new, lower depth.

Underpinning can also be used to lower the ground to construct a basement underneath an existing structure.

Several preliminary works have to be carried out before underpinning begins:

- Obtain local authority approvals for the proposed work.
- Ensure that the owners of adjoining properties are aware of the situation. Obtain permission to check their properties and provide any temporary shoring, etc.

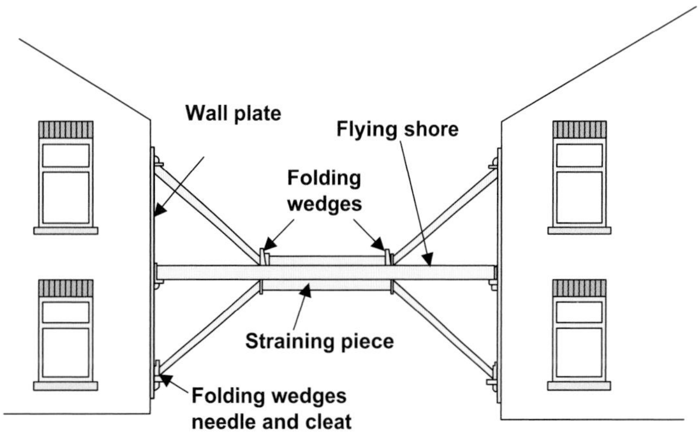

FIGURE 14.12
Flying shoring

- Fix 'tell-tales' over existing cracks to the property and any neighbouring property if involved.
- Fix temporary shoring and/or props to the building and remove any excessive loads from the building.
- Check for any services that enter the building and identify and mark them.

CARRYING OUT THE WORK

Underpinning should only be carried out in small planned bays (Figure 14.13). Each bay should be approximately 1 or 1.5 m in length, depending on the type of foundation being underpinned. If the existing concrete foundation is reinforced then the bays can be between 1.5 and 3 m.

Underpinning could be carried out to a structure that has been built on the natural foundation, and there is no concrete foundation. In this situation the bays should be the shortest dimension of 1 m.

When excavating each trench, allow 1 m working space for the placing of materials, etc. Depending on the type of ground it may be necessary to timber the trench for maximum safety (Figure 14.14).

The new concrete foundation should be laid with steel starter bars left projecting at each end. This is to allow the next concrete section to bond to the previous section.

The brickwork should be built up using engineering quality bricks and 1:3 cement mortar.

As the wall is being built the back of the wall should be backfilled and compacted with weak concrete fill.

The ends of each section need to be toothed to allow the adjoining section to be bonded.

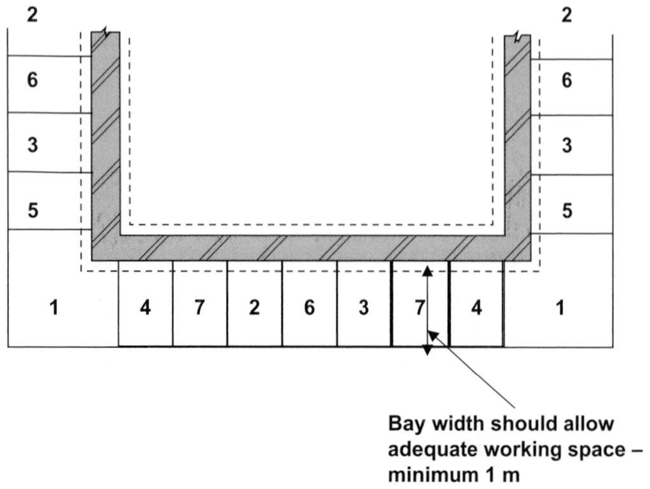

FIGURE 14.13
Planning underpinning

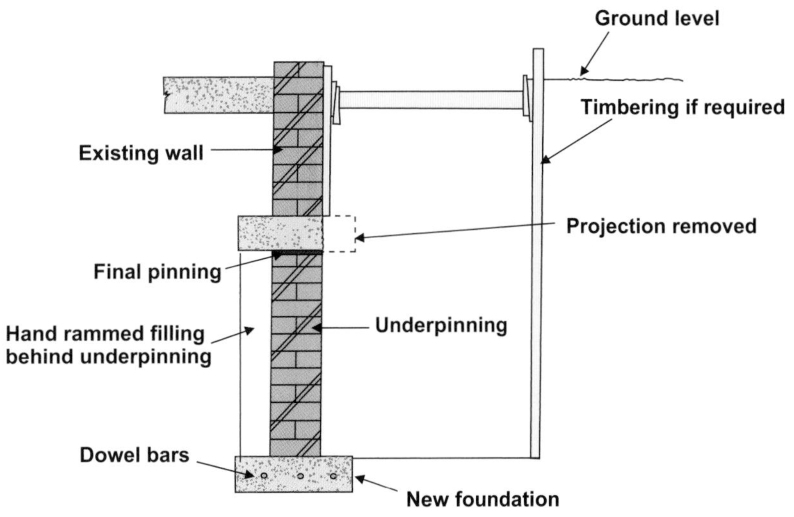

FIGURE 14.14
Underpinning section

The last course should be left until the new brickwork has set and then it should be pinned up to the existing concrete foundation with a 1:3 semi-dry mix of rapid-hardening cement (Figure 14.15). Once it has set the projection of the existing foundation should be carefully removed. This will prevent any load from the building being transferred to the new backfilling.

Each bay is started according to the schedule until all areas have been completed. Ensure that each section bonds to the next with the steel dowels and brick toothings.

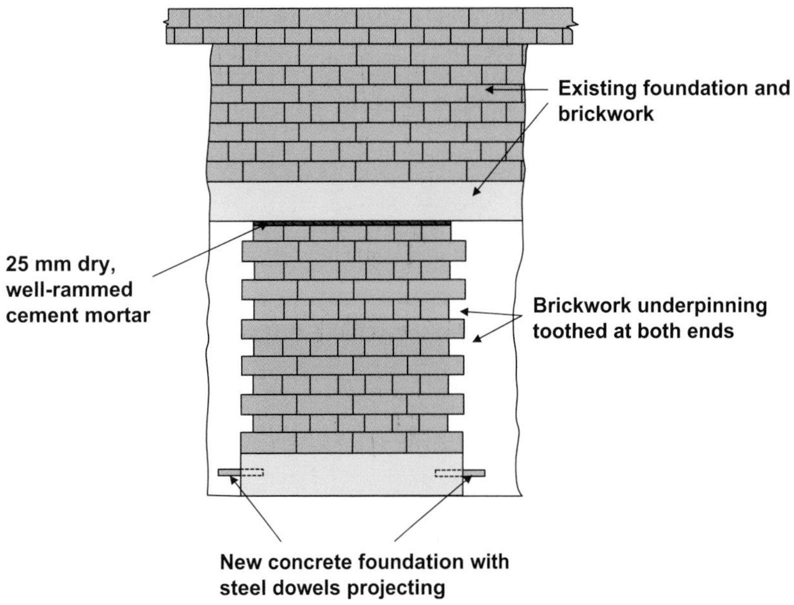

FIGURE 14.15
Elevation of section of underpinning

When all the work has been completed check the rest of the property and any adjoining property for damage. After this inspection the trenches can be backfilled.

Replacing or inserting a damp-proof course

As mentioned before, many older properties did not have a horizontal DPC built into the structure 150 mm above ground level. In addition, some of the older methods may have become fragile and allowed moisture to penetrate the building by rising damp.

In these extreme cases it is necessary to chop out and insert two courses of blue bricks. A modern approach is to inject the affected areas with liquid silicone to provide a DPC barrier.

The procedure for inserting a new DPC has a similar pattern to underpinning. Approximately 1 m of the wall at a time should be worked on and it is essential to plan the sequence before commencing (Figure 14.16).

Usually the existing wall will be a one-brick solid wall. Two courses need to be removed and replaced with DPC or blue bricks laid in cement mortar.

The internal plasterwork and skirting, and in the kitchen wall tiling, will need to be removed.

Under normal conditions there will be no need to support the structure. The building should be surveyed, window and door bracings fitted, and tell-tales fixed over any existing cracks.

The two courses should be cut out using a club hammer and cold chisel or using mechanical methods (Figure 14.17). The two courses have to be cut out through the entire thickness of the wall.

It is normal to start at the corner of a building. Once a section has been cut out it can be cleaned of all dust, wetted down and two courses of blue bricks laid in cement mortar.

English bond is usually used, but any bond can be used in accordance with the rest of the wall.

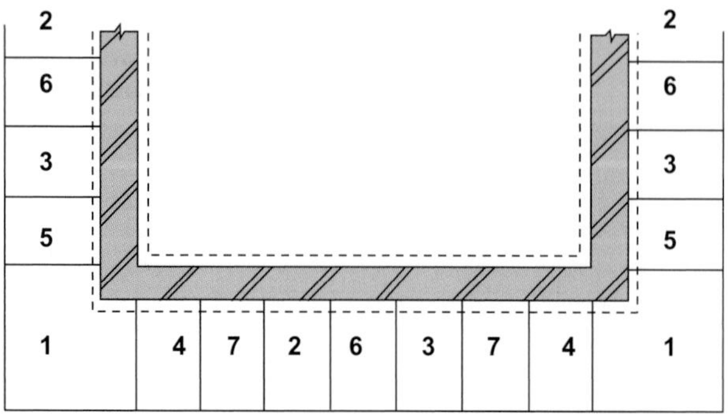

FIGURE 14.16
Sequence for inserting a damp-proof course

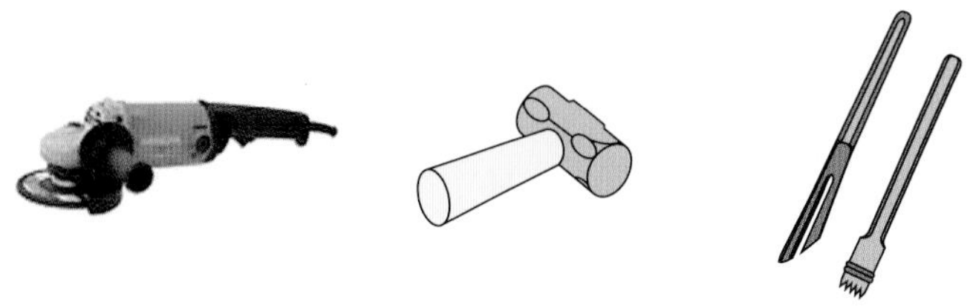

FIGURE 14.17
Equipment required

The new brickwork is allowed to set, normally overnight, and then it is wedged up to the existing wall with slates and a stiff cement mortar rammed into the joint. Figure 14.18 shows the sequence of work. During the cutting out of the two courses other bricks may be damaged or fall. These will need to be replaced and wedged.

Another method is the injection of a liquid silicone into the affected brickwork. Two or three holes are drilled into each brick and the silicone is pressure injected to create a waterproof layer. This can be applied both vertically and horizontally.

Forming an opening and fitting a lintel

There will be occasions when a wall may require removing to make two rooms into one, or part of a wall removing to form a door or window opening. Very often the walls being affected will be load-bearing walls. The removal of non-load-bearing walls will follow a different procedure.

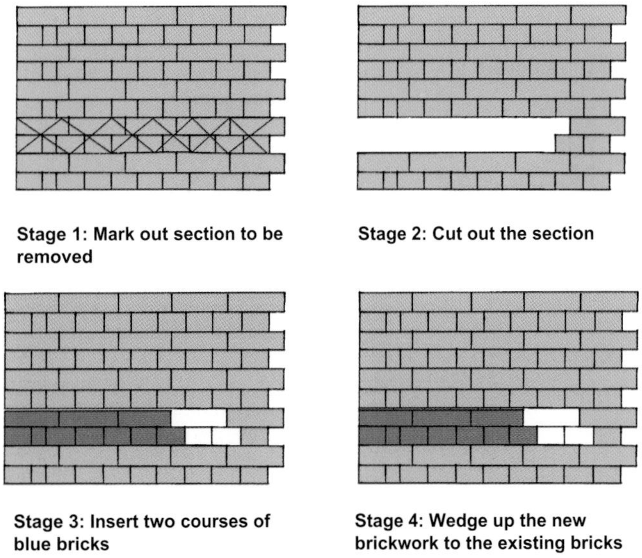

Stage 1: Mark out section to be removed
Stage 2: Cut out the section
Stage 3: Insert two courses of blue bricks
Stage 4: Wedge up the new brickwork to the existing bricks

FIGURE 14.18
Stages in repairing a damp-proof course barrier

Assuming an opening is required to fit a new door or window, the structure must be surveyed and any support put into place. This is one of the most common reasons for using temporary supports. The type of support used will be dead shoring (see Figure 14.10).

Again it is important to plan the sequence of work is beforehand. All floor areas must be cleared of any loading.

Once the dead shores are in place the wall can be removed.

The same tools are required as for cutting out brickwork: club hammer, cold and bolster chisels or mechanical methods.

Then, either the whole area, or just sufficient to fit the new lintel, can be removed.

Whichever method is chosen a lintel is required. This can be either concrete or steel and should overhang the opening by 150 mm at each end. The lintel is fitted and any brickwork above it that may have been disturbed is replaced and wedged using slate and semi-dry cement mortar.

Once the brickwork has set any other removal can take place and all rubbish removed. Figure 14.19 shows the preparatory work.

Note

It is important that this type of work is not rushed. Constant care and attention is vital at all times.

Fitting replacement doors and windows

Openings are formed in walls to allow door and windows to be fitted. Openings could also be produced as access or decorative detail without a frame being fitted.

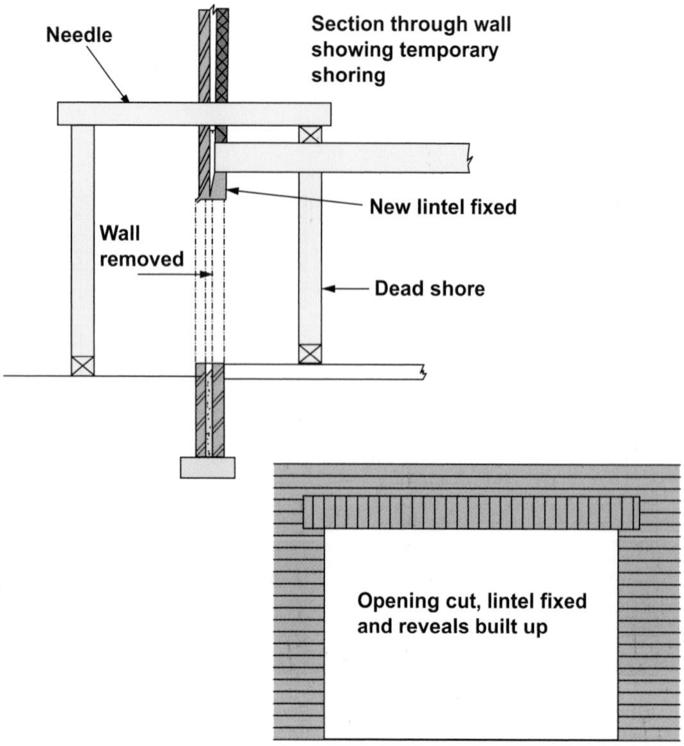

FIGURE 14.19
Sequence for forming a new opening

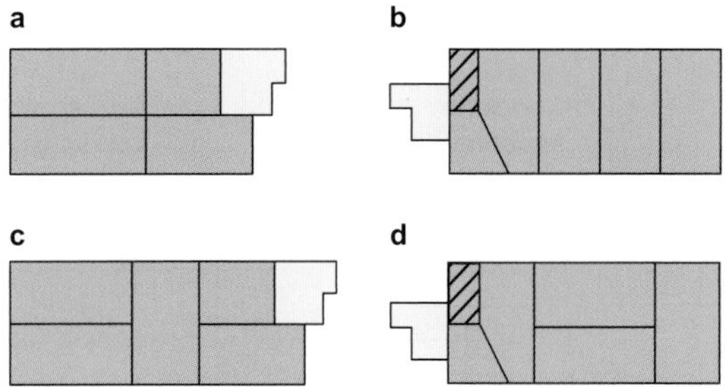

FIGURE 14.20
Square and reveal fitting: (a) English bond with recessed reveal; (b) English bond with recessed reveal built up; (c) Flemish bond with recessed reveal; (d) Flemish bond with recessed reveal built up

Over time door and window frames become defective and beyond repair, so will need replacing.

The original frame can easily be removed and should not require any support for the structure above.

Old frames may have been built set back into recessed reveals. They could be refitted into the same reveal but usually they are fitted to the front of the wall and the reveal is built up. See Figure 14.20 for examples of both.

The replacement frame should have been measured to allow easy fitting into the existing brickwork.

The method of fixing the frame depends on personal preference. Either bed joints are chopped out and wooden pads driven in, or the bricks are drilled and plugs inserted. Whichever method is preferred, the frame can be lifted into place and securely fixed.

The following items need to be considered when building in new frames:

- If they are new wooden frames ensure that the horns are cut and splayed back before positioning the frame.
- Before setting the frames in their correct position it is good practice to check the diagonal dimensions to prove the squareness of the frame.
- Check the drawing for the position of the frame and ensure that the rebates are facing the right way.
- Set up frames on an even bed. Plumb and level and brace securely (Figure 14.21).

Maintaining and extending masonry walls

There will be times when it is necessary to alter or extend existing brick or block walls. Sometimes provision can be made in new work to allow for extensions to be made at a later date.

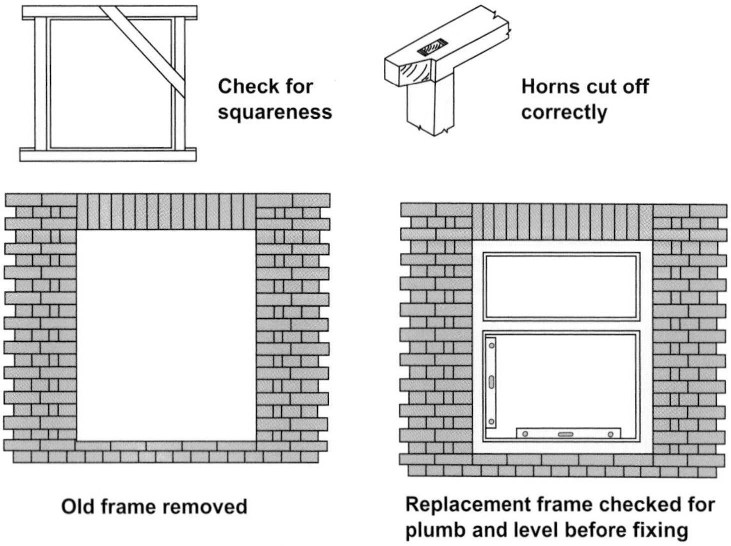

FIGURE 14.21
Building replacement frames in position

Various applications include:

- block bonding
- toothings
- indents
- wall connectors.

BLOCK BONDING

Block bonding is used when you are required to tie a new wall to an existing wall but where appearance is of no importance.

Three types of work are shown:

- block bonding a cross-wall (Figure 14.22)
- extending the length of a wall (Figure 14.23)
- thickening an existing wall (Figure 14.24).

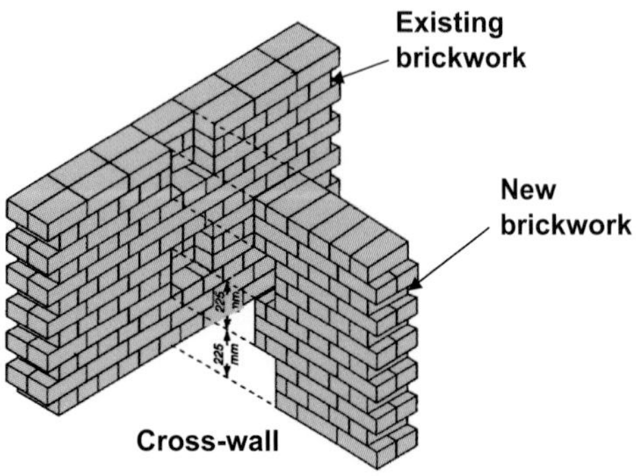

FIGURE 14.22
Block bonding between walls at right angles

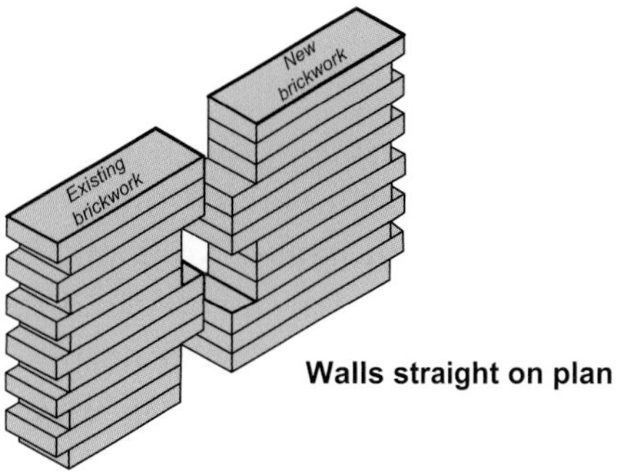

FIGURE 14.23
Block bonding to lengthen the wall

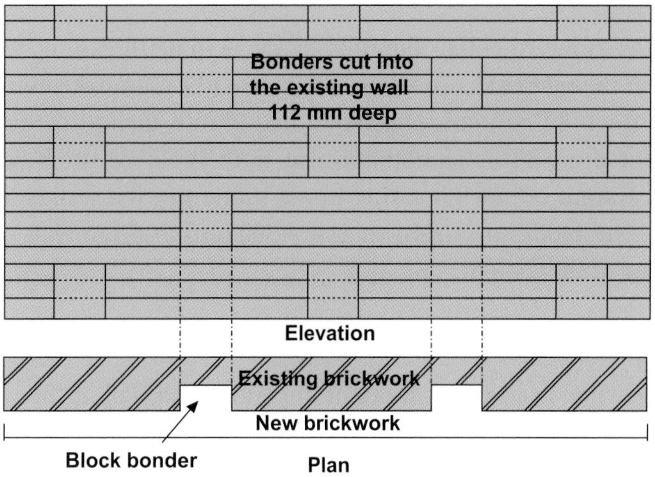

FIGURE 14.24
Block bonding to increase the thickness of a wall

In Figures 14.22 and 14.23 indented toothings could be cut and used for bonding purposes. In some circumstances this is a necessity, but block bonding should be used wherever possible, for two reasons:

- A clear-cut hole and a more adequate bond are obtained.
- The existing brickwork is not always of the same gauge as the new brickwork and attempts to bond on alternate courses are impractical.

In Figure 14.24 block bonding is the most practical solution. Note that the blocks have been placed in a diagonal pattern, which gives greater surface bond than would be the case if the blocks were placed one above the other.

In the illustrations, block bondings have been cut away every 225 mm. This is the usual practice, but the bondings can be extended to 300 mm without loss of stability.

In preparation for work of this type, the bricklayer cuts away the necessary brickwork with a cold chisel and club hammer. There are bricklayers who consider cutting away to be unskilled labour, but this is not the case, for the work requires considerable skill.

Quite often an apprentice employed on this work can be seen attempting to force the chisel, which has been entered into the face of the wall, by using the club hammer with both hands. No doubt expecting to see large pieces of brickwork fall away as a result, he is disappointed when the brickwork splays and fractures in the wrong position, or the chisel disappears to its head and is difficult to extract. The art of cutting away requires that:

- The chisel should never be forced.
- A start should be made with a small hole, which should be cleared as cutting proceeds and kept symmetrical as it is gradually enlarged.
- The cutting edge of the chisel should be watched, and not its head.

Figure 14.25 shows the block bonding of a wall in its thickness, when faced with 50 mm facing bricks. Common bricks are not manufactured in 50 mm sizes, and it would be too expensive to use 50 mm facings throughout the thickness of the wall. A combination of both types of brick is therefore needed. A system of block bonding in the section of the wall is adopted to effect this, the blocks occurring at every 300 mm and the wall being level at multiples of this height.

TOOTHINGS

Toothing is when bricks are left out at the end of the wall so the wall can be extended at a later time. The bricks are left out in such a way that the bond will continue and not be interrupted when the wall is extended. An example of toothing is shown in Figure 14.26.

Care has to be taken when cutting out and building into toothings.

When producing the toothing all perpends have to be kept plumb. Always start to chop out from the top down as this helps to prevent the projecting stretcher from breaking off.

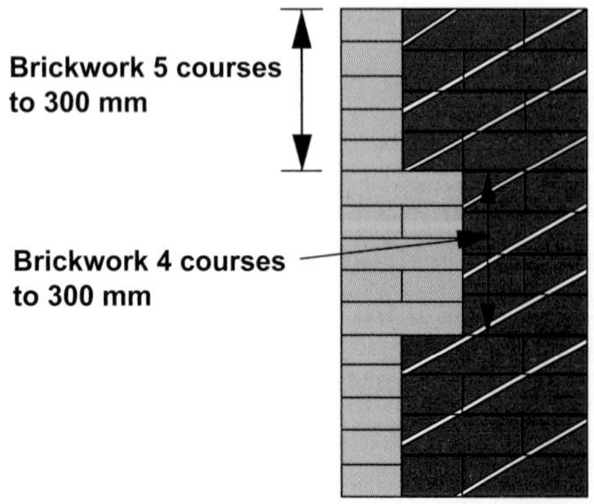

FIGURE 14.25
Block bonding when two different sized bricks are used

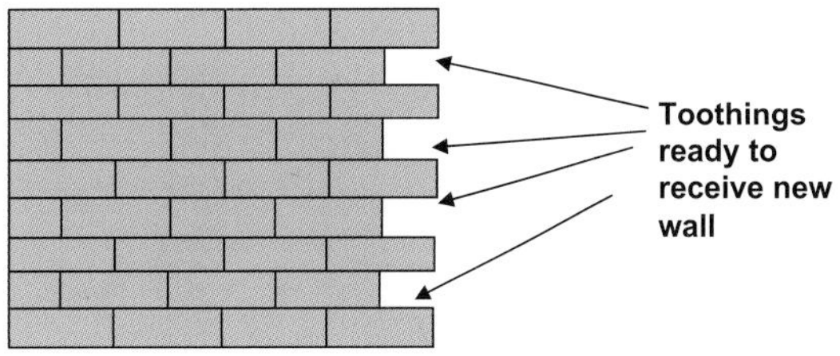

FIGURE 14.26
Toothing cut ready for wall to be extended

When extending it is important to make sure all joints are full. If not this will cause a weak spot in the wall and could cause problems at a later date.

INDENTS

Indents are similar to toothings in that they are left out as the wall is built to allow another wall to be built in later. The difference is that an indent is produced in the middle of a wall so that the extension is built at right angles to the existing wall (Figure 14.27). This makes a T junction.

PROPRIETARY WALL CONNECTORS

Proprietary wall connectors are available to connect a new wall at right angles to an existing wall. This replaces the need to cut toothings into the existing wall to form a tie for the new wall.

A 75 mm wide galvanized metal plate is secured to the existing wall with galvanized screws. The length of the fixing plate depends on the height of the extended wall. The fixing plate has many slots at close centres to allow the new wall to be secured with special brackets (Figure 14.28).

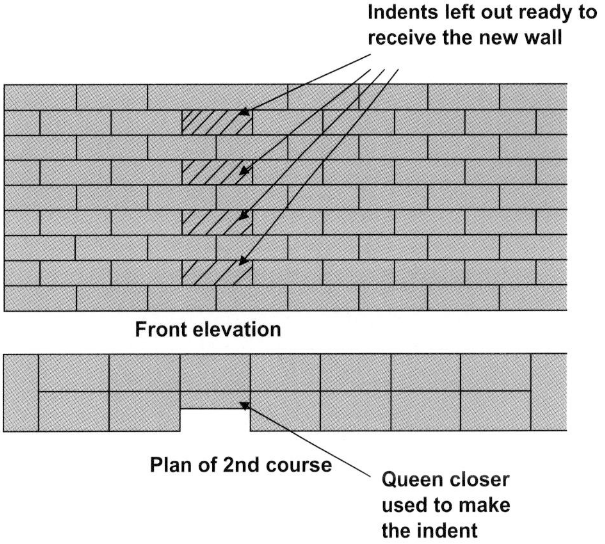

FIGURE 14.27
Indents cut into existing wall

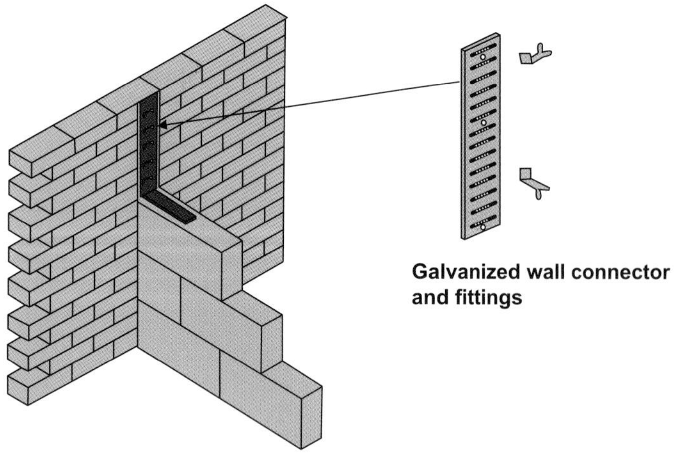

FIGURE 14.28
Proprietary wall connector used to bond a block wall to a brick wall

Chopping out and replacing defective bricks

As with all repairs, it is essential to determine the cause of the problem.

If cracks have appeared in the mortar then repointing may be sufficient, as described next. If the cracks are in the brickwork then further investigation is required to determine the cause.

Settlement can be caused by movement of the building. In this case it is necessary to fix a tell-tale to the building (Figure 14.29). A tell-tale can be a piece of glass fixed over a crack by epoxy resin on each side of the crack. If any more movement takes place then the glass will crack. It is then essential to obtain professional advice before going any further. If no further movement takes place the bricks can be safely removed and replaced with similar bricks and repointed.

Bricks may also require replacing owing to spalling. This could be caused by exposure to severe weather. A poor choice of brick can allow water to enter the face of the brick and if it freezes the face of the brick can be blown off.

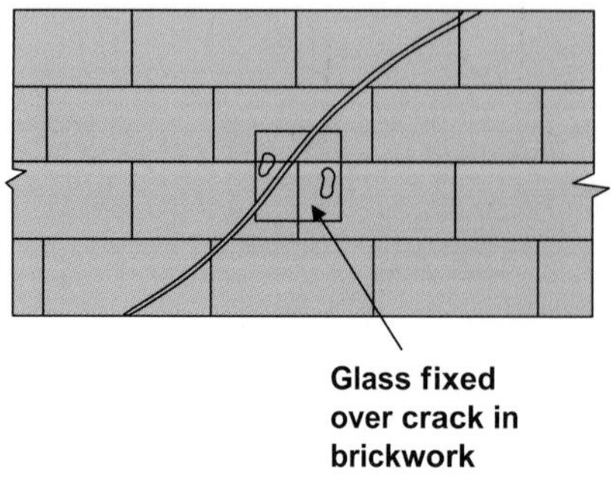

FIGURE 14.29
Tell-tale fixed over a crack in brickwork

When the cause of the defect has been removed the repair can take place.

Bricks can be removed using a club hammer and cold chisel. In severe cases a skill saw or an angle grinder may be used.

The bricks and mortar should be the same as the original if possible, to maintain the correct appearance. Ensure that all the joints are fully packed to prevent cracks appearing after the work has dried out.

Repointing

Repointing is the process of repairing the joints between the bricks.

Chimney stacks and parapet walls are usually the first parts of a brick building that need repointing after 20 or 30 years of exposure to wind, rain and frost.

The mortar should be of the same materials and mix as the original if possible. If the mix is too strong or too weak it will cause problems with movement and the same problem will recur.

The joints should be raked out to a depth of 12–15 mm and all work dusted down to remove any surplus material. The work area should then be dampened to reduce the amount of water taken from the mortar. This would cause the mortar to set too quickly and it may become crumbly.

It is usual to start repointing at the top of the wall and work downwards.

All cross-joints should be completed, first followed by the bed joints. Ensure that all mortar joints are pressed firmly into place (Figure 14.30).

OTHER DEFECTS IN BRICKWORK

Defects in brickwork can arise from the following causes:

- sulphate attack on mortar and renderings
- frost action
- corrosion of iron and steel, which is enclosed by brickwork
- crystallization of salts.

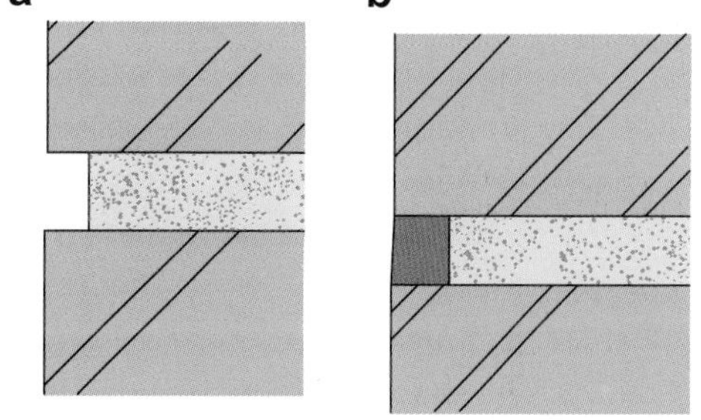

FIGURE 14.30
Repointing brickwork: (a) joints raked out ready for repointing; (b) joints repointed

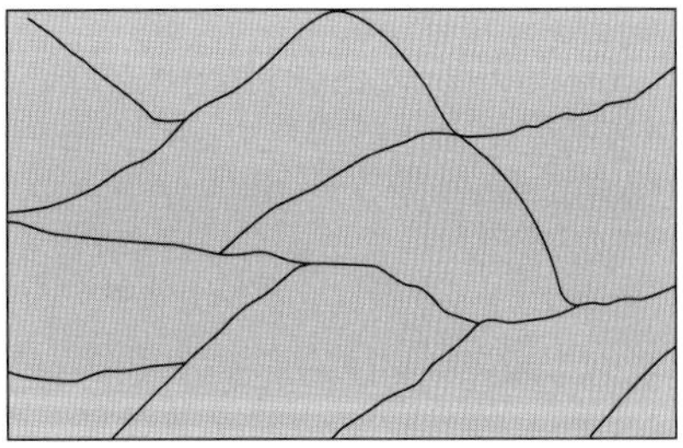

FIGURE 14.31
Crocodile cracking due to sulphate attack

Sulphate attack

When sulphates attack mortar, it expands and causes cracking of the brick-work. Cracking on the face of rendering is known as crocodile cracking. This is caused by the shrinkage of rendering as it dries (Figure 14.31).

Where there is condensation in an old chimney flue, which is not lined, the moisture dissolves the sulphur-containing flue gases, which penetrate the flue walls. This can lead to chimney stacks tending to lean as the brickwork on the wet side gradually expands (Figure 14.32).

The cause of all these failures is the chemical reaction between sulphates in the bricks and the lime present in the mortar.

In these situations it is essential to use bricks with low sulphur content and sulphur-resistant cement.

Frost attack

Frost action can cause the surface of porous mortar joints to crumble. It can also cause the face of porous bricks to spall.

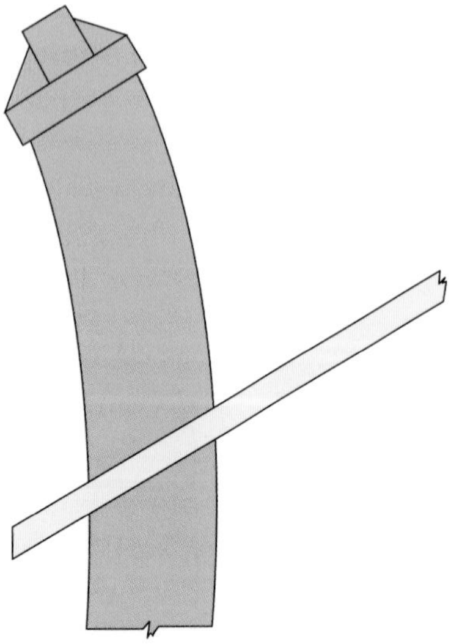

FIGURE 14.32
Chimney leaning due to sulphate attack

Apart from cutting out and replacing the bricks, the surface can be treated with an application of a water sealer.

Corrosion of iron and steel

Unprotected iron and steel will corrode when in contact with damp moist air, especially in the presence of sulphates.

Before building in gate brackets, etc., they should be thoroughly coated with a rust inhibitor, primed and painted with bitumen paint.

All iron and steel prone to corrosion should be kept back from the face of the brickwork.

Crystallization of salts

Lime staining

Many buildings become disfigured by the crystallization of salts on the face of walls (Figure 14.33). Lime staining is a deposit of lime left on the face of brickwork when the water in which it was dissolved evaporates. Lime staining is difficult to remove and may require expensive treatment by experts.

Efflorescence

Efflorescence is also a deposit of soluble salts left on the surface of brickwork when the water in which they were dissolved evaporates (Figure 14.34).

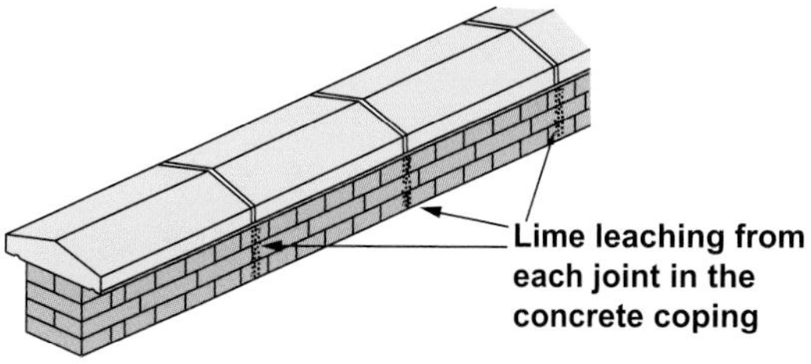

FIGURE 14.33
Lime staining

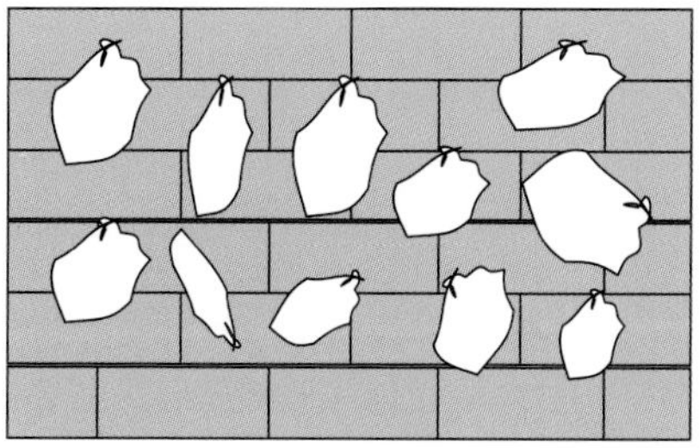

FIGURE 14.34
Efflorescence

This is the most common deposit on the face of bricks but it does not affect the wall as it usually disappears after a few months.

The soluble salts are usually found in clay bricks and sands used for mortars. It is essential not to wash off the efflorescence as it will only dilute it and drive it back into the wall. Brush it off with a soft, dry brush.

Multiple-choice questions

Self-assessment

This section of the book is designed to allow you to check your level of knowledge. The section consists of revision questions for this chapter. The questions are all multiple choice and have four possible answers. The answers are to be found at the end of the book.

The main type of multiple-choice question will be the four-option multiple-choice question. This will consist of a question or statement, known as the stem, followed by a choice of four different answers, called the responses. Only one of these responses is the correct answer; the others are incorrect and are known as distracters.

You should attempt to answer the questions by choosing either (a), (b), (c) or (d).

Example

The person employed by the local authority to ensure that the Building Regulations are observed is called the:

(a) clerk of works
(b) building control officer
(c) council inspector
(d) safety officer

The correct answer is the building control officer, and therefore (b) would be the correct response.

Repairing and maintaining brickwork

Question 1 In dead shoring, the member that penetrates through the wall is called a:

(a) pin
(b) needle
(c) shore
(d) strut

Question 2 Identify the members shown:

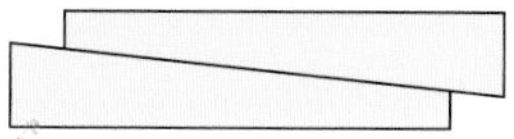

(a) packings
(b) wall plates
(c) folding wedges
(d) timber cleats

Question 3 Which is the recommended method for removing efflorescence from the face of brickwork?

(a) dry brushing

(b) washing down

(c) wet brushing

(d) scraping

Question 4 Identify the item of equipment shown:

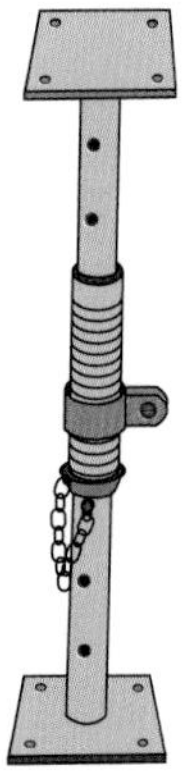

(a) needle

(b) pin

(c) dead shore

(d) steel adjustable prop

Question 5 The protection of existing buildings to enable them to be usable is known as:

(a) refurbishment

(b) renovation

(c) preservation

(d) conservation

Question 6 Water that penetrates the building through faulty flashings around chimney stacks is known as:

(a) penetrating damp

(b) rising damp

(c) percolating damp

(d) efflorescence

Question 7 Name the operation when the existing foundations are taken down to a lower level:

(a) underpinning

(b) shoring

(c) timbering

(d) retaining walls

Question 8 Identify the operation shown for extending a wall:

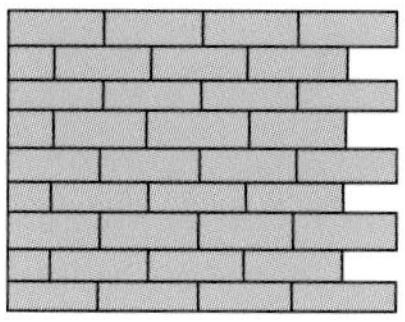

(a) indents

(b) toothings

(c) block bonding

(d) raking back

CHAPTER 15

Answers to Multiple-Choice Questions

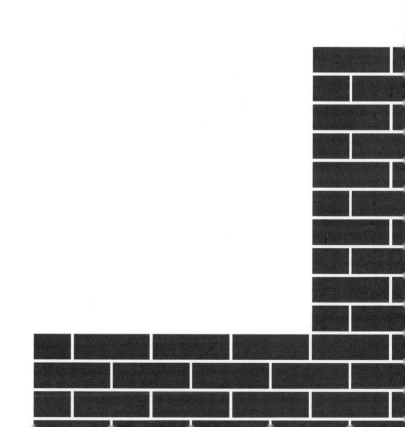

Chapter 1. The Construction Industry

1 (a); 2 (b); 3 (b); 4 (d); 5 (d); 6 (a); 7 (c); 8 (c)

Chapter 2. Health and Safety in the Construction Industry

1 (d); 2 (c); 3 (c); 4 (a); 5 (a); 6 (d); 7 (b); 8 (a)

Chapter 3. Programming and Resources for Work

1 (a); 2 (a); 3 (c); 4 (d); 5 (c); 6 (b); 7 (a); 8 (b)

Chapter 4. Working Relationships

1 (a); 2 (b); 3 (d); 4 (b); 5 (c); 6 (a); 7 (b); 8 (d)

Chapter 5. Working Methods

1 (b); 2 (a); 3 (d); 4 (c); 5 (a); 6 (a); 7 (b); 8 (d)

Chapter 6. Setting Out

1 (a); 2 (d); 3 (b); 4 (c); 5 (a); 6 (c); 7 (b); 8 (d)

Chapter 7. Chimneys, Flues and Fireplaces

1 (a); 2 (d); 3 (a); 4 (c); 5 (a); 6 (b); 7 (d); 8 (b)

Chapter 8. Arches

1 (a); 2 (b); 3 (c); 4 (d); 5 (c); 6 (a); 7 (b); 8 (c)

Chapter 9. Curved Walls on Plan

1 (c); 2 (c); 3 (a); 4 (b); 5 (a); 6 (c); 7 (b); 8 (b)

Chapter 10. Ramped Brickwork

1 (c); 2 (b); 3 (a); 4 (c); 5 (b); 6 (c); 7 (a); 8 (a)

Chapter 11. Splayed Brickwork

1 (b); 2 (c); 3 (a); 4 (d); 5 (b); 6 (c); 7 (a); 8 (d)

Chapter 12. Reinforcement

1 (a); 2 (d); 3 (d); 4 (c); 5 (b); 6 (a); 7 (c); 8 (b)

Chapter 13. Decorative Brickwork

1 (b); 2 (d); 3 (d); 4 (a); 5 (c); 6 (a); 7 (b); 8 (c)

Chapter 14. Repairing and Maintaining Masonry

1 (b); 2 (c); 3 (a); 4 (d); 5 (d); 6 (c); 7 (a); 8 (b)

Index